Corrosion for Everybody

AF665006

A. Groysman

Corrosion for Everybody

Springer

A. Groysman
ORT Braude College of Engineering
Karmiel, Israel

Every effort has been made to contact the copyright holders of the figures which have been reproduced from other sources. Anyone who has not been properly credited is requested to contact the publishers, so that due acknowledgement may be made in subsequent editions.

ISBN 978-94-007-9130-5 ISBN 978-90-481-3477-9 (eBook)
DOI 10.1007/978-90-481-3477-9
Springer Dordrecht Heidelberg London New York

© Springer Science+Business Media B.V. 2010
Softcover re-print of the Hardcover 1st edition 2010
No part of this work may be reproduced, stored in a retrieval system, or transmitted in any form or by any means, electronic, mechanical, photocopying, microfilming, recording or otherwise, without written permission from the Publisher, with the exception of any material supplied specifically for the purpose of being entered and executed on a computer system, for exclusive use by the purchaser of the work.

Printed on acid-free paper

Springer is part of Springer Science+Business Media (www.springer.com)

To my great and wise wife Olga for support, endurance, understanding and assisting in creating, discussing and writing this book, and also to my lovely children Sasha and Anat

What another would have done as well as you, do not do it.
What another would have said as well as you, do not say it;
Written as well, do not write it.
Be faithful to that which exists nowhere but in yourself –
and thus make yourself indispensable.

André Gide (1869–1951)
French writer, humanist, and moralist

Contents

Preface

There are some subjects that most people are always able to discuss among themselves, such as politics, sport, medicine, weather, economics, education, art, and, of course, corrosion. Everyone has observed rust on cars, on pipes, on roofs, and on benches in gardens. Everyone has seen greenish patina on bronze sculptures, statues and ancient coins. You remember, of course, how unpleasant it was to be without water because of corrosion holes in water pipes and when repairs were needed. Thus, there is no one who has not come across this nasty phenomenon related to degradation and deterioration of metals and the environment. Corrosion occurs from the moment when metal comes into contact with the environment (or vice versa). Prehistoric people doubtless observed this phenomenon with the seven metals (gold, silver, copper, tin, iron, lead, and mercury), known to them. Probably the ancient metallurgists were the first who observed corrosion during processing and manufacturing of metals and alloys. But they could not explain the interaction of metals and environment because they had no knowledge of chemistry. The laws of chemistry were discovered later, at the end of the 18th century. Thus, we may say that people began to study corrosion about 200 years ago when the main principles of chemical reactions were defined and established by the French chemist Antoine Lavoisier, and chemists after him.

The discovery of many chemicals (especially inorganic acids) by alchemists in the 12th and 13th centuries, the discovery of new metals in the 18th and 19th centuries, and the main principles of electrochemistry in the 19th century, stimulated the study of corrosion phenomena. Corrosion was taught as a science in England only in 1924. Today one can study the discipline of corrosion in about 300 universities throughout the world. The understanding of corrosion as a phenomenon and as a physico-chemical reaction with the environment began in the 1930s. The development of industry involving processes operating under harsh conditions also gave a powerful incentive to investigations into corrosion. Many books have been written about corrosion in many languages during the past 70 years. Many special corrosion journals are published every month in many countries and a lot of conferences are held every year. Societies dedicated to corrosion have been organized in many countries. There is an information "explosion" about corrosion and preventive anti-corrosion methods. In spite of these events, corrosion phenomena still occur every day in all enterprises, in buildings, in houses, in different equipment and structures all around us. Many specialists and non-specialists want to know why stainless steel corrodes. Why corrosion holes are formed in water pipes? Why silver jewelry becomes black? And many other things.

When I was a child, books such as *Entertainment Physics* and *Entertainment Mathematics* were read. In the former USSR between the 1960s and 1990s the journal *Chemistry and Life* was popular reading. Any person, without special education and knowledge in chemistry and in physics, could enjoy reading this literature and understand many things and phenomena concerning scientific subjects.

On the one hand, corrosion seems a very simple phenomenon, as we observe its destructive results: rust, pits, cracks, and holes. On the other hand, corrosion has many faces, and in order to distinguish and understand all of them, specialists have to know chemistry, electrochemistry, physical chemistry, metallurgy, physics, electricity, material science, microbiology, ecology, economics, and even psychology.

Today, many people of different professions are interested in corrosion: archeologists dealing with ancient coins and artifacts made of copper, bronze, and iron, sculptors who create their works of art from various metals and alloys, architects who design buildings and flats, doctors who deal with metallic implants, prosthetic devices and dental alloys, advocates dealing with the complaints of their clients regarding the corrosion problems at the plants where they work or the buildings in which they live. Therefore I have tried to write about corrosion in such a manner that it would be important and interesting for both specialists and non-specialists in the corrosion field.

As a rule, corrosion causes negative emotion in people. Many of us see it as a nasty phenomenon. Unfortunately, people do not see the other side of the coin. I would like to show corrosion phenomena from a non-traditional point of view at the end of this book, namely, the relationship between corrosion, history, philosophy, and art. I would also like to show that the subject of corrosion is interesting and can be enjoyed and understood by studying the subject. I wrote this book with these feelings in mind, and I hope that you will enjoy it. Now I invite you to journey into the *world of corrosion.*

Alec Groysman

Chapter 1
Corrosion Mechanism and Corrosion Factors

1.1 Definition of Corrosion

First of all, we have to define the term "corrosion" in an interesting way. When you observe the rust spots on a car, the question is why was the rust formed? It was formed because of the reaction between iron and the environment. This may be the atmosphere, water, soil, chemicals, etc. We call rust a "corrosion product" formed on iron surfaces.

Can we always observe a corrosion product (substance, for example, rust) as a result of a reaction between a metal and its environment? No! Sometimes you observe pits in a metal surface, but you do not see the corrosion product in the pits (Figure 1.1).

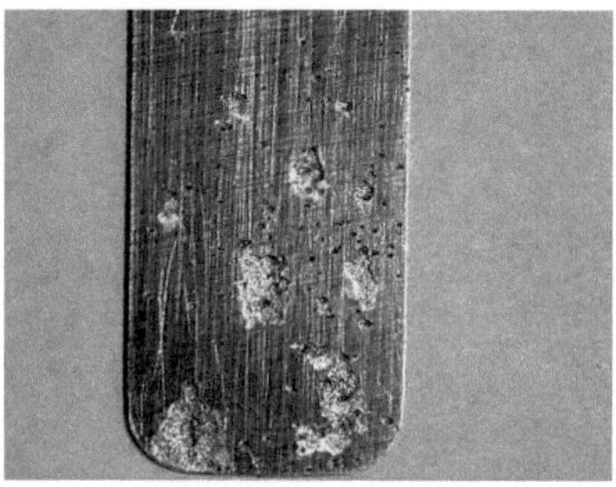

Fig. 1.1 Pits on a metal surface.

Why were the pits formed? They were formed as a result of the reaction between a metal and the environment, but we do not see the corrosion product.

Another example – cracks in a tube (Figure 1.2). We do not observe the corrosion product as a substance in the cracks or in the vicinity of the cracks. Why were

A. Groysman, *Corrosion for Everybody*, DOI 10.1007/978-90-481-3477-9_1,
© Springer Science+Business Media B.V. 2010

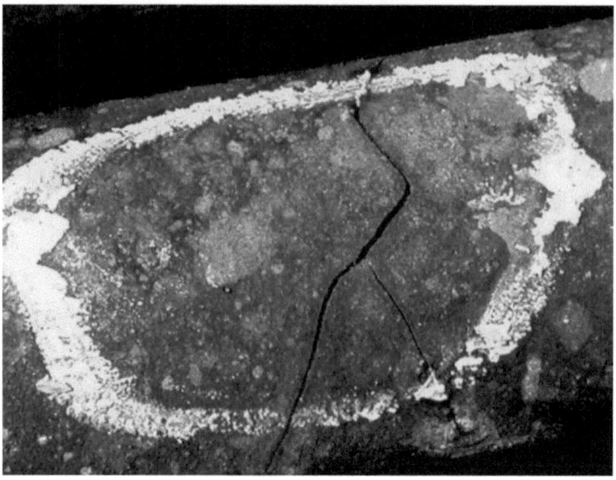

Fig. 1.2 Cracks in Admiralty brass tube.

they formed? They were formed as a result of a reaction between a metal and its environment.

What is the common feature among these three examples?

Metals respond to a specific imposed stimulus of the external environment. This may be chemical, physical (mechanical), or mixed chemical and mechanical. The stimulus evokes some type of response and determines the nature of the deterioration of the material. Thus, the deteriorative characteristics of materials indicate their chemical resistance, or activity. Chemical influence results in a chemical reaction occurring on a metal surface or inside the metal, and we are able to observe corrosion products, pits, or cracks. Mechanical influence (for example, erosion) causes the removal (physical thinning and wastage) of metal. Erosion is the removal of surface material by means of turbulence of solid particles in a gas or a liquid. Sometimes the influence of the environment may be mixed: chemicals may contain solid particles or cause conditions for combined chemical and physical removal of metals. As a result of such a response and the reactions with the environment, metals lose their pleasant appearance, their thickness, and mechanical, thermal, and optical properties.

Manufactured substances around us (medicines, food, fuels, paints, chemicals, etc.) are usually produced in metal equipment. Corrosion products formed as a result of a reaction with these substances (environment) deteriorate them. If pinholes and cracks are formed in metal apparatus as a result of its corrosion, dangerous chemicals (fuels, gases, acids, etc.) may affect the atmosphere, water, soil, and even threaten people's lives and the animal kingdom.

Now we are able to define the term "corrosion":

Corrosion is an interaction between a material, usually a metal, and its environment that results in deterioration of the material, and the environment.

People began to use the term "corrosion" to describe the reaction between metals and the environment at the end of the 17th century, the period between alchemy and chemistry. The term "corrosion" is derived from the Latin word *corrodere* which means "to eat away". This approach also lets us talk about "corrosion" of non-metallic materials: polymers, ceramics, glasses, and composites. Many people use the term only in relation to metals, and feel that it is better to refer to "destruction, destroying, degradation, or deterioration" of non-metallic materials. Today more and more specialists talk about corrosion of non-metallic materials too. This is correct if we follow the above-mentioned definition.

We have to mention that not all reactions of substances with the environment and not every destruction of materials may be called "corrosion". Dissolution of sugar and salt in water is not corrosion, and burning (oxidation) of wood or paper is not corrosion. But dissolution of metals in hydrochloric acid, oxidation of some metals in a furnace or in the atmosphere, is corrosion.

Another note concerns the term "rust". Some people call corrosion products formed on all metals "rust". Historically, it is a brown-red corrosion product formed only on iron and iron alloy surfaces. For deterioration of other metals, we have to use the term "corrosion product", and not "rust". For example, white corrosion products are formed on aluminum and zinc, green on nickel, blue and blue-green on copper, pink on cobalt, etc.

And one last note. We have to differentiate between a metal and an alloy. A metal is a material composed only of one chemical element, for example, iron, nickel, gold, zinc, copper, etc. An alloy is a material having metallic properties and composed of two or more chemical elements of which at least one is a metal. For example: carbon steel consists of iron and carbon, brass consists of copper and zinc, and so on.

Now we have to explain why we have gold, silver, platinum, and some other pure metals, in nature (they are called "noble"). The reason is that they do not usually react with the environment. We cannot find pure iron, aluminum, zinc, or titanium in nature. "Non-noble" metals are only present in ores, that is in compounds. We call these compounds (metallic ores) corrosion products of metals.

In order to understand the different behavior of "noble" and "non-noble" metals, we have to have a knowledge of metals and non-metals.

1.2 Metals and Non-Metals

Of course, you are familiar with Mendeleev's Periodic Table of the chemical elements (Figure 1.3). This table shows that 112 elements are familiar to man today. These elements are divided into metals and non-metals. If we draw a boundary-line (staircase) between the elements as shown in Figure 1.3, all chemical elements in the Periodic Table left of the staircase are metals. The elements which are placed in the right upper corner of the staircase are non-metals. Knowledge of all 112 elements was only achieved over a long period of time. Seven metals (gold, silver,

1 H 1.01																		2 He 4.00
3 Li 6.94	4 Be 9.01												5 B 10.81	6 C 12.01	7 N 14.01	8 O 16.00	9 F 19.00	10 Ne 20.18
11 Na 22.99	12 Mg 24.31												13 Al 26.98	14 Si 28.09	15 P 30.97	16 S 32.07	17 Cl 35.45	18 Ar 39.95
19 K 39.10	20 Ca 40.08		21 Sc 44.96	22 Ti 47.87	23 V 50.94	24 Cr 52.00	25 Mn 54.94	26 Fe 55.85	27 Co 58.93	28 Ni 58.69	29 Cu 63.55	30 Zn 65.39	31 Ga 69.72	32 Ge 72.61	33 As 74.92	34 Se 78.96	35 Br 79.90	36 Kr 83.80
37 Rb 85.47	38 Sr 87.62		39 Y 88.91	40 Zr 91.22	41 Nb 92.91	42 Mo 95.94	43 Tc (99)	44 Ru 107.07	45 Rh 102.91	46 Pd 106.42	47 Ag 107.87	48 Cd 112.41	49 In 114.82	50 Sn 118.71	51 Sb 121.76	52 Te 127.60	53 I 126.90	54 Xe 131.29
55 Cs 132.91	56 Ba 137.33	57-70	71 Lu 174.97	72 Hf 178.49	73 Ta 180.95	74 W 183.84	75 Re 186.21	76 Os 190.23	77 Ir 192.21	78 Pt 195.08	79 Au 196.97	80 Hg 200.59	81 Tl 204.38	82 Pb 207.2	83 Bi 208.98	84 Po (209)	85 At (210)	86 Rn (222.01)
87 Fr (223)	88 Ra (226)	89-102	103 Lr (262)	104 Rf (261)	105 Db (262)	106 Sg (263)	107 Bh (264)	108 Hs (265)	109 Mt (268)	110 Uun (269)	111 Uuu (272)	112 Uub (277)	113 Uut	114 Uuq (289)	115 Uup	116 Uuh (289)	117 Uus	118 Uuo (293)

57 La 138.91	58 Ce 140.12	59 Pr 140.91	60 Nd 144.24	61 Pm (145)	62 Sm 150.36	63 Eu 151.96	64 Gd 157.25	65 Tb 158.63	66 Dy 162.5	67 Ho 164.93	68 Er 167.26	69 Tm 168.93	70 Yb 173.04
89 Ac (227)	90 Th 232.04	91 Pa 231.04	92 U 238.03	93 Np (237)	94 Pu (244)	95 Am (243)	96 Cm (247)	97 Bk (247)	98 Cf (251)	99 Es (252)	100 Fm (257)	101 Md (258)	102 No (259)

Fig. 1.3 Mendeleev's Periodic Table of the Chemical Elements. (The number in each square gives the atomic number of the element – the number of protons in the nucleus. The conventional symbol of each element is given.)

copper, iron, mercury, tin, and lead) and two non-metals (carbon and sulphur) were known at the time of the writing of the Old Testament of the Bible (approximately 11th century BC). For nearly 3000 years, from about the 11th century BC to the middle of the 18th century AD, only five more chemical elements were isolated: arsenic, antimony, phosphorus, platinum, and bismuth. Sixteen elements (among them 13 metals) were discovered in the 18th century and 50 elements (among them 38 metals) were also discovered in the 19th century. Two of the naturally occurring elements (metals) and then another 17 man-made metals were discovered at the beginning of the 20th century.

Metals have a strong tendency to give up electrons to non-metals. Therefore, we can call metals "male" and non-metals "female". Some elements, boron, silicon, germanium, arsenic, antimony, tellurium, and polonium are at the margin between metals and non-metals, and are called "semimetals" or "metalloids". They may accept or donate electrons, therefore, could be called "bisexual". Thus 80, or three-quarters, of all known chemical elements are metals which are characterized by high electrical and thermal conductivity, ductility and malleability, high reflectivity of light (when polished), non-transparent and lustrous (metallic shine). "Malleable" means that metals can be hammered, pounded, or pressed into different shapes without breaking. "Ductile" means that metals can be drawn into thin sheets or wires without breaking. All metals are solid materials at room temperature, with the exception of mercury, gallium and cesium which are liquids. Gallium and cesium can be melted in a man's hand (melting point 30°C). All properties of metals are ex-

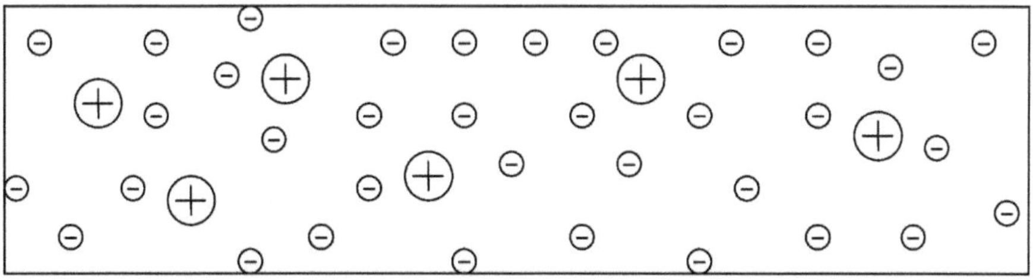

Fig. 1.4 Schematic illustration of metallic bond. ⊕ – nucleus (positive ions); ⊖ – valence electrons, which do not belong to one specific metallic atom.

plained by the metallic bond, a special type of chemical bond that exists only in metals and their alloys (Figure 1.4).

When identical metallic atoms get together in a piece of a metal, the bonding between them is realized by means of "collectivized" electrons which do not belong to one specific metal atom. Certainly in the inanimate world we may observe realization of the "communist" idea about collectivization among people. Thus a piece of a metal is a collection of positive ions floating in a "sea" of electrons. Without these electrons, the positively charged metal ions would repel one another. A free "electron gas" or "electron cloud", acts as a "glue" to hold the metal ions together. Free electrons in metals have high mobility.

We have known about this for only about 100 years, after the discovery of the electron by the English physicist J.J. Thomson in 1897, and the suggestion of German physicist Paul Drude in 1900, that the electrons in metals may move freely. Nowadays, we recognize that the "flow" of electrons in a metal is realized by a pushing mechanism as in a "domino" fall.

What else is common for all metals? Metals are prone to release the electrons when they meet non-metals under particular conditions and, as a result, be oxidized. For example, neutral iron atoms (Fe) may turn into iron cations (Fe^{2+}), releasing two electrons (e^-):

$$\mathrm{Fe} \rightarrow \mathrm{Fe}^{2+} + 2e^-. \tag{1.1}$$

This process cannot occur without a partner, some other atom or a particle able to obtain the electrons released in the reaction (1.1). Such a partner may be non-metal atoms (oxygen, sulphur or chlorine) which obtain electrons lost by neutral metal atoms. There are 18 non-metals in the Periodic Table, taking part in the process of electrons' gain, and their properties are opposite to those of metals. They rarely have metallic luster. They exist in three physical states at room temperature: gas (oxygen O_2, chlorine Cl_2, etc.), liquid (bromine, Br_2), and solid (iodine I_2, sulphur S_8, etc.). Non-metallic solids are neither malleable nor ductile. They are non-conductors of electricity and heat (except graphite C). The most important characteristic for non-metals: they are able to gain electrons. Thus, neutral oxygen atom (O) may gain up to two electrons and turn into negative ions O^{2-} called anions:

$$\mathrm{O} + 2e^- \rightarrow \mathrm{O}^{2-}. \tag{1.2}$$

When metals release their electrons they oxidize and corrode. Therefore *corrosion is a chemical oxidizing process*. For example, when iron meets oxygen, the first gives electrons to oxygen, and two stable particles are formed: iron cations (Fe^{2+}) and oxygen anions (O^{2-}), which are energetically more stable than separate iron and oxygen atoms. Thus iron oxide (FeO) composed of these two stable ions is a corrosion product of iron, that is, the result of a chemical reaction between iron and oxygen. "You may be virtuous alone, but only together you may sin". In order for corrosion to occur, two sinners must be present: a metal and the environment.

Why do corrosion processes occur? Why are some metals (for example, gold and platinum) stable in water, in the atmosphere, in soil, and exist as pure metals in the earth, but others (for example, iron and zinc) corrode under similar conditions, and do not exist as pure metals, existing only in the ores of the earth? Undoubtedly there is no absolute resistant metal under all conditions. All metals, even pure ones existing in the earth, can corrode under some particular conditions. Can we predict if a particular metal will corrode under specific conditions? We shall answer these questions in the next section.

1.3 Prediction of Corrosion of Metals

All the above questions are similar to the main question for chemists: why do some elements and molecules react under particular conditions and others do not? Chemists found the answer using two very important tools: thermodynamics and kinetics. Thermodynamics is a field of science that studies the changes and relationships between heat, work, and energy. It uses the energetic state of the chemical substances under equilibrium, and allows the prediction of the possibility of a chemical process. We may write any chemical reaction:

$$aA + bB \leftrightarrow cC + dD, \tag{1.3}$$

where A and B are two substances (they are named reactants); for example, in our corrosion case, iron and water (or copper and water, or iron and hydrochloric acid, etc.) before their contact; C and D are corrosion products, i.e. the substances formed after reaction has been completed (rust, metal oxides, sulphides, etc.); a, b, c, and d are quantities (in moles) of each participant in chemical reaction. In order to predict the possibility of a chemical reaction (1.3), we have to know the energetic states for all participants (iron, copper, water, hydrochloric acid, etc.) before their contact, and for the products (possible corrosion products: rust, copper hydroxide, etc.) after the reaction has been completed. We may explain the energetic states of substances in chemical reactions by means of analogy with the "tendency to the minimum of potential energy" in mechanics. Thus, the tendency for the motion of some body from high to low level is defined by the differences in the levels of the body (heights) in the initial and final states. This tendency does not depend on the body's path. The motion of the body is completed after the gravitation potential reaches its minimum.

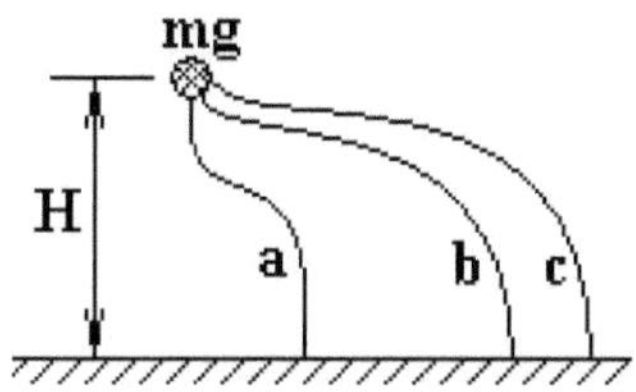

Fig. 1.5 The loss of gravitation potential energy (E) equals the work of the body's fall, and does not depend on its path (**a**, **b** or **c**): $E = mgH$ (m – mass of the body; g – acceleration of gravity; H – height above the ground).

The loss of the gravitation potential energy (for the mass unit) equals the work of the body's fall, and does not depend on its path (Figure 1.5). We may say that gravitational energy is a form of potential energy in mechanics. What happens in the world of atoms and molecules?

There is a *chemical energy* for the chemical reactions with participation of atoms and molecules, similar to the *gravitational energy* in mechanics.

We are all familiar with other forms of potential energy. For example, the energy stored between the plates of a charged capacitor is an *electrical potential energy*. A *nuclear energy* is also a form of potential energy and is stored in nuclei. *Chemical energy* is a capacity of substances to do work or to evolve heat by undergoing a change of composition, and similar to mechanics, electricity and nuclear physics, may be regarded as a potential energy resulting from the mutual forces between atoms and molecules. As a body in mechanics possesses *gravitation potential*, any substance taking part in a chemical reaction possesses a *chemical potential*. The latter decreases in any spontaneous chemical process. The American scientist Josiah Willard Gibbs (1839–1903) was the first to suggest in 1876 the use of *chemical potential* for the identification of the direction of the chemical (among them corrosion) processes. Therefore, today this *chemical potential* is called *Gibbs energy*, and is designated as "G" for pure substances. Changes in Gibbs energy of substances (participants in chemical reaction) are used to judge whether a certain chemical process can occur spontaneously under certain conditions (constant temperature and pressure). Such changes do not depend on the "chemical path" but depend only on the initial and final energetic states (a *chemical*, or *Gibbs energy*) of substances. The sign in Gibbs energy changes for a certain chemical process defines its possible spontaneous proceeding, without external influence, i.e., naturally.

Any chemical process (corrosion as well) may occur if $\Delta G^\circ_T < 0$ ("green light" in the traffic lights). The "o" and "T" show that chemical potentials of all participants were taken for calculations under standard conditions: pressure 101.325 Pa (1 atm) and any constant temperature T (usually 298 K, but one may calculate at any other temperature T).

The corrosion process cannot occur if $\Delta G^\circ_T > 0$ ("red light" in the traffic lights).

The traffic lights allow a third possible situation – "yellow light". The chemical language for the reaction (1.3) is $\Delta G^\circ_T = 0$. This means that the sum of chemical potentials (Gibbs energies) of reactants equals the sum of chemical potentials of

Table 1.1 Gibbs energy changes for some corrosion reactions.

Reaction	*ΔG^o_{298}, kJ/mol*
$4Fe_{(s)} + 3O_{2(g)} \rightarrow 2Fe_2O_{3(s)}$	-742.2
$Fe_{(s)} + 2H_2O_{(l)} \rightarrow Fe(OH)_{2(s)} + H_{2(g)}$	-15.7
$4Fe_{(s)} + 6H_2O_{(l)} + 3O_{2(g)} \rightarrow 4Fe(OH)_{3(s)}$	-1397.2
$4Au_{(s)} + 6H_2O_{(l)} + 3O_{2(g)} \rightarrow 4Au(OH)_{3(s)}$	+24.0
$Cu_{(s)} + 2H_2O_{(l)} \rightarrow Cu(OH)_{2(s)} + H_{2(g)}$	+100.9
$2Cu_{(s)} + 2H_2O_{(l)} + O_{2(g)} \rightarrow 2Cu(OH)_{2(s)}$	-272.5
$Cu_{(s)} + 2HCl_{(aq)} \rightarrow CuCl_{2(aq)} + H_{2(g)}$	+65.5
$3Cu_{(s)} + 4HCl_{(aq)} + O_{2(g)} \rightarrow 2CuCl_{2(aq)} + H_{2(g)} + Cu(OH)_{2(s)}$	-199.8
$Zn_{(s)} + 2HCl_{(aq)} \rightarrow ZnCl_{2(aq)} + H_{2(g)}$	-147.0
$Fe_{(s)} + 2HCl_{(aq)} \rightarrow FeCl_{2(aq)} + H_{2(g)}$	-84.9

Note: the symbols "s, l, aq", and "g" relate to the "solid, liquid, aqueous", and "gaseous" phases of substances in the above-mentioned reactions.

products in the reaction (1.3). In other words, reactants and products are in chemical equilibrium. This equilibrium is not static, but dynamic, and two reactions (in right and left directions) occur at equal rates in opposite directions. Any changes in the conditions (temperature, concentrations of the participants, pressure) may result in Gibbs energy changes and may also lead to the process occurring in direct or opposite direction. Chemists evaluate the Gibbs energy changes (ΔG) of any chemical reaction (1.3) according to the formula:

$$\Delta G^\circ = \sum n_i \cdot \Delta G^\circ_{f(\text{products})} - \sum n_i \cdot \Delta G^\circ_{f(\text{reagents})}, \tag{1.4}$$

n_i is the number of moles of species participating in the reaction (1.3).

One can find ΔG°_f (Gibbs energy formation of substance) values for many substances in chemical textbooks and encyclopedias. Some examples of corrosion reactions and changes of Gibbs energy (ΔG°_{298}) are shown in Table 1.1.

The data of ΔG°_T for the above-mentioned corrosion reactions show that iron (Fe) may corrode in air, in water with and without dissolved oxygen, and in acid because $\Delta G^\circ_T < 0$. Copper (Cu) is resistant to deoxygenated water ($\Delta G^\circ_T > 0$), but may corrode in water containing dissolved oxygen ($\Delta G^\circ_T < 0$). Gold (Au) is resistant to water ($\Delta G^\circ_T > 0$). Copper is resistant to hydrochloric acid (HCl) if there is no dissolved oxygen in the acid ($\Delta G^\circ_T > 0$), but zinc (Zn) and iron (Fe) will corrode in the same acid ($\Delta G^\circ_T < 0$). But if dissolved oxygen is present in the hydrochloric acid, copper may corrode ($\Delta G^\circ_T < 0$). One can read more about the thermodynamics of oxidation of iron and carbon steels in water in Appendix A. It is important to emphasize several points related to the prediction of corrosion of metals by means of thermodynamic Gibbs energy values:

(a) The negative value of Gibbs energy changes (ΔG°_T) points out only the possibility of corrosion reaction and not its obligatory occurrence. We cannot even decide about the probability of the occurrence of a corrosion reaction based only on Gibbs energy values.

(b) The value of the Gibbs energy change (ΔG_T°) does not show the reaction rate. We are not able to predict if a certain corrosion process will occur slowly or rapidly. For example, based on two values, −1397.2 kJ/mol for the corrosion of iron in water (containing dissolved oxygen), and −84.9 kJ/mol for the dissolution of iron in hydrochloric acid, we are not able to say which corrosion reaction will occur faster. We may even be confused, because you know that iron dissolves faster in hydrochloric acid than in water. However, the Gibbs energy change for the corrosion of iron in water 16 times more negative than that for the corrosion of iron in hydrochloric acid. The negative value of the Gibbs energy change shows only how far this particular reaction is from equilibrium. There is no way to calculate theoretically and to predict the rate of a corrosion reaction. In order to determine the rate of a corrosion process, we have to carry out an experiment.

(c) Fortunately, the rates of many corrosion reactions are very low. For example, in spite of the thermodynamic possibility of corrosion of iron and aluminum in air, in water and in soil, they are widely used in these media. Why? Because kinetic restrictions exist for the realization of the thermodynamic possibility of corrosion of metals such as iron, aluminum, magnesium, chromium, nickel, and titanium. If these pure metals or their alloys are in media containing oxygen (atmosphere, water, or soil), they react immediately with oxygen, and a tenacious dense film of metal oxide is formed on the metal's surface which protects them from subsequent corrosion (except iron). That is, kinetic restrictions prevail over thermodynamic possibility. Iron and its alloys in contact with oxygen also give tenacious oxide films, but they are very porous and do not protect against the ingress of water and oxygen, and the subsequent corrosion reaction with water and oxygen. Certainly we may predict the possibility of corrosion of metals under particular conditions by means of thermodynamic principles, but only kinetics (dynamics of chemical reactions) will define the real rate of corrosion processes and may predict how long a metallic object will resist environmental attack.

We shall discuss this in the next section.

1.4 Corrosion Kinetics

Corrosion kinetics is a discipline dealing with the definition of corrosion rate, influence of various factors on corrosion rate, and determination of the corrosion mechanism. In other words, kinetics deals with the dynamics of chemical reactions (including corrosion). If we know the mechanism of a corrosion reaction, we may be able to follow and to control its rate and its direction. In any chemical reaction, the substances originally present disappear, and substances that were not initially present appear. The properties that were initially observable are no longer observed, and properties not originally observable are now noted. In corrosion, a substance called a metal, with its unique properties (dense, solid material with high electrical and thermal conductivity, ductile and malleable, lustrous), and water, and oxygen with their unique properties, all of which were capable of detection before corrosion occurred, cannot be detected after corrosion. Instead, after corrosion of iron in water, which is described by the reaction

$$4Fe_{(s)} + 6H_2O_{(l)} + 3O_{2(g)} \rightarrow 4Fe(OH)_{3(s)}, \tag{1.5}$$

only the properties of rust (reddish-brown ferric hydroxide) can be detected.

Corrosion object (metal) disappears after full dissolution (after the corrosion process has finished). If iron comes into contact with hydrochloric acid,

$$Fe_{(s)} + 2HCl_{(aq)} \rightarrow FeCl_{2(aq)} + H_{2(g)}, \tag{1.6}$$

only the properties of the salt ferrous chloride and hydrogen gas can be detected.

If iron comes into contact with oxygen at ambient temperatures, magnetite Fe_3O_4 can be formed:

$$6Fe_{(s)} + 4O_{2(g)} \rightarrow 2Fe_3O_{4(s)}. \tag{1.7}$$

Only the properties of Fe_3O_4 may be detected.

Corrosion reactions may be fast or slow, complicated or relatively simple, and the modes by which they proceed are the subject of corrosion kinetics.

As in thermodynamics, we may compare the occurrence of a corrosion reaction with mechanics. A car goes from one town to another over a distance of 60 km. It is clear that it may travel for half an hour, for an hour, and even for two hours. Driving time depends on inner and outer factors. Inner factor: engine capacity. Outer factors: the state of the road (highway or simple road, weather – sunny, rainy, or foggy), and the state of the driver (tired or rested, object of the trip – business or tourist).

A similar situation occurs with corrosion reactions. Fast or slow – this depends on inner and outer factors. Inner factors: type of metal and its properties, such as metallurgy (crystallography, amorphous, inclusions, heterogeneity), presence of surface film, mechanical properties (presence of stresses), and thermal treatment. Outer factors: type of environment (water, acid, alkali, salt, soil, atmosphere, etc.), temperature, flow rate, and others (presence of microorganisms). Not only does the corrosion rate depend on inner and outer factors but so does the corrosion mechanism.

How to define corrosion rate? This is the main question of corrosion kinetics.

1.4.1 Measuring the Corrosion Rate

There are various methods of measuring the corrosion rate based on the type of corrosion reaction and corrosion products formed on metal surfaces. First of all, we have to define what we mean by "corrosion rate". The corrosion rate is the amount of corrosion occurring in unit time. We may measure it in different units according to the reactions (1.5), (1.6) or (1.7). If we weigh a sample of iron before contact with the environment and after some immersion period (not very long – we shall define this period later), we observe that rust is formed after contact with water as in reaction (1.5). After mechanical cleaning of the iron sample from rust, we weigh it again and determine the weight decrease. If an iron strip is immersed in hydrochloric acid (see reaction (1.6)), we do not observe corrosion products, because iron turns into soluble salt (ferrous chloride), and the iron surface will be always clean during its immersion in an acid up to complete iron dissolution and its disappearance. Therefore, we need not clean the iron surface from corrosion products after a period of immersion in the hydrochloric acid, and after weighing of a sample of the remaining iron, we determine its weight decrease. Now we may define the corrosion rate as a weight change (in this case weight loss) per unit area A per unit time T:

$$k = (M_i - M_f)/(A \cdot T), \tag{1.8}$$

k is the corrosion rate, gram/(cm^2· hour); M_i and M_f is the initial and final mass of a piece of iron, gram; A is the area of iron sample, cm^2; T is the period of iron-sample immersion in solution in hours.

The advantages and disadvantages of this method of measuring the corrosion rate will be described in Section 5.2.1. When we weigh a piece of iron after contact with oxygen gas in reaction (1.7) after some period, we determine an increase of weight because of the formation of magnetite (Fe_3O_4) on the iron surface. It is not simple (but possible!) to clean the iron surface from this dense magnetite film without removing the pure iron under the film. Therefore, the corrosion rate of metals in gases is usually defined as a gain of weight per unit area per unit time. It is very interesting to mention that this corrosion reaction of metals with oxygen at high temperatures was used by the French chemist Antoine Laurent Lavoisier in the 18th century for the definition of the main law of chemistry: the Law of Mass Conservation. Thus, corrosion of metals by oxygen at high temperatures helped the evolution of science and the change from alchemy to chemistry.

The corrosion reactions (1.5) and (1.6) show that we may also measure a decreasing of dissolved oxygen concentration in water for reaction (1.5), and increase of ferrous ion concentration in hydrochloric acid, or the volume of the hydrogen gas formed in reaction (1.6). Then we may recalculate these data for the quantity of corroded iron.

1.4.2 Duration of Corrosion Test

It is very important to define the duration of a corrosion test. For example, we would like to know the corrosion resistance of iron in hydrochloric acid, in water, and in gasoline. It can be immersed for a second or for an hour in the acid. We shall not find a change in the weight of the iron strip in the first case and our conclusion that iron is resistant to hydrochloric acid will be false. We shall not find any iron strip remaining in the second case. The period of full dissolution of an iron strip in acid as any chemical process depends on the following kinetic parameters: initial weight of iron and area contact with acid, its concentration, temperature, and agitation. It

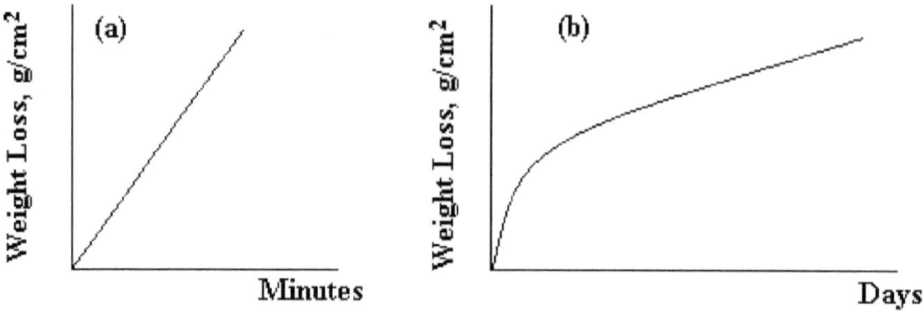

Fig. 1.6 Kinetic curves for corrosion of iron in hydrochloric acid (a) and in water (b).

is clear that there is some period between a second and an hour for measuring the corrosion rate of iron with acid. This means that we have to sink several iron strips in acid and to take out one strip every 5, 10 or 20 minutes. We have to weigh these iron strips after immersion periods and to determine weight loss per unit area of iron strip (gram per cm^2) (Figure 1.6a). Unlike acid, more time is needed for measuring the corrosion rate of iron in water, and immersed periods must be increased to 1, 2, 3 up to 15, and even 30 days. This period for iron is defined by water corrosiveness (chemical content of water) and environmental conditions (temperature and agitation). As a result, we build kinetic curve (graph) "weight loss per unit area (corrosion loss) – immersion period" (Figure 1.6b).

The form of a kinetic curve depends on inner factors (metal type, surface conditions) and outer factors (environment). The kinetic curve as shown in Figure 1.6 is typical for the corrosion of iron in acids (curve "a"), and in water, aqueous solutions, air, and soil (curve "b"). Figure 1.6 shows that iron corrodes in acids with a constant rate to full disappearance. Corrosion of iron in water slows down as corrosion products (rust) are formed on iron surface and protect (but not fully) it from further corrosion. Unlike acid and water, if an iron strip would be immersed in gasoline for several months and even years, we would not find any changes in the weight of iron. Iron is resistant to pure gasoline. If you want to define the corrosion rate of copper, aluminum, and their alloys in water, the weighing of copper and aluminum strips will also not show marked changes in weight, because copper and aluminum are resistant to general corrosion in water. In such cases, some more complicated methods of analytical chemistry are used (see Chapter 5).

Creation of a kinetic curve "weight loss – time" is an initial stage in the determination of the corrosion rate of metals. Using Equation (1.8), we may calculate the average corrosion rate for any immersion period. Engineers usually want to know the thickness loss of the equipment. It is very simple to recalculate the "weight loss" into "thickness loss", or "corrosion penetration per unit time", knowing the density d of the metal.

$$k_p = (k \cdot 10 \cdot 8760)/d, \tag{1.9}$$

where k_p is the penetration of corrosion, mm/year; k is the corrosion rate defined in Equation (1.8), gram/(cm^2·hour); d is the density of metal, gram/cm^3; 10 is the conversion of cm into mm; 8760 is the conversion of hours into years.

We discussed the relatively simple "weight loss method". There is another way for the determination of the corrosion rate of metals in water and in an aqueous solution of electrolytes – the electrochemical methods, which are more complicated, less precise and less reliable, but are very fast. We can define the corrosion rate of metals in electrolytes in 10 to 20 minutes by means of the electrochemical methods which are described in Section 5.4.

1.5 Corrosion Mechanisms

What do we mean by "mechanism of corrosion reaction"? We mean the behavior of a metal, or the way a metal reacts with an environment. This behavior may be simple and consist of one stage, for example, corrosion of iron in the oxygen atmosphere at high temperature. The corrosion reaction may be more complicated and consist of two and more stages, for example, when iron comes into contact with water or with hydrochloric acid. If two dissimilar metals, iron and zinc, contact together in salt water, or metal is under stress in some environment, the corrosion reaction may be more complicated.

In spite of many different metals (about 80) and many different environments (certainly more than 80!), all corrosion reactions are described by two mechanisms: with and without the formation of electric current and electric potential at a metal surface. The first mechanism (without the formation of electric current) is realized if metals come into contact with non-electrolytes – substances which do not conduct electric current. The second mechanism (with the formation of electric current and potential) is realized if metals come into contact with electrolytes – substances which conduct electric current. In order to understand the difference between these two mechanisms, we should be familiar with them both.

1.5.1 Electrolytes and Non-Electrolytes

Electrolytes. Metals are good conductors of electricity because of "free" electrons, which characterize a metallic bond. Some solutions (salts, acids and alkalies dissolved in water or other polar liquids) and liquids may also conduct electric current. There are no "free" electrons in such liquids like in metals or in alloys. There are positively and negatively charged particles called "ions" in these liquids, and they are responsible for the conductance of electric current. *Electrolytes are the substances whose water solutions conduct electric current on account of free ions.* Pure water is not an electrolyte because of the poor dissociation of water molecules.

Non-electrolytes. *Non-electrolytes are the substances that do not dissociate into ions and do not conduct electric current being dissolved in some solvent.* Example: Sugar or acetone dissolved in water. They are dissolved as molecules in water, do not dissociate into ions, and therefore do not conduct electric current.

Metals behave in a different way (from an electrochemical point of view) when in contact with electrolytes and non-electrolytes.

1.5.2 Corrosion of Metals by Non-Electrolytes

Non-electrolytes exist in gaseous (O_2, Cl_2), liquid (Br_2) and solid (sugar) states. If metals come into contact with any dry gas or liquid non-electrolyte, for example, with oxygen in a dry atmosphere (without water vapors), or with dry gaseous chlorine (Cl_2), dry liquid bromine (Br_2), or dry liquid sulphur (S_8), corrosion (chemical reaction) occurs in one stage. Metals give their outer electrons to non-metals (O_2, Cl_2, Br_2 or S_8) and are oxidized in one reaction:

$$Fe_{(s)} + 3/2\ Cl_{2(g)} \rightarrow FeCl_{3(s)}, \tag{1.10}$$

$$Fe_{(s)} + 3/2\ Br_{2(l)} \rightarrow FeBr_{3(s)}, \tag{1.11}$$

$$8Fe_{(s)} + S_{8(l)} \rightarrow 8FeS_{(s)}. \tag{1.12}$$

Corrosion, or chemical reaction (1.7), is an example of such processes. The metals' oxidation by oxygen at high temperature is a widespread corrosion process occurring in furnaces and engines. Some metals react with non-electrolyte organic solvents in a similar way. For example, the Grignard reagents are prepared by *corrosion* of magnesium in organic halide under a nitrogen atmosphere:

$$CH_3I_{(l)} + Mg_{(s)} \rightarrow CH_3MgI_{(l)}. \tag{1.13}$$

Metals contacting with non-electrolytes do not have electric potential at their surfaces.

Usually, such corrosion reactions occur under dry conditions: without water or, more precisely, without electrolyte. Characteristic feature of this corrosion mechanism is the absence of electric current and electric potential on the metal surface during corrosion.

1.5.3 Corrosion of Metals in the Presence of Electrolytes

If we put a piece of metal into a polar solvent (water or alcohol), or electrolyte solution (salt, acid, or alkali dissolved in water), the metal surface acquires electric

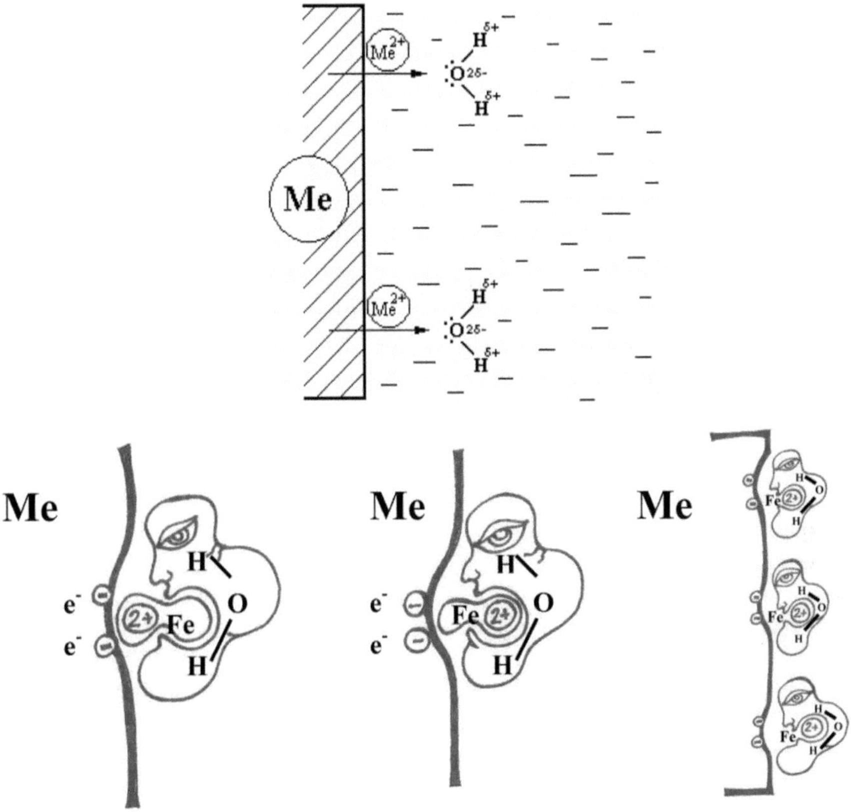

Fig. 1.7 Influence of polar water molecules (H_2O) on the metal (Me) atoms in a crystal. Water "eats" ferrous cations Fe^{2+} (Me^{2+}). $\mathbf{e}^-$ denotes the electrons. δ^- and $2\delta^+$ are the charges on H and O atoms, respectively, of H_2O. The lower pictures are by the artist Vladimir Frenkel.

potential. Why? Polar molecules of water have a surplus (redundant) negative charge on the oxygen atom and positive charges on the hydrogen atoms.

Electrical forces arising among water molecules and metal atoms compel metal ions to abandon their places in the crystal and to move into the solution as cations (Figure 1.7).

Water molecules do not permit these metal cations, which left their places in the crystal and were currently disposed at the metal–water boundary, to return into crystal. More precisely, some metal cations return into a crystal, and there is equilibrium between metal atoms in the crystal and metal cations disposed in the solution at the border with a metal. These metal cations form a positive layer in the surroundings of water molecules at the metal–water boundary. Electrons (e^-) are not able to move into the solution, and they remain in the outer metal surface close to the positive metal cations' boundary layer arranged in the water. As one can see in Figure 1.8,

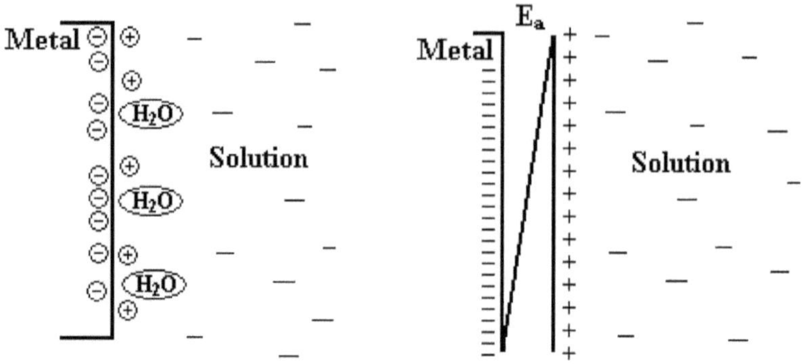

Fig. 1.8 The *electric double layer* (capacitor) formed at the metal surface in water. ⊖ – electrons; ⊕ – metallic cations. E_a – absolute electric potential.

a capacitor is formed. This is similar to a device for the concentration of electric charges.

The formed capacitor at the metal–water boundary is called an *electric double layer* – a term suggested by the German physical chemist Hermann Ludwig Ferdinand von Helmholtz in 1879. Therefore, this layer is called the *Helmholtz layer*. What is the main characteristic of the *electric double layer* in the capacitor? According to the law of neutrality, the absolute charge value Q_m formed by the remaining electrons at the metal surface must be equal to the charge Q_s formed by metal cations that formed in the solution close to the metal surface:

$$|-Q_m| = |Q_s|. \tag{1.14}$$

This charge is proportional to the electric potential E_a formed in the electric double layer:

$$Q_m = C \cdot E_a, \tag{1.15}$$

where C is the coefficient of proportionality, or capacity of a capacitor.

The electric potential E_a is an *absolute electric potential* formed on the metal surface in the solution of the electrolyte. It is impossible to measure or to calculate an absolute electric potential E_a. All 80 metals found in nature have their own electric potentials when immersed in an electrolyte with the same metal cations. Such electric potentials are similar to the different fingerprints of every person around the world. If we cannot measure the absolute electric potential on the metal surface, how can we differentiate the metals? We should choose a relative reference point for comparison.

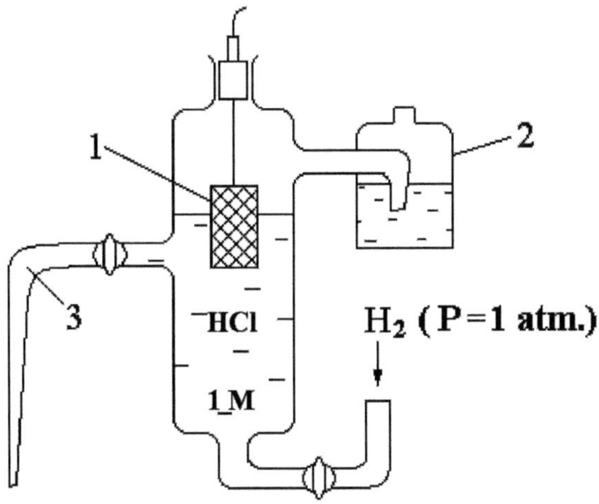

Fig. 1.9 The standard hydrogen reference electrode (SHE). (1) Platinum; (2) water plug; (3) electrolytic key. 1M HCl – Hydrochloric acid of concentration 1 mol/liter.

1.5.4 Reference Electrodes

In order to measure the electric potential formed in the electric double layer at the metal–electrolyte boundary, we have to take any electrode with some electric constant potential and accept it as a conventional reference value. Scientists took a piece of platinum (inert metal in many media!) and placed it in a glass cell with an aqueous solution of hydrochloric acid (HCl) of concentration 1 mol/liter. Hydrogen gas (H_2) at pressure 1 atm was introduced into this cell (Figure 1.9).

The following reaction occurs at the platinum surface:

$$2H_3O^+_{(aq)} + 2e^- \leftrightarrow H_{2(g)} + 2H_2O_{(l)}. \tag{1.16}$$

Usually the platinum electrode is covered by black dispersed platinum in order to accelerate the action of reaction (1.16) reaching equilibrium. One half of a platinum surface must be immersed in an aqueous solution of HCl and half of an electrode is in a H_2 gas (see Figure 1.9). As a result of reaction (1.16), the surface of the platinum acquires some electric potential E_{H_3O+/H_2}. Nobody knows this potential value, and it is accepted that the electric potential E_{H_3O+/H_2} is equal to zero at any temperature. This is the standard hydrogen reference electrode (SHE). Thus, if we connect it to any metal immersed in an electrolyte solution, we may measure and compare all absolute electric potentials formed on the surfaces of different metals (Figure 1.10).

Zinc electrode is an example of the electrode possessing negative electric potential regarding the SHE. But copper, silver, gold, platinum, and palladium are metals with electric potentials higher than the potential of the SHE. We do not know the real electric potentials on the metals' surfaces, but we know them regarding the con-

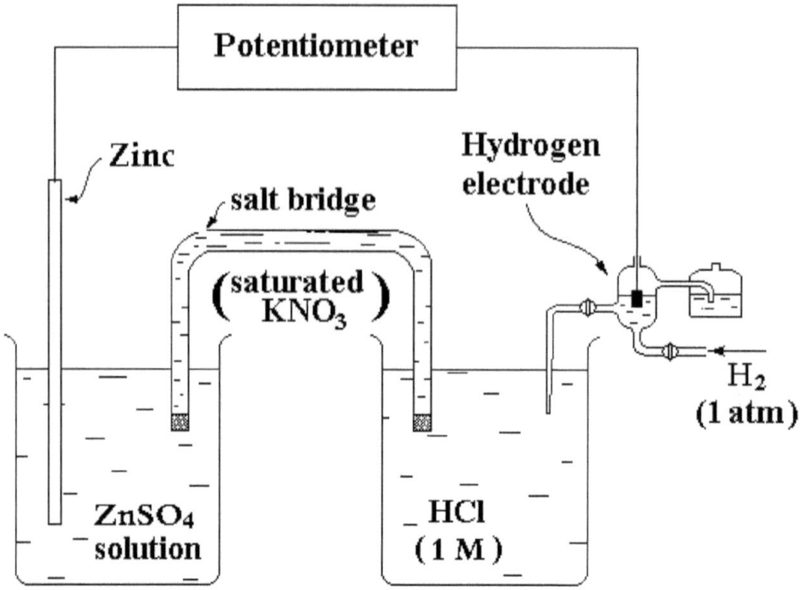

Fig. 1.10 The measuring of the electric potentials on the metals' surfaces by means of the standard hydrogen reference electrode.

ventional standard electrode and, therefore, different electrode potentials of various metals may be compared. The SHE is not convenient for use, because it is not simple and is not safe to use with hydrogen gas and hydrochloric acid as a platinum surface is very sensitive to the impurities contained in the gas and in the acid solution (electrode potential on the platinum surface may change). Therefore, safe reference electrodes were invented. The silver–silver chloride and calomel reference electrodes found wide use in laboratory corrosion measurements, especially in solutions containing chlorides. The main requirement of the reference electrodes is a constant electric potential formed on the electrode's surface. All the above-mentioned reference electrodes comply with this demand. The copper–copper sulphate reference electrode is also very simple and convenient to use. This electrode consists of a copper strip immersed in saturated cupric sulphate ($CuSO_4$) solution. A *saturated* solution is a solution with maximum concentration of a dissolved substance, and its residue is in solid state, and cannot be dissolved in water under appropriate conditions. The Dead Sea is an example of a saturated solution of many salts, because of the salt crystals observed in the water. The following reaction occurs on a copper electrode surface in a saturated solution of cupric sulphate ($CuSO_4$):

$$Cu^{2+}_{(aq)} + 2e^- \leftrightarrow Cu_{(s)}. \tag{1.17}$$

The reversible electrode potential of this reaction is plus 0.337 V to the SHE and does not change with time. We may carry this electrode in the pocket and do not

Table 1.2 Reference electrodes used in corrosion measurements.

Reference Electrode	*Me/Me^{n+} system*	*Electrolyte*	*Potential E_{SHE} at 25°C (V)*	*Temperature Dependence (mV/°C)*	*Application*
$Cu\text{-}CuSO_4$ (Copper-copper sulfate)	Cu/Cu^{2+}	Saturated $CuSO_4$	+0.316	0.97	Soils, water
Ag-AgCl (Silver-silver chloride)	Ag/Ag^{+}	Saturated KCl	+0.202	1.0	Saline and fresh water
Saturated Calomel	Hg/Hg_2^{2+}	Saturated KCl	+0.242	0.65	Water, laboratory
1 M Calomel	Hg/Hg_2^{2+}	1 M KCl	+0.290	0.24	Laboratory

Note: $E_{\mathrm{SHE}} = 0$ Volt (electrode potential of standard hydrogen electrode).

need to worry. Of course, it must be tightly closed in order to prevent any leakage through the membrane. The copper–copper sulphate reference electrode is widely used for measuring the electrode potentials of equipment under cathodic protection. This electrode is not used in sea water because of the possible deterioration of the copper sulphate solution with chloride anions. In these cases, the silver–silver chloride reference electrode is used for measuring the electrode potentials of steel structures, equipment, and ships under cathodic protection in sea water. Some reference electrodes, their electrode potentials regarding the SHE, temperature dependence, and application are shown in Table 1.2.

1.5.5 Electromotive Force Series

If we have some reference electrode, we may compare all metals according to their standard electrode potentials (electric "fingerprints"). All metals are arranged in the electromotive force series related to the SHE which is accepted as zero (Table 1.3). These electrode potentials are called *reversible* and are different from *irreversible* ones (see Appendix B). The metals in the lower part of this table (lithium, potassium, calcium, sodium, magnesium, aluminum, etc.) have negative electrode potentials regarding SHE, which means that they are not stable in water solutions and are prone to oxidizing (corrosion). The metals in the upper part of this table (gold, platinum, palladium, silver, etc.) have positive electrode potentials regarding SHE, which means that they are stable in water solutions and are not prone to oxidation. Some gases (helium, neon, argon, krypton, xenon, and radon) are called "noble" because they are stable and do not react with other elements or substances under

Table 1.3 Electromotive force series.

Electrode reaction	*Standard potential at 25°C, Volts versus SHE*
$Au^{3+}_{(aq)} + 3e^- \leftrightarrow Au_{(s)}$	1.50
$Pt^{2+}_{(aq)} + 2e^- \leftrightarrow Pt_{(s)}$	1.200
$Pd^{2+}_{(aq)} + 2e^- \leftrightarrow Pd_{(s)}$	0.987
$Hg^{2+}_{(aq)} + 2e^- \leftrightarrow Hg_{(s)}$	0.854
$Ag^{+}_{(aq)} + e^- \leftrightarrow Ag_{(s)}$	0.800
$Hg_2^{2+}{}_{(aq)} + 2e^- \leftrightarrow 2Hg_{(s)}$	0.789
$Cu^{+}_{(aq)} + e^- \leftrightarrow Cu_{(s)}$	0.521
$Cu^{2+}_{(aq)} + 2e^- \leftrightarrow Cu_{(s)}$	0.337
$2H^{+}_{(aq)} + 2e^- \leftrightarrow H_{2(g)}$	(Reference 0.000)
$Pb^{2+}_{(aq)} + 2e^- \leftrightarrow Pb_{(s)}$	- 0.126
$Sn^{2+}_{(aq)} + 2e^- \leftrightarrow Sn_{(s)}$	- 0.136
$Ni^{2+}_{(aq)} + 2e^- \leftrightarrow Ni_{(s)}$	- 0.250
$Co^{2+}_{(aq)} + 2e^- \leftrightarrow Co_{(s)}$	- 0.277
$Tl^{+}_{(aq)} + e^- \leftrightarrow Tl_{(s)}$	- 0.336
$In^{3+}_{(aq)} + 3e^- \leftrightarrow In_{(s)}$	- 0.342
$Cd^{2+}_{(aq)} + 2e^- \leftrightarrow Cd_{(s)}$	- 0.403
$Fe^{2+}_{(aq)} + 2e^- \leftrightarrow Fe_{(s)}$	- 0.440
$Ga^{3+}_{(aq)} + 3e^- \leftrightarrow Ga_{(s)}$	- 0.53
$Cr^{3+}_{(aq)} + 3e^- \leftrightarrow Cr_{(s)}$	- 0.74
$Cr^{2+}_{(aq)} + 2e^- \leftrightarrow Cr_{(s)}$	- 0.91
$Zn^{2+}_{(aq)} + 2e^- \leftrightarrow Zn_{(s)}$	- 0.763
$Mn^{2+}_{(aq)} + 2e^- \leftrightarrow Mn_{(s)}$	- 1.18
$Zr^{4+}_{(aq)} + 4e^- \leftrightarrow Zr_{(s)}$	- 1.53
$Ti^{2+}_{(aq)} + 2e^- \leftrightarrow Ti_{(s)}$	- 1.63
$Al^{3+}_{(aq)} + 3e^- \leftrightarrow Al_{(s)}$	- 1.66
$Hf^{4+}_{(aq)} + 4e^- \leftrightarrow Hf_{(s)}$	- 1.70
$U^{3+}_{(aq)} + 3e^- \leftrightarrow U_{(s)}$	- 1.80
$Be^{2+}_{(aq)} + 2e^- \leftrightarrow Be_{(s)}$	- 1.85
$Mg^{2+}_{(aq)} + 2e^- \leftrightarrow Mg_{(s)}$	- 2.37
$Na^{+}_{(aq)} + e^- \leftrightarrow Na_{(s)}$	- 2.71
$Ca^{2+}_{(aq)} + 2e^- \leftrightarrow Ca_{(s)}$	- 2.87
$K^{+}_{(aq)} + e^- \leftrightarrow K_{(s)}$	- 2.93
$Li^{+}_{(aq)} + e^- \leftrightarrow Li_{(s)}$	- 3.05

Note: SHE – standard hydrogen electrode.

usual conditions. By analogy with the "noble" gases, metals possessing high positive electric potentials are stable to oxidation (corrosion), and are called "noble" metals. The more the "desire" of a metal to oxidize (to get away from the crystal), the greater the negative charge will be on the metal surface (because of electrons remaining on the metal surface during the exit of metal cations from the crystal into the solution). The more the "desire" of the metal cations to reduce (to return to the crystal), the greater the positive charge will be on the metal surface because of the neutralization of the electrons by the metal cations approaching, or returning from, the solution, and additional attaining of positive charges.

The more a metal is placed in the upper part of the electromotive force series, the more its resistance to corrosion. It does not mean that the "noble" metals do not corrode at all, or "non-noble" metals must corrode in any medium. We know that aluminum and magnesium are widely used in the aircraft and spacecraft industry, building, machinery, automobiles, utensils, etc. Vice versa, gold, platinum, palladium and silver dissolve well (corrode!) in some solutions, for example, in *aqua regia* ("royal water" – a mixture of concentrated nitric and hydrochloric acids), and in cyanide solutions. This means that the electromotive force series has thermodynamic significance and therefore foresees only the possibility of the oxidation (corrosion) of metals, and does not predict a corrosion rate and mechanism. Once more, thermodynamics points out the possibility of corrosion, but there are sometimes kinetic limitations for the realization of thermodynamic chance.

1.5.6 Electrochemical Corrosion Mechanism (Corrosion by Electrolytes)

Electrochemical corrosion is a process occurring between a metal and the electrolyte environment not in one electrochemical reaction, and the corrosion rate depends on the electric potential on the metal surface.

Why does electrochemical corrosion not occur in one electrochemical reaction? Let us place an iron strip electrode in a ferrous sulphate solution and, separately, a copper strip electrode in a cupric sulphate solution (Figure 1.11). These solutions do not contain dissolved oxygen.

Each metallic electrode immersed in the solution of its salt is called a *half-cell*. We link the two *half-cells* by means of a salt bridge (in order to form an electric closed circuit) and a potentiometer (in order to measure the electric difference potential between the two *half-cells*). What will happen with iron and copper electrodes if they came into electric contact? Let us examine the reversible potentials for two possible reactions occurring on the metal surfaces (see Table 1.2).

$$Fe^{2+}_{(aq)} + 2e^- \rightarrow Fe_{(s)}, \quad E^\circ_{Fe_2+/Fe} = -0.44\ V, \qquad (1.18)$$

$$Cu^{2+}_{(aq)} + 2e^- \rightarrow Cu_{(s)}, \quad E^\circ_{Cu_2+/Cu} = +0.337\ V. \qquad (1.19)$$

The electric potential of reaction (1.18) is less than that of (1.19). Therefore, the iron electrode is less "noble" than the copper one when contacting together in an electrolyte solution. The iron will oxidize, or corrode, according to reaction:

$$Fe_{(s)} \rightarrow Fe^{2+}_{(aq)} + 2e^-. \qquad (1.20)$$

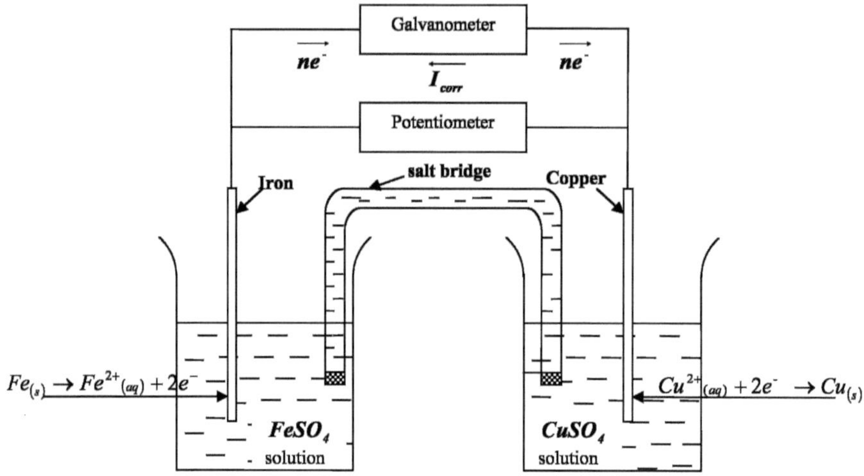

Fig. 1.11 An electrochemical corrosion cell consists of iron and copper electrodes. I_{corr} is the corrosion current.

This reaction is called *anodic*, and a corroded (oxidized) electrode is called an *anode*. The iron ions retreat from the anode into the solution and the anode corrodes. The electrons released in reaction (1.20) are accumulated on the anode's surface and result in a negative charge formation on the surface. These electrons are "looking for" a place with higher electrical potential. Such a place is a copper electrode with the potential +0.337 V. Therefore, accumulated electrons "flow" from the iron anode through the galvanometer to the copper electrode (see Figure 1.11). An electric current, called a corrosion current, appears in this process. The electrons do not "flow" in the solution! What happens at the copper electrode? The Cu^{2+} cations in the aqueous sulphate solution move toward the copper electrode and react with the free electrons released in the anodic process (1.20) and that are coming from the iron anode:

$$Cu^{2+}_{(aq)} + 2e^- \rightarrow Cu_{(s)}. \tag{1.21}$$

This reaction is called *cathodic*. The copper electrode which is the place for the reduction of Cu^{2+} cations by the electrons to copper neutral atoms is called a *cathode* and does not corrode.

This example shows that neutral atoms of iron are oxidized and iron electrode corrodes in the anodic process, and copper cations are reduced to the neutral copper atoms in the cathodic process. Now we see that corrosion of metals in the electrolytes is a process including two reactions, anodic and cathodic, which take place simultaneously. The anodic reaction cannot occur without a cathodic one. Thus, electrochemical corrosion occurs because of different electrode potentials on dissimilar metal surfaces in the electrolyte solution.

One may ask if a copper strip would be immersed in pure water and not in a cupric sulphate solution, that is there would be no copper cations Cu^{2+} taking part in the cathodic process, what would happen? Other particles exist in aqueous solution and are able to gain electrons on the copper surface: dissolved oxygen (O_2) and hydronium cations (H_3O^+). But we shall discuss later their participation in a cathodic "absorption" of electrons.

The combination of two dissimilar metals (electrodes) in a general electrolyte as observed in Figure 1.11, is called a *galvanic cell, voltaic cell, electrolytic cell*, or an *electrochemical cell*, or *voltaic battery*. Strictly speaking, the term *battery* should only be used when two or more unit cells are connected together. Such cells are important in the electrochemical corrosion. Therefore, sometimes electrochemical corrosion is called *galvanic*, or *electrolytic*.

Luigi Galvani, Professor of Anatomy in Bologna (Italy), discovered in 1780 that when a muscle of a dissected frog touches two dissimilar metals in contact, a contraction of the muscle occurs. This discovery led the Italian physicist Alessandro Volta in 1797 to use this corrosion phenomenon in the construction of *batteries* (generators of electric currents). Today, everyone uses clocks, cars, etc., and no one think that these common goods work because of corrosion of a less "noble" metal which comes into contact with a more "noble" metal in the electrolyte (see Chapter 6). Electrochemical corrosion occurs only in electrolyte solutions, mainly in water containing electrolytes, in soil, in a high humid atmosphere, and in many chemicals. There are no electrode potentials on metal surfaces in the absence of electrolytes, namely, in dry air, in dry liquid bromine, in dry organic solvents (acetone, benzene, toluene, gasoline, gas oil, fuel oil, crude oil, etc.). However, in the presence of water in air and in such liquids as bromine and organic solvents, metals may acquire the electric potentials, and electrochemical corrosion would occur. "Critical" water concentration must exist in organic solvents and in liquid bromine for corrosion to occur.

In the above-mentioned example (see Figure 1.11) iron corrodes, and copper does not. Will iron always corrode? It depends on the couple. If we take an iron in a ferrous sulphate solution as the first half-cell, and zinc in a zinc sulphate solution as the second half-cell, the electromotive force series gives (see Table 1.2):

$$E^\circ_{Fe^{2+}/Fe} = -0.44\ \text{V} \quad \text{and} \quad E^\circ_{Zn^{2+}/Zn} = -0.763\ \text{V}.$$

The electrode potential of iron is more positive than that of zinc in this galvanic cell. Zinc will be an anode and will corrode according to the reaction

$$Zn_{(s)} \rightarrow Zn^{2+}_{(aq)} + 2e^-. \tag{1.22}$$

The electrons releasing in the anodic reaction (1.22) flow through the galvanometer to the iron electrode and reduce the ferrous cations Fe^{2+} coming up from the solution to the iron surface:

$$Fe^{2+}_{(aq)} + 2e^- \rightarrow Fe_{(s)}. \tag{1.23}$$

Now an iron electrode is a cathode, has a positive charge, and does not corrode in this couple.

Fig. 1.12 Corroded anchor made of carbon steel. (For a full color version of this figure, see the Color Section)

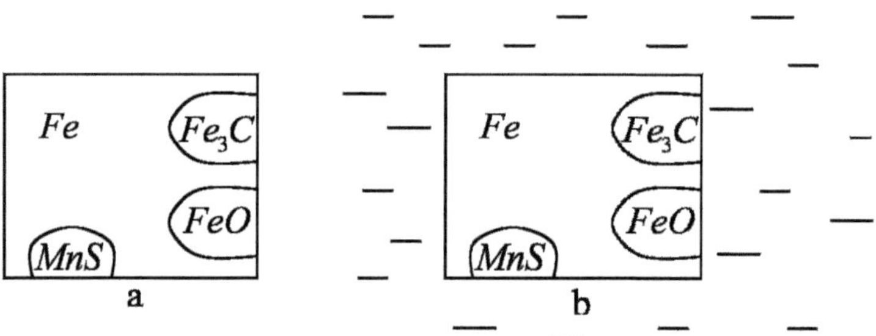

Fig. 1.13 Carbon steel surface; (a) in a dry atmosphere; (b) in the presence of liquid electrolytes (solution of NaCl in water).

1.5.7 Electrochemical Mechanism on a Single Whole Alloy or Metal

We have observed many times the corroded artifacts made of carbon steel that did not come into contact with other alloys and metals (Figure 1.12). Why and how does a single whole alloy or metal corrode?

All alloys consist of different components. For example, carbon steel contains iron (Fe) and iron carbide (Fe_3C). Various iron oxides (FeO, Fe_2O_3, or Fe_3O_4) are formed on the surface of steel (Figure 1.13).

If ironware is in a dry atmosphere, nothing takes place on its surface. If some electrolyte (aqueous solution of acid, alkali or salt) appears on steel surface, different components of steel (iron, iron carbides, iron oxides, or such impurities as Fe_2SiO_3, FeS, MnS) acquire various electric potentials. Usually iron has less electric potential value than its compounds iron carbides, iron oxides or ferrous salts.

For example, iron and iron carbide in hydrochloric acid have electric potentials of −0.5 V and +0.4 V respectively. This is a good "battery" (0.9 Volt) which appears in the presence of an electrolyte! This means that the ironware surface acquires the electrochemical heterogeneity. A similar situation occurs not only on the surface of the carbon steel alloy, but on the "pure" iron surface too. It is impossible to produce a metal of 100% purity. For example, if we manufacture a metal of 99.99999% purity, this means that there are one thousand million million (or one with 15 zeros) impure atoms (carbon in the main) in one gram of iron. This value is enough for the formation of the electrochemical heterogeneity on the "pure" iron or at any other "pure" metal surface. The idea of electrochemical heterogeneity belongs to the Swiss scientist De la Rive (1837) who investigated dissolution of zinc and its alloys in hydrochloric acid, and observed that the more the contaminates that are in zinc alloy, the greater the dissolution rate of zinc in the acid. De la Rive suggested that such contaminates were micro-galvanic cells and probably resulted in acceleration of corrosion of zinc in the acid. We have to emphasize that electrochemical heterogeneity only influences the corrosion rate. *The cause of corrosion is thermodynamic instability of metals, and not their heterogeneity. Even if the metals would be of 100% purity, they would corrode in most media because $\Delta G < 0$ for these reactions* (see Section 1.3).

Thus, pure iron on a carbon steel surface has less electric potential value than its compounds and turns into anodes in the presence of electrolytes, which corrodes and gives Fe^{2+} cations to the solution (Figure 1.14a). *An anode is an electrode where corrosion occurs.*

The electrons released in the anodic process (corrosion reaction!) flow to the iron carbide (or iron oxides, or some other contaminates) with more positive electric potentials (Figure 1.14b). These regions are the *cathodes*, and electrons reaching them are looking for some particles needed by them on the cathodic sites. There are many particles in solutions and on metallic surfaces, which desire to have the electrons liberated, and are then able to devour them (Figure 1.14c). These particles may be

(a) neutral molecules (for example, dissolved O_2, or Cl_2):

$$O_{2(g)} + 2H_2O_{(l)} + 4e^- \rightarrow 4OH^-_{(aq)}, \qquad (1.24)$$

$$Cl_{2(g)} + 2e^- \rightarrow 2Cl^-_{(aq)}, \qquad (1.25)$$

(b) cations:

$$2H_3O^+_{(aq)} + 2e^- \rightarrow H_{2(g)} + 2H_2O_{(l)}, \qquad (1.26)$$

$$Fe^{3+}_{(aq)} + e^- \rightarrow Fe^{2+}_{(aq)}, \qquad (1.27)$$

$$Cu^{2+}_{(aq)} + e^- \rightarrow Cu^+_{(aq}, \qquad (1.28)$$

(c) dissolved molecules and ions together:

$$O_{2(g)} + 4H_3O^+_{(aq)} + 4e^- \rightarrow 6H_2O_{(l)}, \qquad (1.29)$$

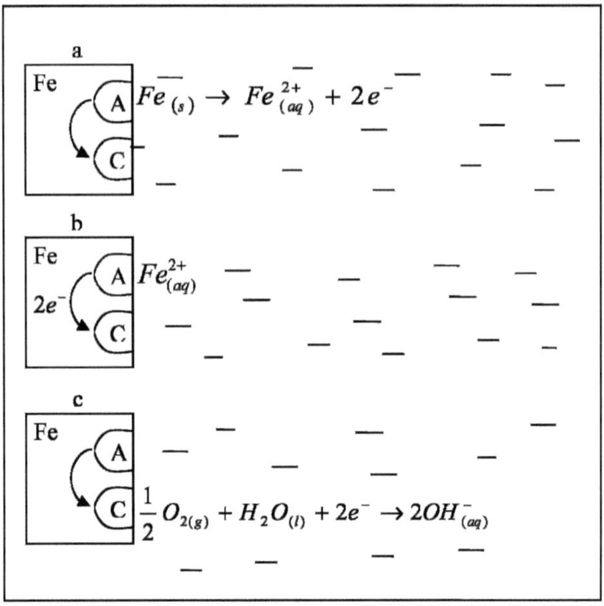

Fig. 1.14 Corrosion on a single whole carbon steel; (a) anodic process; (b) flow of electrons from an anode "A" to a cathode "C"; (c) cathodic process.

(d) oxide or hydroxide films on the metal surface:

$$Fe_3O_{4(s)} + 2e^- + H_2O_{(l)} \rightarrow 3FeO_{(s)} + 2OH^-_{(aq)}, \tag{1.30}$$

$$Fe(OH)_{3(s)} + e^- \rightarrow Fe(OH)_{2(s)} + OH^-_{(aq)}. \tag{1.31}$$

All reactions (1.24–1.31) are called *cathodic*, and they take place on the metal surface (not in the solution!) on the sites with the higher electric potentials than *anodic* zones. *A cathode is an electrode where a reduction of some species occurs.*

Oxygen contained in an atmosphere in a quantity of 21%, dissolves in all kinds of water. The solubility of oxygen in water is not great, about 0.0008%, but this quantity is enough for the existence of many kinds of fish, and as we now know in order for corrosion to occur.

When metal equipment is in contact with water, soil or atmosphere of high humidity, dissolved oxygen in water plays an important role in the cathodic reaction (1.24). This means that the cathodic process in neutral and alkali aqueous solutions on a metal surface is the reduction of dissolved oxygen, and hydroxide anions OH^- are formed on the *cathodic* sites. Scientists observed this phenomenon at the end of the 18th century – alkalization of water when iron was immersed and corroded – but they could not explain it.

Acids are used for cleaning metal surfaces before the application of a coating. Hydronium cations H_3O^+ are also desirous that electons are liberated in the anode

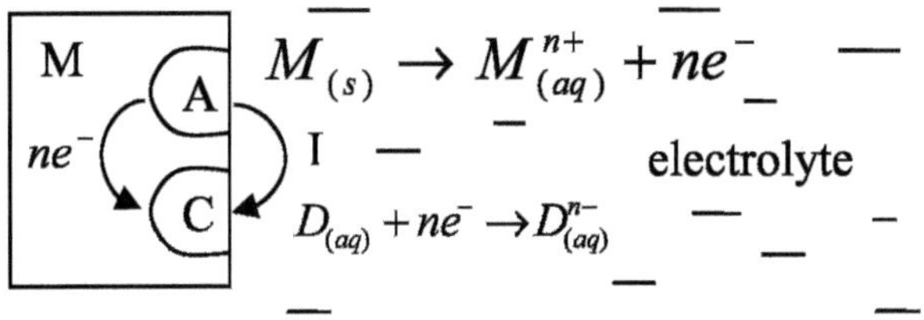

Fig. 1.15 Electrochemical corrosion mechanism. I – ionic current path; A – anode; C – cathode; M – metal; D – species (O_2, H_3O^+, Fe^{3+}) taking part in the cathodic process; ne^- – n mol of electrons.

process. Therefore, if metal is in contact with acid, the cathodic process is reaction (1.26) or (1.29) (as oxygen can dissolve in acidic solutions too). Anodic reaction of metal corrosion cannot exist alone without its pair – cathodic reaction. Here we reveal a very important way to prevent corrosion in neutral and alkali aqueous solutions – removing the main cathodic "actor" dissolved oxygen. We now understand that electrochemical corrosion occurs not in one reaction, but in two – anodic and cathodic, and they are interrelated. To conclude, we note that electrochemical corrosion consists of three processes (Figure 1.15):

1. Anodic, or oxidizing, process – passing of metal cations into solution and releasing of electrons on the metal surface:

$$M_{(s)} \rightarrow M^{n+}_{(aq)} + ne^-, \tag{1.32}$$

   M – metal; $M^{n+}_{(aq)}$ – metallic cations in solution.
2. Flow of the electrons released from the anode A (low electric potential site) to the cathode C (high electric potential site).
3. Cathodic, or reduction, process – any process that receives electrons by neutral or charged species D (see reactions (1.24–1.31)):

$$D_{(aq)} + ne^- \rightarrow D^{n-}_{(aq)}. \tag{1.33}$$

Elimination of one of these processes will decrease corrosion of a metal. The stage of electrons' flow from the anode to the cathode on a metal's surface is faster than the occurrence of anodic (1.32) and cathodic (1.33) reactions. Therefore, electrochemical corrosion occurs in two stages: anodic dissolution of the metal and cathodic reduction of some species existing in the solution on the cathodic sites of a metal surface close to the anodic zone on a single, whole metal surface (see Figure 1.15). The corroding metal is similar to the short-circuited electrochemical cell illustrated in Figure 1.11. Thus, many such "micro-cells" exist on a metal surface. The energy is produced in the "corrosion (electrochemical) cell" because of the corrosion process – electrochemical reactions occurring on the metal surface. Unfortunately, we cannot use the energy produced in the corrosion reaction on the single whole metal. This energy is dissipated, but if we couple two different metals in a common electrolyte, we can use this energy. Telegraph services in England in the beginning of 1830s needed stable sources of electrical energy. The English chemist John Frederic Daniell (1790–1845), a friend of Michael Faraday, took copper instead of iron in the galvanic couple shown in Figure 1.11, and created his famous "Daniell cell" with a voltage of 1.1 Volt. Zinc corroded in a "copper–zinc" couple, and gave a stable source of electric energy. Thus, corrosion of zinc helped Daniell to solve the problem of the telegraphic needs in electric batteries in England in 1836.

A very interesting and important question is how does the presence of electrolytes in water influence corrosion? Physicists long before the discovery of the electron in 1897 by Joseph John Thompson accepted that an electric current I is a flow of positive charges (see Figure 1.15).

This flow is opposite to the electron flow (ne^-). It was revealed that the law discovered by the German physicist Georg Simon Ohm (1827) for a solid conductor could be used for electrolytes:

$$I = E/R, \tag{1.34}$$

I – electric current; E – electric potential difference between anode and cathode; R – sum of electrical resistance of electrolyte (R_e) and metal (R_m).

$$R_e + R_m = R. \tag{1.35}$$

As electrical resistance of an electrolyte is much more than that of a metal ($R_e \gg R_m$), R_e is only used in (1.34). If corrosion products are formed on the metal surface, or there are some other films, R is the sum of electrical resistance of corrosion products (R_{products}), of films (R_{film}), and of an electrolyte (R_e):

$$R = R_{\text{products}} + R_{\text{film}} + R_e. \tag{1.36}$$

The electric current appearing on the metal surface and in the electrolyte solution between an anode and a cathode is equivalent to corrosion current I. If an electrical resistance of electrolyte (R_e) near a metal surface is a large value, corrosion current I is negligible. When we add salt (or any other electrolyte) to water, electrical resistance of the electrolyte solution (R_e) diminishes, and corrosion current I increases according to (1.34).

However, we should not think that the more salt we add, the more corrosion will occur. We may add so much salt to water that all dissolved oxygen (the main cathode "actor" in neutral aqueous solutions) will take out the solution, and corrosion will stop, but we shall discuss this phenomenon later (see Section 1.8).

Now we know how to decrease or even to prevent corrosion – to remove electrolytes, or to form isolated films on the metal surface and to increase electrical resistance of a medium near the surface of a metal in such way and, according to (1.34), to diminish corrosion current I. Films on the metal surface may consist of non-solubles in water salts, such as phosphates $Fe_3(PO_4)_2$, sulphides FeS, carbonates $CaCO_3$, oxides Fe_3O_4, or corrosion inhibitors and organic polymers (coatings).

If we measure the quantity of ions forming on the surface unit during some period in the anodic reaction (1.32), we may measure the corrosion rate. The great English scientist Michael Faraday showed us in the 1830s how to calculate the quantity m of corroded metal by means of measuring the electric current I:

$$I \cdot t = (n \cdot F \cdot m/M), \tag{1.37}$$

I – the corrosion current; t – the period of current flow; n – the number of moles of electrons involved in the metal dissolution; F – Faraday's constant; M – the molecular weight of the metal; m – the mass of corroded metal. Calculations were carried out many times according to formula (1.37), and thus an electrochemical mechanism of corrosion of metals in electrolyte solutions was proved.

Some important notes.

(a) In any *chemical reaction*, oxidation of one substance is accompanied without fail by reduction of another one (*corrosion by non-electrolytes*). If, in any *chemical reactions*, oxidation and reduction are combined in one process, in *electrochemical reactions* (corrosion by electrolytes), they represent different independent processes.
(b) We may emphasize that the fewer the imperfections and impurities in the metal, the less the corrosion rate would be. This means that the more a purity of a metal, the more its corrosion resistance. Many impurities exist at the grain boundaries of metal crystals. The amorphous, or glassy, metals (without a crystal structure) have no grain boundaries. Therefore, amorphous metals are more resistant to corrosion than conventional metals with crystal structures.
(c) If we would produce 100% pure metal, it would be resistant to corrosion. We have to mention that corrosion is a reaction between a metal and the environment. The heterogeneity of the environment (for example, differences in concentrations of dissolved oxygen, pH, and temperature gradient on the metal's surface), or mechanical stresses on a pure metal may result in the formation of anodic and cathodic zones on the metal surface (electrochemical heterogeneity), resulting in corrosion.

Some features of electrochemical corrosion.

(a) The differentiation into two electrochemical processes, anodic and cathodic, taking place simultaneously on the surface of metal.
(b) We may influence the kinetics of the electrochemical corrosion by means of changes of the electric potential of the metal (see Appendix C). If the electric potential increases, the anodic reaction (1.32) increases, and the cathodic reaction (1.33) decreases, respectively. If the electric potential decreases, the cathodic reaction (1.33) increases and the anodic reaction (1.32) decreases, respectively. This means that corrosion, or anodic reaction, may be suppressed by means of decreasing the electrode potential of the metal. This is the main principle of cathodic protection (see Section 4.2).
(c) We may define the anodic and cathodic sites on the metal surface. If a metal is immersed in water, anodic zones are acidic because of cations being formed, which may hydrolyze and form acid (H^+_{aq}):

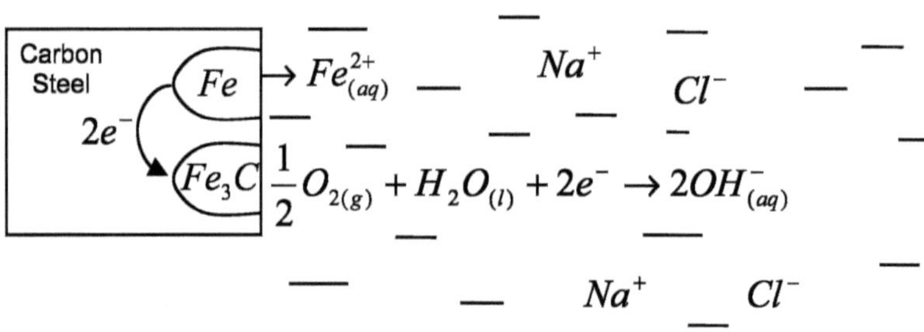

Fig. 1.16 Electrochemical corrosion of carbon steel in water and aqueous salt solutions. Iron (Fe) is an anode; cementite (Fe_3C) is a cathode.

$$Fe^{2+}_{(aq)} + H_2O_{(l)} \rightarrow Fe(OH)^{+}_{(aq)} + H^{+}_{(aq)}. \tag{1.38}$$

Cathodic zones in neutral solutions are alkaline because of the formation of hydroxide anions according to (1.24).

(d) If a metal is immersed in an acidic solution, hydrogen gas bubbles are formed on the cathodic zones according to reaction (1.26). Metallic neutral atoms are dissolved and appropriate metallic cations are formed on the anodic sites. If the cathodic isles are surrounded by anodic sites, after dissolution of anodes, cathodes are also removed. New anodes and cathodes appear. Thus, anodic and cathodic sites interchange all the time, and corrosion spreads gradually on all surfaces of the metal.

How do iron and carbon steel corrode in water and aqueous solutions of electrolytes (salts, acids, and alkalies)? This will be the theme of the next two sections.

1.6 Corrosion of Iron in Water and Aqueous Salt Solutions

Iron or carbon steel immersed in water or aqueous salt solutions will corrode according to the electrochemical mechanism. Heterogeneity of carbon steel or non-uniformity of the environment results in the formation of anodes with negative electric potential and cathodes with positive electric potential on the metallic surface (Figure 1.16).

Iron oxidizes on the anodic sites, according to reaction (1.20), and the releasing electrons flow to the cathodic sites possessing higher electric potential. What molecules and ions exist in water close to the cathodic regions and are capable of receiving the liberated electrons? Dissolved oxygen (O_2) at the concentration of 8 ppm (or 8×10^{-4} wt%), cations H^+ at a concentration of 10^{-7} mol/L (or 10^{-8} wt%) in neutral aqueous solutions, and cations Fe^{3+} and Fe^{2+} at a concentration of about 0.1 to 0.5 ppm (10^{-5} wt%) may take part in the cathodic reactions (at pH = 7):

$$O_{2(g)} + 2H_2O_{(l)} + 4e^- \leftrightarrow 4OH^-_{(aq)}, \qquad E^\circ = +0.815\ \mathrm{V}, \tag{1.39}$$

Table 1.4 Influence of real concentrations of redox species on electric potentials E for reactions (1.39–1.41) at pH = 7 at 25°C.

Reaction	Concentration of redox species, mol/L						E°, V	E, V
	P_{O2},atm	$OH^-_{(aq)}$	$H_3O^+_{(aq)}$	P_{H2},atm	$Fe^{3+}_{(aq)}$	$Fe^{2+}_{(aq)}$		
(1.39)	0.21	10^{-7}	-	-		-	+0.815	+0.805
(1.40)	-	-	10^{-7}	$5\cdot10^{-7}$		-	-0.414	-0.225
(1.41)	-	-	-	-	$8.9\cdot10^{-6}$	$8.9\cdot10^{-6}$	+0.771	+0.771

Note: P_{O2} = 0.21 atm and $P_{H2} = 5\cdot10^{-7}$ atm are the partial pressures of O_2 and H_2 in air contacting an iron electrode dipped in an aqueous solution.

$$2H_3O^+_{(aq)} + 2e^- \leftrightarrow H_{2(g)} + 2H_2O_{(l)}, \qquad E^\circ = -0.414\text{ V}, \tag{1.40}$$

$$Fe^{3+}_{(aq)} + e^- \leftrightarrow Fe^{2+}_{(aq)}, \qquad E^\circ = +0.771\text{ V}. \tag{1.41}$$

There is an electric potential E° for every electrochemical reaction. We have to emphasize that if one anodic process [$Fe_{(s)} \rightarrow Fe^{2+}_{(aq)} + 2e^-$] occurs with electric potential −0.44 V, and several cathodic reactions (in this case, three) with several electric potentials, the most possible cathodic reaction will take place when the difference between potentials of cathodic and anodic reactions, $E^\circ_{(cathodic)} - E^\circ_{(anodic)}$, will be the largest. The more positive the $E^\circ_{(cathodic)}$ value, the more possible cathodic process would occur. Therefore, among three possible cathodic processes (1.39–1.41), the reduction of oxygen is the most probable since reaction (1.39) has the highest potential and oxygen concentration is more than the concentrations of H^+ and Fe^{3+} cations in water. For a more precise decision, we should not use the standard potential E°, but real electric potentials E calculated according to the Nernst equation:

$$E = E^\circ + \frac{RT}{nF}\ln\frac{[O_x]}{[Red]}. \tag{1.42}$$

The real potentials E calculated according to (1.42) and the real concentrations of redox species are summarized in Table 1.4.

Oxygen molecules move to the cathodic regions where electrons are "waiting" for them, and reaction (1.39) occurs on the metal surface (not in the solution!) with the formation of hydroxyl anions OH^- in the vicinity of the anodic sites. Cations Fe^{2+} formed in the anodic process "meet" the products of the cathodic reaction, OH^- anions, at the border between anodic and cathodic sites, and ferrous (II) hydroxide is formed (see Figure 1.16):

$$Fe^{2+}_{(aq)} + 2OH^-_{(aq)} \rightarrow Fe(OH)_{2(s)}. \tag{1.43}$$

Cations Fe^{2+} and the hydroxide $Fe(OH)_2$ of a beautiful blue-green color may only exist in the absence of oxygen (and other oxidizers). Dissolved oxygen molecules immediately come to the ferrous (II) hydroxide and oxidize it to ferric (III) hydroxide of a reddish-brown color:

$$Fe(OH)_{2(s)} + O_{2(g)} + 2H_2O_{(l)} \rightarrow 4Fe(OH)_{3(s)}. \tag{1.44}$$

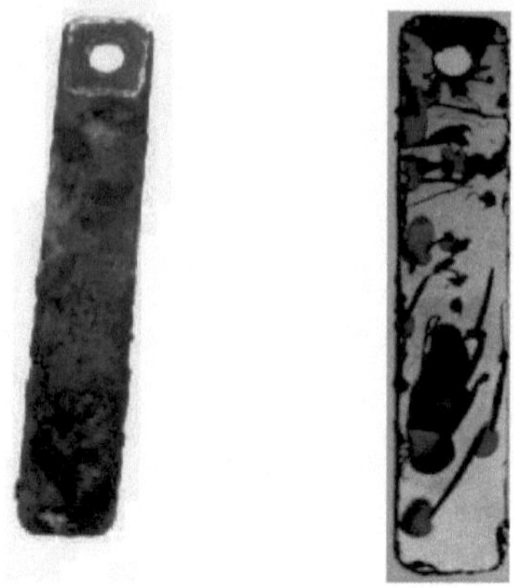

Fig. 1.17 Carbon steel after 7 days immersion in cooling water (a) and in water with the addition of sodium peroxocarbonate ($NaBO_3$) (b). The "beautiful" rust is formed on the steel's surface in a peroxocarbonate solution. (For a full color version of this figure, see the Color Section)

Of course, we are familiar with this compound – rust formed on carbon steel equipment and structures in various kinds of water (sea, ocean, river, lake, potable water, cooling water, etc.), in soil, and in the atmosphere in the presence of water vapor (see Figure 1.12). If we put a piece of carbon steel in water or in an aqueous electrolyte solution, we should observe the rust formation at the borders between the anode and cathode (Figure 1.17). The anodic and cathodic sites are not constant places on the metal surface, because the conditions both on the metal surface and in the solution are changing all the time. Iron at the anodic sites dissolves, and cathodic inclusions appear at the places of the anode. Cathodic components disappear because they, as islands, are circled by iron anodes which dissolve, and cathodes throw out the metal crystal – anodes appear instead of cathodes and vice versa. Rust may cover the anodic sites and so they become cathodic. As a result of such interchanges between anodic and cathodic zones, all the carbon steel surface will be covered by rust (see Figures 1.12 and 1.17).

We may ask if rust is a pure compound of ferric (III) hydroxide, $Fe(OH)_3$. This amorphous brown-red compound changes with time: polymerization and dehydration occur. Intermediate products are formed coloring in dirty-green colors, simultaneously containing Fe^{2+} and Fe^{3+}. Rust $Fe(OH)_3$ is not dissolved in water and forms a colloidal solution. After dehydration, *Goethite* [α-FeO(OH)] in the form of mutually intersecting round reddish-brown platelets, or needle-like brown *Lepidokrokite* [γ-FeO(OH)], or mixed oxides are formed: $FeO{\cdot}Fe_2O_3{\cdot}nH_2O$ (Figures 1.17 and 1.18):

$$Fe(OH)_3 \xrightarrow{-H_2O} \alpha\text{-FeO(OH) (Goethite) or } \gamma\text{-FeO(OH) (Lepidokrokite)}. \quad (1.45)$$

Fig. 1.18 Scanning electron microphotograph of rust formed in warm water: Lepidokrokite [γ-FeO(OH)] ($\times$ 10,000).

Some physical properties of corrosion products of iron are shown in Table 1.5.

Iron (II) oxide FeO is a gray-black compound, iron (III) oxide Fe_2O_3 is brown-red, and iron (II, III) oxide Fe_3O_4 is a black-grayish ferromagnetic compound. The first oxide FeO is well dissolved in mineral acids. The two others (Fe_2O_3 and Fe_3O_4) are poorly dissolved.

In many cases, various components of media (anions, cations, non-dissolved oxides and salts, organic molecules, and microorganisms) are incorporated in rust. Therefore if we say "rust", we mean a mixture of ferric and ferrous hydroxides and oxides with such impurities as chlorides (Cl^-), sulphates (SO_4^{2-}), carbonates (CO_3^{2-}), calcium (Ca^{2+}), magnesium (Mg^{2+}), etc. If rust is formed in the atmosphere, various pollutants may be incorporated in the rust, for example, sand (SiO_2), carbon particles (C), sulphur (S_8), etc. All these impurities incorporating in rust depend on the type of environment.

Corrosion resistance of pure iron and various kinds of carbon steels in water and neutral aqueous salt solutions is a very fascinating problem. What type of carbon steel may be chosen from the corrosion point of view? Carbon steel is an alloy containing iron and carbon at concentrations from 0.008 to 2 wt%. Carbon dissolves in iron, forming iron carbide Fe_3C called *cementite*.

In other words, carbon steel is a mixture of pure iron and iron carbide. Pure iron is a very soft metal, softer than copper. The addition of carbon improves the mechanical properties of iron. Does the carbon content influence the corrosion resistance of steel in water and in neutral aqueous solutions? The cathodic process in water is the reduction of dissolved oxygen according to reaction (1.39). When all oxygen molecules arranging near the cathode take part in (1.39), surplus of electrons at the cathode will wait for new oxygen molecules. If there is no agitation in a solution, dissolved oxygen molecules do not diffuse so quickly compared with the reduction rate on the cathodic sites. Usually, the corrosion rate is limited by cathodic reduction of oxygen or by the rate of diffusion of neutral oxygen molecules to the cathodes on a steel surface, or both. "Cathodic" inclusions (iron carbide and iron oxides) do not

Table 1.5 Physico-chemical properties of corrosion products of iron.

Chemical formula	Name	Density, g/cm^3	Color	Electro-conductive properties	Thermal behavior
FeO	Wüstite	5.4–5.7	Gray-black	Very low electro-conductivity, 10 $Ohm^{-1}\cdot m^{-1}$, Semiconductor	Decomposes to Fe and Fe_3O_4 below 570°C. Melts at 1371–1424°C
α-Fe_2O_3	Haematite	5.25	Brick-red to black	Insulator	Decomposes to Fe_3O_4 at 1457°C
γ-Fe_2O_3	Maghemite	4.88	Brown	Semiconductor to insulator	Transforms to α-Fe_2O_3 above 250°C
Fe_3O_4	Magnetite	5.20	Black-grayish	Good electro-conductivity, 10^4–10^5 $Ohm^{-1}\cdot m^{-1}$, Conductor	Melts at 1597°C
$Fe(OH)_2$	Ferrous Hydroxide	3.40	White-blue	Insulator	Decomposes* to Fe_3O_4 and H_2 at ~100°C
α-FeOOH	Goethite	4.20	Yellow, reddish-brown	Insulator	Dehydrates to α-Fe_2O_3 at ~200°C**
γ-FeOOH	Lepidokrokite	3.97	Orange	Insulator	Dehydrates to γ-Fe_2O_3 at ~200°C**

*$Fe(OH)_2$ is unstable in contact with traces of O_2 at ~20°C and transforms into α-FeOOH, γ-FeOOH or Fe_3O_4 depending on the conditions.
**The presence of water causes conversion to α-Fe_2O_3 at lower temperature.

influence the rate of oxygen reduction. The limiting stage of corrosion in this case is the rate of oxygen diffusion in a solution. This means that the nature of a cathode on a steel surface (also, kinds of carbon steel) does not influence the corrosion rate of carbon steel in water and in neutral aqueous solutions. Anodic reaction of iron dissolution cannot occur without its "couple" cathodic reaction. Therefore, if we remove oxygen from water or aqueous solution (by boiling or blowing with inert gas), corrosion will slow down and may even be stopped. We must not forget that H_3O^+ and Fe^{3+} ions are present in water and may take part in cathodic reactions (see (1.40) and (1.41)), but their concentrations in water are usually very small in order for a cathodic reaction to occur with their participation and, as a rule, do not influ-

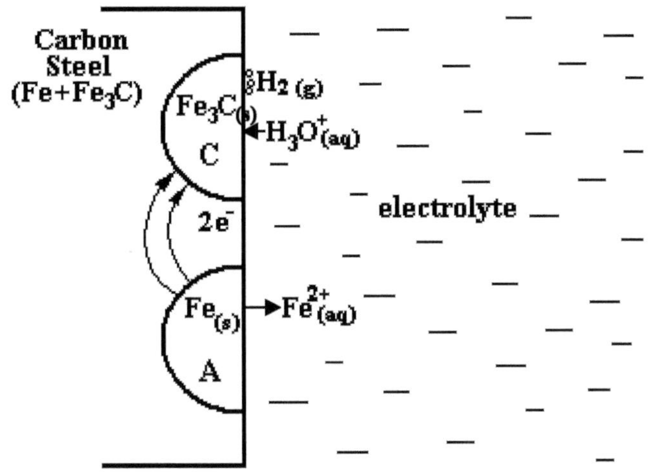

Fig. 1.19 Electrochemical corrosion mechanism of carbon steel in acid. "A" (Fe) – anode; "C" (Fe_3C) – cathode.

ence the corrosion of steel in neutral aqueous solutions. Only large concentrations of Fe^{3+} may influence the corrosion of iron (see Appendix A).

1.7 Corrosion of Iron and Carbon Steels in Acids

Acids are widely used in many branches of industry: processes of synthesis in the chemical and petrochemical industry, for chemical cleaning before galvanizing, phosphatizing, passivation, or other coating use. If carbon steel is immersed in acid, the situation is similar when it comes into contact with water. Because of the heterogeneity of carbon steel, various phases acquire different electric potentials in contact with acid. The electric potentials of iron in acid is -0.4 to -0.5 V and of iron carbide (Fe_3C) is $+0.37$ V, respectively, to the SHE.

The iron phase is an anode related to iron carbide Fe_3C (cathode), and will corrode (Figure 1.19). The electrons liberating in the anodic reaction "flow" to the cathode with higher electric potential and are looking for species which are able to receive them. What are these species? They are cations H_3O^+ (at a concentration of 1 mol-ion/l, if pH = 0), dissolved oxygen O_2 (at a concentration of about 2.5×10^{-4} mol/liter (in reality, this concentration depends on the acid concentration and temperature), and cations Fe^{3+} (at a concentration of about 2×10^{-5} mol/liter) which are formed as a result of oxidation of Fe^{2+} by dissolved O_2. Let us write three possible cathodic reactions on the carbon steel surface in acid:

$$O_{2(g)} + 4H_3O^+_{(aq)} + 4e^- \leftrightarrow 6H_2O_{(l)}, \qquad E^\circ = +1.229\ \text{V}, \tag{1.46}$$

$$2H_3O^+_{(aq)} + 2e^- \leftrightarrow H_{2(g)} + 2H_2O_{(l)}, \qquad E^\circ = 0\ \text{V}, \tag{1.47}$$

$$Fe^{3+}_{(aq)} + e^- \leftrightarrow Fe^{2+}_{(aq)}, \qquad E^\circ = +0.771\ \text{V}. \tag{1.48}$$

Corrosion reaction may occur if $E^\circ_{(\text{cathodic})} > E^\circ_{(\text{anodic})}$. The electric potentials of all three cathodic reactions (1.46–1.48) are more positive than that of the anodic dissolution of iron (−0.44 V). Therefore, when carbon steel is in contact with acid, all three cathodic processes may occur. In spite of the most positive electric potential of oxygen reduction (1.46) and high positive electric potential of ferric ions reduction (1.48), the concentration of hydronium cations H_3O^+ is so large that their reduction is the prevalent cathodic process for the corrosion of carbon steel in acids, and we observe the evolution of hydrogen gas. The total reaction of the dissolution of iron in hydrochloric acid (in the absent of dissolved O_2):

$$Fe_{(s)} + 2HCl_{(aq)} \rightarrow FeCl_{2(aq)} + H_2(g). \tag{1.49}$$

If iron or carbon steel is immersed in acid with dissolved O_2, cations Fe^{2+} are oxidized by dissolved oxygen to cations Fe^{3+} and the latter are very greedy (even more than H_3O^+) for electrons. Therefore, the cathodic reaction (1.48) plays a very important role, after some period of iron being immersed in acid. Dissolved oxygen in acid may reach the cathodic sites on metal surfaces and take part in the cathodic process (1.46). Thus, one anodic process (iron oxidation) may correspond to several parallel cathodic reactions taking part at the same time on the cathodic sites. If alloy consists of several metals, for example, stainless steel 316 consists of iron (74%), chromium (18%) and nickel (8%), and is immersed in acid, all three metallic components may dissolve. In this case, three anodic reactions occur at the same time. Electrons released in all anodic processes take part in all cathodic processes. Cathodic inclusions (for example, Fe_3C) in carbon steel influence significantly the reaction (1.47). The higher the number of cathodic sites on carbon steel surfaces, the more violently the reduction of hydronium cations H_3O^+ occurs. Therefore, the carbon steel type influences the cathodic processes and their corrosion rate, respectively. Black carbon deposits may appear on carbon steel's surface during corrosion in acid because of the following reaction:

$$Fe_3C_{(s)} + 6H_3O^+_{(aq)} \rightarrow 3Fe^{2+}_{(aq)} + C_{(s)} + 3H_{2(g)} + 6H_2O_{(l)}. \tag{1.50}$$

This is the reason why we observe black powder on a carbon steel surface after immersion in acids. This reaction helped Swedish and French chemists to define the difference between iron and carbon steel in the end of the 18th century (see Section 6.1).

1.8 Corrosion Factors

If we know the electrode potentials of metals in some electrolyte, we may predict whether metal would corrode or not. The electromotive force E is the difference between electric potentials of cathodic and anodic reactions:

$$E = E_{\text{cathodic}} - E_{\text{anodic}}. \tag{1.51}$$

This value is a thermodynamic parameter and relates to Gibbs energy by the equation:

$$\Delta G = -nFE, \tag{1.52}$$

ΔG – the change of Gibbs energy of corrosion reaction; n – the number of mol electrons taking part in the corrosion reaction; F – Faraday's constant.

We also want to know the corrosion rate and the corrosion form, in order to prevent corrosion failure. Both depend on various factors changing with time. The majority of metals are not resistant to the environment: water, atmosphere, soil, and industrial streams. A metal which is not resistant under some conditions may be resistant under other conditions. Contrarily, resistance of metals is different in the same media. For example, aluminum is not resistant to caustic solution (it dissolves even faster in it than in acidic solution), but iron and chromium are highly resistant to the same caustic solution at pH = 9 to 12. But if caustic solutions are very concentrated (pH above 13.5), caustic corrosion and caustic embrittlement of iron may occur (especially at temperatures above 90°C). Here are some other examples. Titanium is dissolved in hydrochloric acid. However, if we add 0.2% of palladium to titanium, this alloy will be resistant to the same hydrochloric acid. Copper is resistant to hydrochloric acid which does not contain dissolved oxygen and copper corrodes in the same acid if dissolved oxygen is present in the acid.

These examples show that there are internal and external factors influencing the corrosion of metals. The internal factors relate to the metal properties, namely to the metallurgical properties of a metal: type of the metal (chemical content, imperfections, and structure), heat treatment, presence of stresses, etc. The external factors relate to the properties of the environment and conditions: type of chemicals and their concentrations, temperature, pressure, flow rate, presence of stray electric current, presence of dissimilar metals, etc. There is an additional factor – the boundary between a metal and the environment. Two electrochemical reactions, anodic oxidation of a metal and cathodic reduction of some species, make up the corrosion process proceeding at the metal–environment boundary. Corrosion products are formed at this boundary. Of course, the kinetics of corrosion process will also depend on this boundary which is defined both by a metal and by the environment. This means that the presence of oxides (Al_2O_3, Cr_2O_3, Fe_3O_4, TiO_2), hydroxides [$Zn(OH)_2$, $FeO(OH)$], salts [$PbSO_4$, $CaCO_3$, $Fe_3(PO_4)_2$], mixed compounds [$ZnCO_3 \cdot Zn(OH)_2$], or some other substances (oil, surfactants, microorganisms) may influence the corrosion rate and its mechanism (Table 1.6). To sum up, the behavior of a metal depends on both internal and external factors, as well as

Table 1.6 Internal and external factors influencing corrosion of metals.

Internal factors	Metal–environment boundary	External factors
Metal type: chemical content, imperfections, structure, heat treatment, stresses	Corrosion products and other species on the metal's surface	Chemical type of the environment (components), their concentration, temperature, flow rate, pressure

on the conditions at the boundary metal–environment. We may often change these factors. Sometimes we are not able to change or influence them. Of course, we may reduce stresses in metallic equipment and structures under particular conditions by using special heat treatment.

Usually we receive equipment with particular metallurgical properties, and we cannot change them. Therefore, we shall concentrate on the external factors and consider their influence on corrosion of metals, namely, the influence of chemical content of the electrolyte solutions. We should know this influence in order to predict the corrosion behavior of metals under particular conditions.

1.8.1 Influence of pH

We cannot talk about this influence if we do not know what pH means. The pH value shows the activity (concentration) of H^+ (H_3O^+) cations, that is the environment is neutral, acidic, or alkaline, and may range between -1 and 15. The lower the pH value, the more acidic the solution is. Pure water is neutral with pH = 7. Distilled water is slightly acidic with pH = 5.5 because of ingressing carbon dioxide (CO_2) from the air and formation of weak carbonic acid (H_2CO_3). Acids are the substances releasing H^+ cations in aqueous solutions. This small particle H^+ cannot exist alone without any electrons, and is immediately captured by water molecule H_2O forming H_3O^+ cation. For simplicity, we may write H^+ in chemical reactions, but should keep in mind that H_3O^+ takes part in chemical reactions, not H^+. The presence of inorganic acids (hydrochloric HCl, sulphuric H_2SO_4, nitric HNO_3, phosphoric H_3PO_4, boric H_3BO_3, etc.) and organic acids (formic HCOOH, acetic CH_3COOH, citric $HOC(CH_2COOH)_2COOH$, etc.) in aqueous solutions diminish their pH (< 7). Inorganic acids are stronger than organic ones, and may diminish the pH to 0, and even less. Organic acids are weaker and the pH of their aqueous solutions is not less than 2.3 to 4.

Alkalies are substances releasing OH^- anions in water. For example, caustic solutions of sodium (NaOH) and potassium (KOH), ammonia (NH_4OH), baking soda ($NaHCO_3$), and soda (Na_2CO_3) dissolved in water give OH^- anions. The more alkaline a solution is, the higher is its pH (> 7). Note that there are some substances without OH groups, but when dissolved in water they are hydrolyzed and OH^-

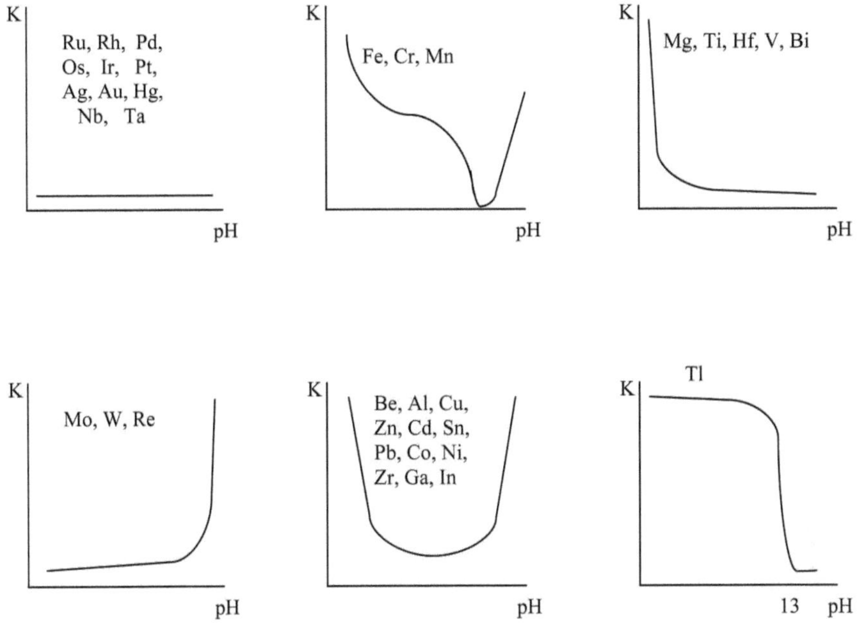

Fig. 1.20 Influence of pH on corrosion resistance of metals. *K* – relative corrosion rate.

anions are released. For example,

$$NaHCO_{3(s)} + H_2O_{(l)} \rightarrow Na^+_{(aq)} + \mathbf{OH^-_{(aq)}} + H_2CO_{3(aq)}, \quad (1.53)$$

$$Na_2CO_{3(s)} + H_2O_{(l)} \rightarrow 2Na^+_{(aq)} + \mathbf{OH^-_{(aq)}} + HCO^-_{3(aq)}. \quad (1.54)$$

Therefore, the aqueous solutions of such salts (for example, soda) are alkaline. Other salts ($CaCl_2$, $MgCl_2$ and NH_4Cl) dissolved in water are hydrolyzed and solutions become acidic:

$$CaCl_{2(s)} + 2H_2O_{(l)} \rightarrow \mathbf{H_3O^+_{(aq)}} + Ca(OH)^+_{(aq)} + 2Cl^-_{(aq)}. \quad (1.55)$$

All metals may be divided into five groups according to their relationship (corrosion resistance) to pH values (Figure 1.20):

1. The pH value does not influence the corrosion resistance of gold (Au), silver (Ag), platinum (Pt), palladium (Pd), rhodium (Rh), ruthenium (Ru), mercury (Hg), tantalum (Ta), niobium (Nb), osmium (Os) and iridium (Ir). They are resistant at pH = 0 to 14.
2. Iron (Fe), chromium (Cr) and manganese (Mn) corrode at a high rate in acidic solutions (pH $<$ 4) and in very strong alkaline solutions (pH $>$ 13.5) at temperatures above 80°C. They also corrode, but at a low rate at neutral pH (6 to 8), and are resistant at pH = 9 to 13.

3. Magnesium (Mg), titanium (Ti), hafnium (Hf), vanadium (V) and bismuth (Bi) corrode at a high rate at low pH (in acidic solutions) and are resistant to neutral and alkaline solutions (high pH).
4. Molybdenum (Mo), tungsten (W) and renium (Re) are resistant to acids and neutral solutions, but corrode in alkaline solutions.
5. The last group is large: beryllium (Be), aluminum (Al), copper (Cu), zinc (Zn), cadmium (Cd), tin (Sn), lead (Pb), cobalt (Co), nickel (Ni), zirconium (Zr), gallium (Ga) and indium (In). They corrode both in acidic and alkaline solutions and, therefore, are called the *amphoteric* metals (from the Greek word *amphoteros* – *both* – the ability of some substances to manifest acidic or alkaline properties depending on conditions). These metals are resistant only to neutral solutions. It is obvious that a region of pH where high resistance exists, is different for various *amphoteric* metals. Aluminum is resistant at pH = 4.5 to 8.3, zinc is resistant at pH = 6.5 to 12. Temperature and presence of various ions in a solution may drastically change the pH region of high resistance of the *amphoteric* metals. Strictly speaking, iron and chromium also relate to the *amphoteric* metals, as they dissolve at low and very high pH (above 13.5); the higher temperature, the lower the pH value when iron dissolves in alkaline solutions.

One may ask if lead and iron corrode in acidic solutions, but we know that sulphuric acid is stored in carbon steel tanks cladding with lead, and carbon steel tanks are used for transportation and storage of concentrated sulphuric and nitric acids. Why? Not only does the pH value influence the corrosion rate of metals. Acid type (that is anion type in the acid) also influences the corrosion resistance of metals (Figure 1.21).

For instance, lead begins to corrode in sulphuric acid, but corrosion stops after formation of non-soluble salt deposits ($PbSO_4$) on lead surface. This deposit which stops corrosion may be called a passive film, and depends on the concentration of sulphuric acid. If the acid concentration is high (above 80%) the lead sulphate salt $PbSO_4$ forming on the lead surface will dissolve in sulphuric acid with the formation of dissolved salt $Pb(HSO_4)_2$. Therefore, lead is used for the protection of carbon steel from corrosion in sulphuric acid at concentrations under 70%. Today lead is not used for this purpose because of its ecological and safety problems. Concentrated sulphuric (above 60%) and nitric (above 40%) acids favor the formation of passive iron oxide layers on iron and carbon steel surfaces (see Appendix D). Such layers protect iron alloys from corrosion in sulphuric and nitric concentrated acids. Any dilution of sulphuric and nitric acids may also cause the destruction of passive iron oxides and give rise to intensive corrosion of iron. Phosphoric acid (H_3PO_4) at low concentrations (50 to 200 ppm) and some organic acids inhibit corrosion of iron and carbon steel in neutral solutions of electrolytes (see Section 4.3).

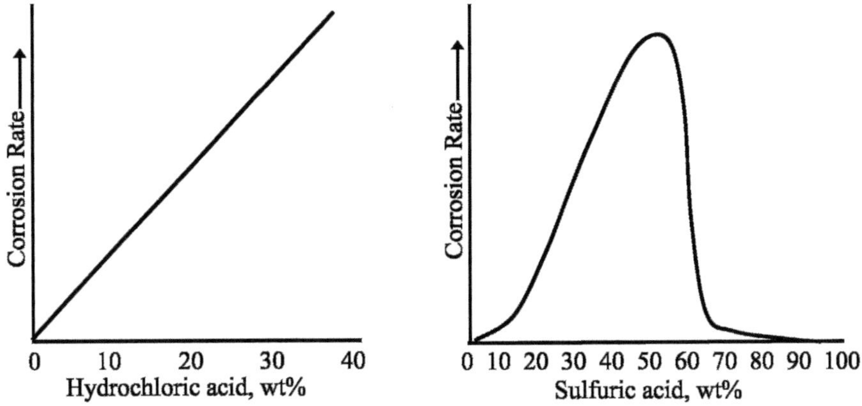

Fig. 1.21 Effect of type of acid and its concentration on the corrosion rate of iron immersed in aqueous solutions of two inorganic acids at room temperature.

1.8.2 Influence of Dissolved Salts in Water on Corrosion

The chemical content of potable (drinking, or tap) waters, cooling waters, seas, oceans, rivers, lakes, and underground waters is different. The common denominator for all these water types is their nearly neutral pH (usually between 5.5 to 8.3), presence of inorganic and organic substances, and dissolved gases. Slight changes of the chemical content of species in water may cause drastic changes of the corrosion rate of metals. There is little effect of pH values from 5.5 to 8.3 on the corrosion of metals. Metal cations formed in anodic dissolution of metal react with hydroxide anions OH^- formed in a cathodic reduction of dissolved oxygen in aqueous neutral solutions, and corrosion products in the form of hydroxides are formed. They are rust [ferric hydroxide – FeO(OH)] or zinc hydroxide $Zn(OH)_2$ which are not dissolved in water. These corrosion products cover the metal surface, and corrosion proceeds at a constant rate at some stable pH values. The corrosion rate depends on the salt type and its concentration. We differentiate four groups of salts according to their influence on the corrosion of metals.

Acidic and alkali salts. These salts are hydrolyzed with the formation of acids or alkalies, and thus cause a decrease or increase in the pH of the solution. For example, if there is ferric chloride ($FeCl_3$) in water, it hydrolyzes with the formation of hydrochloric acid:

$$FeCl_{3(aq)} + 2H_2O_{(l)} \leftrightarrow \mathbf{H_3O^+_{(aq)}} + 3Cl^-_{(aq)} + Fe(OH)^{2+}_{(aq)}. \qquad (1.56)$$

As a result, the pH of the aqueous solution containing ferric chloride salt decreases to 1–2. Many salts similar to ferric chloride may hydrolyze in water with a drastically decreased pH. For example, calcium chloride ($CaCl_2$), magnesium chloride ($MgCl_2$), ammonium chloride (NH_4Cl), sodium dihydrophosphate (NaH_2PO_4), and

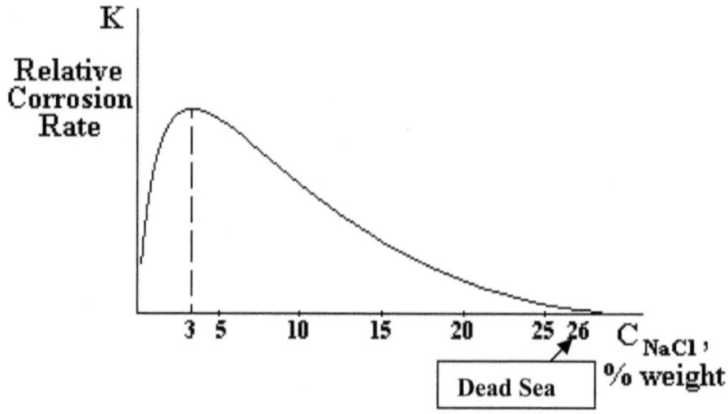

Fig. 1.22 The influence of sodium chloride concentration in water on the corrosion rate of carbon steel.

aluminum sulphate $Al_2(SO_4)_3$. Soda ash (Na_2CO_3) in water hydrolyzes with OH^- formation according to reaction (1.54), and the pH of such a solution will increase to above 8. The following salts hydrolyze with increasing pH: baking soda ($NaHCO_3$), sodium phosphate (Na_3PO_4), disodium hydrophosphate (Na_2HPO_4), potassium cyanide (KCN), and sodium silicate (Na_2SiO_3).

Salts which form deposits on a metal's surface and protect it: brown ferric phosphate [$Fe_3(PO_4)_2$], black ferrous carbonate ($FeCO_3$) and black iron sulphide (FeS) on iron alloys, green-blue patina [$CuCO_3{\cdot}Cu(OH)_2$] and [$CuSO_4{\cdot}Cu(OH)_2$] on copper alloys, white aluminum hydroxide (AlOOH) on aluminum alloys, white mixed corrosion product $Zn(OH)_2{\cdot}ZnCO_3$ on zinc, and gray $PbSO_4$ on lead.

Neutral salts, which added to water cause an increase in the corrosion rate, but with further addition, corrosion reaches its maximum and then decreases (Figure 1.22).

This dependence requires an explanation. Sodium chloride NaCl being dissolved in water does not change the pH of water. The aqueous solutions of such electrolytes are called neutral. If such salt is added to water, corrosion of iron increases because of a decreasing of the electrical resistance of a solution and a proper increasing of corrosion current on the metal surface according to Ohm's law (see formula (1.34)). A cathodic process on the metal surface in neutral aqueous solutions is a reduction of oxygen according to reaction (1.39). If salt or another electrolyte is added to water, the concentration of dissolved oxygen decreases in water. This phenomenon is called *salting out* and concerns all gases dissolved in water with electrolytes (see Appendix E). Other gases salt out together with oxygen if salt is added to water. Only oxygen is interesting here because it is a main "actor" in the cathodic process on the metal surface in neutral solutions of electrolytes. If the oxygen concentration in aqueous solution diminishes, there is no consumer of electrons releasing in anodic reaction of metal dissolution, and corrosion decreases. We may add so much salt to water that nearly all dissolved oxygen will disappear, and metal will stay in the aqueous salt solution without the "cathodic participant". If there is no electron consumer O_2, corrosion decreases and the corrosion rate of carbon steel in a 26 wt% NaCl solution (the Dead Sea!) may even be less than that in the distilled water (see Figure 1.22). Therefore, you will not observe rusted steel equipment immersed in the Dead Sea and you will not find fish there, as dissolved oxygen is absent because of *salting out*.

To sum up, addition of salts to water influences corrosion of carbon steel by means of two factors: decreases electrical resistance of the solution (and thus increases the corrosion current), and oxygen concentration in water. Up to some salt concentration value in water, the first factor prevails over the second. Then with an increase of salt concentration in water, the more prevailing factor is the concentration of dissolved oxygen. Therefore, the dependence "corrosion rate – salt concentration" is described by the curve with a maximum (see Figure 1.22). This is determined by the kind of salt (electrolyte) and temperature. It is good to know that the maximum corrosion rate for carbon steel occurs in a 3 wt% NaCl solution. This is the mean concentration of salts in seas and oceans. Therefore, many corrosion tests are carried out in 3 wt% NaCl solution. This is correct only regarding iron and carbon steel. For other metals and alloys, the maximum corrosion rate may be at another salt concentration.

1.8.2.1 Influence of Substances Forming Chemical Complexes with Metals on Corrosion

Substances forming chemical complexes with metals accelerate their corrosion. These substances may be neutral molecules (ammonia NH_3) or ions (cyanides CN^-). The English scientist Joseph Priestley who was the first to prepare pure ammonia in 1774, did not imagine how many corrosion problems would be caused by his discovery. Ammonia is highly soluble in water, forming an alkaline solution of ammonium hydroxide. Copper and copper alloys are not resistant in the presence of ammonia and dissolved oxygen in water because of the formation of complex:

$$4Cu_{(s)} + 8NH_{3(aq)} + O_{2(g)} + 2H_2O_{(l)} \leftrightarrow 4[Cu(NH_3)_2]^+_{(aq)} + 4OH^-_{(aq)}. \quad (1.57)$$

Therefore, it is very dangerous to use copper condensers with water from bays near agricultural fields where ammonia fertilizers are used.

We call gold and silver "noble" metals because of their high corrosion resistance. Many people think that they are resistant to all media. But if you put them in an aqueous solution of sodium or potassium cyanide (NaCN or KCN), anions CN^- form chemical complexes with these "noble" metals:

$$4Au_{(s)} + 8CN^-_{(aq)} + 2H_2O_{(l)} + O_{2(g)} \rightarrow 4[Au(CN)_2]^-_{(aq)} + 4OH^-_{(aq)}, \quad (1.58)$$

$$4Ag_{(s)} + 8CN^-_{(aq)} + 2H_2O_{(l)} + O_{2(g)} \rightarrow 4[Ag(CN)_2]^-_{(aq)} + 4OH^-_{(aq)}. \quad (1.59)$$

Therefore, gold and silver are not resistant in solutions containing cyanide anions. These two corrosion reactions helped the Scottish chemists John S. MacArthur,

Robert W. Forrest and William Forrest to invent in 1887 the method of extracting gold and silver from their ores by dissolving them in a dilute solution of sodium cyanide or potassium cyanide.

We should keep in mind that cyanide anions are highly toxic. The best "solvent" for gold is "aqua regia" (a mixture of hydrochloric and nitric acids):

$$Au_{(s)} + HNO_{3(aq)} + 4HCl_{(aq)} \rightarrow H[AuCl_4]_{(aq)} + NO_{(g)} + 2H_2O_{(l)}. \quad (1.60)$$

Other "noble" metals, such as platinum, palladium, rhodium, iridium and ruthenium also dissolve in "aqua regia" at heating, and this is one of the main ways to separate "noble" metals from each other. Really, corrosion helps to produce pure "noble" metals. Once more, we have to mention that all in life is relative. There is no material resistant to all media, as well as there is no material non-resistant at all.

1.8.2.2 Influence of Cations Participating in Cathodic Reactions

In addition to the participation of dissolved oxygen (O_2) in neutral and alkaline solutions, and cations H_3O^+ in acidic solutions, cations Fe^{3+} and Cu^{2+} may take part in cathodic processes on the metal surface and accelerate corrosion. It is good to say that Fe^{2+} cations formed in cathodic reaction (1.27) may be oxidized to Fe^{3+} according to the reaction

$$4Fe^{2+}_{(aq)} + O_{2(g)} + 4H_3O^+_{(aq)} \leftrightarrow 4Fe^{3+}_{(aq)} + 6H_2O_{(l)}. \quad (1.61)$$

Thus, regeneration of Fe^{3+} cations proceeds all the time, and corrosion is accelerated. There is a vicious circle! Therefore it is very important to determine and follow such cations in a media which are able to participate in cathodic reactions and accelerate corrosion.

1.8.3 Influence of Electrical Conductance of Media

If corrosion of metals in electrolyte solutions is the electrochemical process, how does the electric conductance of media influence corrosion of metals? The electric conductance of a solution is an ability to conduct electric current by means of ion migration.

Ohm's law is also applicable for the solutions of electrolytes (see (1.34)):

$$V = I \cdot R, \quad (1.62)$$

V – change of electric potential in solution, Volts; I – electric current, Amperes; R – electrical resistance of a solution, Ohms. Electrical resistance of a solution appears if there is a voltage drop, or potential difference V between cathode and anode electrodes (Figure 1.23). Electrical resistance R of a solution is defined by

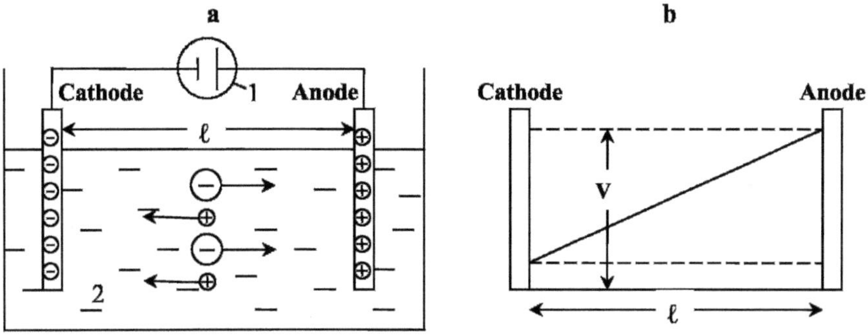

Fig. 1.23 The voltage drop V in the electrolyte solution and appearance of electrical resistance R in solution: (a) cell for electrolysis; (b) voltage drop in the cell. 1 – source of direct electric current; 2 – solution of electrolyte. Cations (⊕) move to a cathode and anions (⊖) move to an anode in solution; ℓ – the distance between a cathode and an anode.

$$R = \rho \cdot \ell / A, \tag{1.63}$$

ρ – the specific electrical resistance, Ohm·m (characteristics of conductor, in our case solution); ℓ – the length of the conductor (the distance between the electrodes, see Figure 1.23); A – a cross-section surface of the conductor (the surface of the electrodes in the solution of electrolytes). Now we may define the electrical conductance L and the specific electrical conductance κ:

$$L = 1/R, \tag{1.64}$$

$$\kappa = 1/\rho, \tag{1.65}$$

κ – the conductance of the solution layer with a volume of 1 m^3 and a thickness of 1 m.

Therefore, $\kappa = (1/R) \cdot (\ell/A)$, and is measured in Ohm·m or Siemens/m (S/m), where Siemens = 1/Ohm. Electrical conductance in solutions is determined by ion mobility and is a very important value for the definition of the corrosiveness (aggressiveness towards metals) of water, aqueous solutions of electrolytes, and other liquids. The more electrical conductance the liquid has, the higher the ability to carry the electric current on the metal surface between anode and cathode sites, and the corrosion current is consequently more. We may write down from Ohm's law: $I = V/R = V \cdot L$. For example, the electrical conductance of fuels is very low, 10^{-12} S/m, and their corrosiveness is consequently very low, close to zero. The electrical conductance of pure water is more, 10^{-6} S/m, and its corrosiveness is more. The electrical conductance of aqueous electrolyte solutions is high, 10^{-3} to 40 S/m, and they are very aggressive towards metals (Table 1.7). Ions H^+ and OH^- are responsible for the electrical conductance of pure water. Various cations and anions are responsible for the electrical conductance if electrolytes are present in water. Which charged species are responsible for low electrical conductance of

Table 1.7 The electrical conductance of various liquids and corrosion rate of carbon steel in them.

Liquid	Specific electrical conductance (κ), S/m	Corrosion rate of carbon steel[a], mm/year
Demineralized water	10^{-4}	0.1
Potable water[b]	0.05	0.2 to 0.3
Cooling water (industrial)[c]	0.2	0.6 to 1
Gasoline	10^{-10}	0.001
Gas oil	10^{-12}	0.0001
Kerosene	10^{-12}	0.0005
Gasoline + 0.02% water	Two-phase system	0.4
Kerosene + 0.02% water	Two-phase system	0.5

[a]Corrosion rate of carbon steel was determined by the weight loss method with intensive agitation during one week at 25°C.
[b]Potable water in Israel.
[c]Cooling water in the chemical plant (without any chemical additives: inhibitors, biocides, and anti-scaling agents).

fuels? Small quantities of water, ions, and organic acids dissolved in fuels are responsible for their electrical conductance. For instance, gasoline usually contains about 80 ppm of water and can dissolve in up to 250 ppm of water. But we shall discuss this in Section 3.5. It is good to say that engineers and scientists have tried for many years to determine the influence of the chemical content of various types of soil on its corrosiveness, but without success. It was accepted to determine the corrosiveness of soils in accordance with their electrical resistance.

We may sum up that electrical conductance of media is very important factor in corrosion processes because of:

(a) Acceleration of the corrosion current if the electrical conductance is high (see Figure 1.15). The lower the electrical resistance R of a medium, the higher the corrosion current I_{corr}:

$$I_{corr} = V/R = (E_k - E_a)/R. \tag{1.66}$$

(b) All parameters of the electrochemical (cathodic) protection (the choice of its type, electric current density, etc.) depend on the electrical resistance of soil (or other media) and type of anodes.
(c) The distribution of stray electric currents also depends on the electrical resistance of soil.
(d) And, of course, the electrochemical measurements, for example, the determination of the electric current as the function of electric potential depends on the electrical resistance of media. During an electric current flow between working electrode and a reference electrode in solution, the Ohmic potential drop, called IR drop, appears. This is a potential drop due to the electrical resistance of the solution layer. The higher the electrical resistance of a solution, the more the IR drop. This mistake in potential measurements must be compensated. Polar-

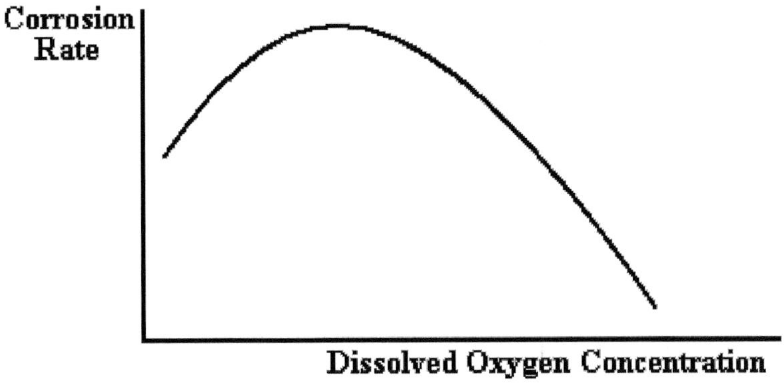

Fig. 1.24 The influence of dissolved oxygen concentration in water of high purity on the corrosion rate of iron.

ization measurements in pure water and non-aqueous solvents, as well as metal electrodes with coatings can include an IR drop.

1.8.4 The Influence of Dissolved Oxygen

Dissolved oxygen plays a very important and complicated role in the corrosion of metals. Oxygen takes part in cathodic processes on the metal surface in neutral, alkaline, and acidic media. Therefore, its presence is necessary for corrosion to occur. If dissolved oxygen is absent in water, corrosion diminishes nearly to zero in neutral and alkaline solutions. If the concentration of dissolved oxygen increases, corrosion accelerates as a result of oxygen participation in the cathodic processes. What would happen if we inject more and more oxygen in water? It was defined that oxygen under some particular conditions (in water of high purity) and high temperature may result in the formation of a passive protective dense film composed of metal oxides on the metal surface, and corrosion would decrease (Figure 1.24). The injection of oxygen in water is one of the corrosion control methods at power stations. If water contains ions (Cl^-, SO_4^{2-}), they can destroy passive films and the effect of passivation will not be obtained.

1.8.5 The Influence of Temperature

We live on the Earth in an environment with broad changes of temperature: from −60 to +50°C. It is cold in winter, and we may dress accordingly. It is hot in summer, and so we may shed our clothing. The biochemical processes in our organism proceed at the constant temperature of 36.6°C. These reactions may even stop if the

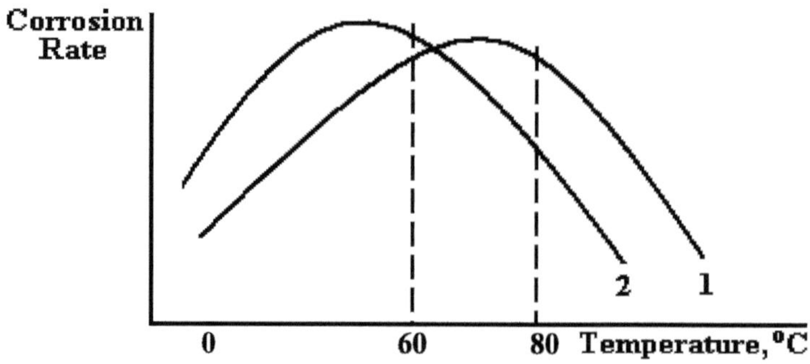

Fig. 1.25 The influence of temperature on corrosion of iron in water (1) and aqueous neutral solution of electrolytes (2).

temperature falls, and may accelerate if the temperature rises. The organism may be destroyed because of high acceleration of biochemical processes. Thus, all chemical processes depend on temperature. Temperature is the main parameter in chemistry to change the rate of chemical reactions. Metals and alloys are used under wide changes of environmental conditions. For example, the rectification of liquid hydrogen (in order to separate hydrogen and deuterium isotopes) proceeds at −249°C. The cracking process at petrochemical plants proceeds at +1,000°C. We shall not discuss corrosion processes at these critical temperatures but we shall concentrate on how corrosion processes are influenced by temperature in the range between 0 and +100°C. How does temperature influence corrosion of metals? Temperature may influence both thermodynamics and kinetics of metallic corrosion. Appendix A shows the thermodynamics of corrosion is governed by temperature. Increase of temperature usually accelerates the anodic and cathodic reactions, diminishes the dissolution of gaseous oxygen in media, and accelerates diffusion of cathodic participants (O_2, H_3O^+, Fe^{3+}, etc.). Temperature also changes the dissolution and transformation of corrosion products forming on metal surfaces. The influence of temperature is complicated, and the prediction of behavior of metals with changes of temperature is not simple. For example, we may expect that increase of temperature will accelerate corrosion of iron in aqueous solutions of electrolytes. Experiments show that the corrosion rate of iron raises at the beginning if the temperature increases, then reaches maximum at about 60 to 80°C, and then diminishes (Figure 1.25).

We must explain this dependence in Figure 1.25. Dissolved oxygen, the cathodic "actor", is responsible for the curves at maximum in both cases. At first, increase of temperature accelerates both reactions on the metallic surface: anodic dissolution of iron and cathodic reduction of oxygen. The rising of temperature also results in a decrease of oxygen solubility in water, and at 100°C nearly all dissolved oxygen escapes (Figure 1.26). This is the main principle of deaeration – removing of dissolved oxygen from water by heating up to 100°C at atmospheric pressure.

Thus, temperature influences corrosion by two factors: accelerates both anodic and cathodic reactions, and diminishes dissolved oxygen concentration (as a result,

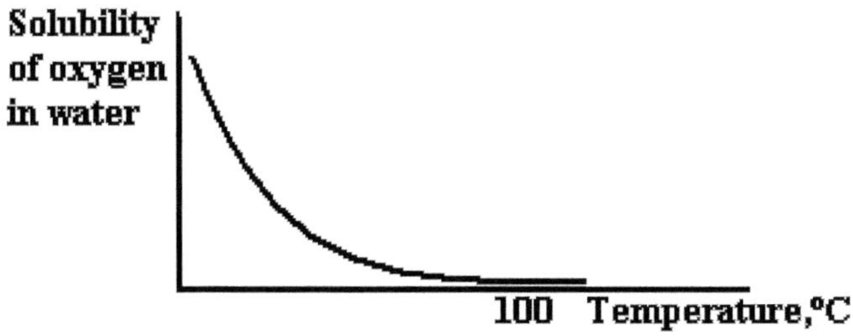

Fig. 1.26 Temperature influences the solubility of oxygen in water.

cathodic reaction). At some temperature these two factors are equal, and corrosion reaches the maximum value. Then at higher temperatures, the decreasing of oxygen concentration in water prevails over the increase of the anodic and cathodic reaction rates. Dissolved oxygen as cathodic participant disappears from the "scene", and corrosion decreases. Oxygen is so important in the corrosion of metals that it may result in acceleration of corrosion in particular cases, and in other cases to diminish and even stop it. The concentration of dissolved oxygen is non-uniform in media in many cases, and causes the formation of *differential aeration cells.*

We shall explain what this is in the next section.

1.9 Differential Aeration Cell

We often observed the uneven spreading of rust inside water pipes, severe shallow pits under deposits on a carbon steel surface in water or in soil, and severe corrosion of various structures and equipment on the air–water boundary. One often hears the question: "Why were pits formed in this place and not 20 cm right or left?" We may reply: "The pits were formed because of the formation of *differential aeration cells* on a steel surface in water and in neutral aqueous electrolyte solutions". The concentration of dissolved oxygen may change due to factors such as flow distribution, electrolyte concentration, temperature, and pressure. If two sites on a carbon steel surface differ in dissolved oxygen concentrations $[O_2]_1$ and $[O_2]_2$, these sites acquire different electric potentials

$$E = E_1 - E_2 = \frac{RT}{nF} \ln \frac{[O_2]_1}{[O_2]_2} \tag{1.67}$$

and a *differential aeration cell* appears (Figure 1.27). A carbon steel surface with a larger dissolved oxygen concentration on it will be a cathode (oxygen participates in reduction process), and will not corrode. A carbon steel surface with a smaller dissolved oxygen concentration on it will be an anode (electric potential will be less than that of the site with a high oxygen concentration), and will corrode.

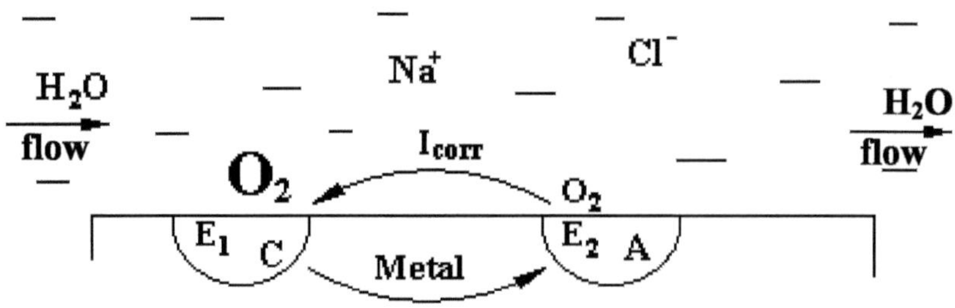

Fig. 1.27 Appearance of the *differential aeration cell* on a metal surface in water. E_1 – region of high oxygen concentration (C – cathode) with electric potential E_1; E_2 – region of low oxygen concentration (A – anode) with electric potential E_2.

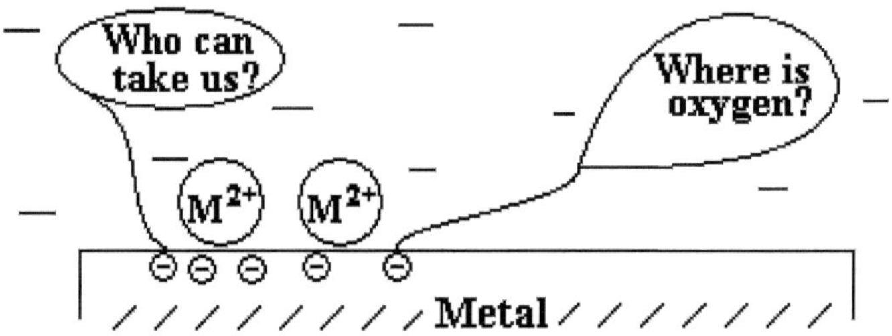

Fig. 1.28 The "limiting" case of the *differential aeration cell.*

This means that a carbon steel surface with a free oxygen way will always be a cathode (not corrodes and will "breathe!") compared to another part of the carbon steel surface with a difficult way for oxygen which will be an anode and will corrode. There is an oxygen deficiency on the anodic sites for "breathing", and it will "suffocate". Diminishing the oxygen concentration or removing the dissolved oxygen as a cathodic participant results in a decrease of corrosion. Now we say that the sites with low oxygen concentration will corrode locally severly because of anode formation at these sites. Is this a contradiction? No! Deaeration (removing of oxygen) is the limiting case of the *differential aeration cell* when an oxygen concentration is equal to zero, and there is no difference in the oxygen concentration at various metal sites (Figure 1.28). We have to remember that H_3O^+ cations are present in water and can take part in the cathodic process and metals may also be oxidized directly by H_2O molecules (see Appendix A). These factors have little influence on the corrosion of iron and carbon steels in water and neutral aqueous solutions of electrolytes. If dissolved oxygen is present in a media, this is the general case of a *differential aeration cell.*

No dissolved oxygen, nearly no corrosion. Any deposit of inorganic or organic material on a metal surface results in the formation of differential aeration cells. Inorganic deposits may be rust, non-soluble salts, scale, oxides, sand, stones, etc. The organic deposits may be oil, microorganisms, etc. The corrosion mechanism as

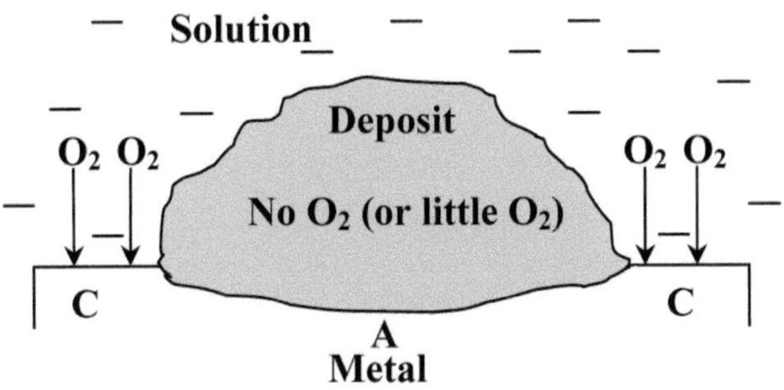

Fig. 1.29 Differential aeration cell forms in the presence of any deposits. C – cathode; A – anode.

a result of the *differential aeration cell* formation does not depend on the kind of deposit (Figure 1.29).

Differential aeration cells are responsible for the pitting corrosion, crevice corrosion, corrosion of structures and equipment at the interface border air–water, air–soil, corrosion in petroleum distillates (gasoline, kerosene, etc.), and corrosion in soils.

If we know all factors that result in corrosion, we are able to control the corrosion situation and to predict the corrosion behavior of metals. The factors can result in diverse corrosion phenomena and various corrosion forms. We have to differentiate them. This will be the theme of the next chapter.

Chapter 2
Corrosion Phenomena

Corrosion mechanism and factors, which we have just analyzed, relate to corrosion phenomena defining by corrosion forms. There are both many metals (about 80) and a lot of environmental types (of course, more than 80!), and we might expect many different corrosion phenomena. All corrosion reactions can be divided according to their mechanism in two types: reactions in the presence of non-electrolytes, and in the presence of electrolytes (electrochemical mechanism). All corrosion phenomena can be also divided into two types: uniform (general) and non-uniform (localized) corrosion. Uniform corrosion is a process when all of a metal surface corrodes, or dissolves. Non-uniform corrosion is a process when only some definite part of a metal surface corrodes, i.e. anodic process concentrates on a small part of the metal's surface. A corrosion phenomenon is a process that occurs on the surface of or inside the metal. The corrosion form is a result of a corrosion phenomenon. It is possible to define the corrosion phenomenon according to the corrosion form. On the other hand, if we know the corrosion phenomenon type, we may expect what corrosion form, or deterioration type, will happen with a metal: uniform thickness loss, pits, cracks, blistering, etc. We have to keep in mind that many corrosion phenomena influence such important functions of metallic equipment as mechanical and electrical properties. All corrosion phenomena and corrosion forms are described by specific mechanisms.

2.1 Uniform, or General, Corrosion

If we put a piece of zinc in a hydrochloric acid solution, or aluminum in a soda solution, we will observe uniform diminishing of the metal's thickness. If we put a piece of carbon steel in a furnace at high temperature, we will observe the formation of even iron oxides film on the steel surface. *Uniform, or general, corrosion is a type of corrosion attack uniformly distributed over a metal surface* (Figure 2.1). Here are some examples of uniform corrosion of metals and alloys.

A. Groysman, *Corrosion for Everybody*, DOI 10.1007/978-90-481-3477-9_2,

© Springer Science+Business Media B.V. 2010

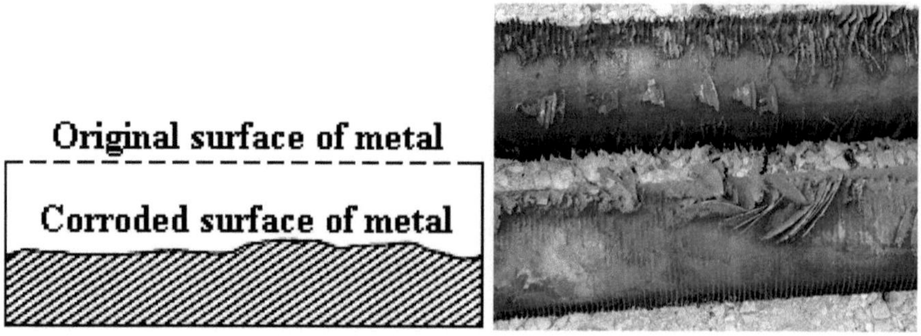

a b

Fig. 2.1 Uniform, or general, corrosion. (a) Scheme of corroded surface of metal; (b) carbon steel tubes after uniform corrosion in the furnace. (For a full color version of this figure, see the Color Section)

1. Metal dissolution in acids (zinc in hydrochloric acid) and in alkali solutions (aluminum in caustic soda solution):

$$Zn_{(s)} + 2HCl_{(aq)} \rightarrow ZnCl_{2(aq)} + H_{2(g)}, \quad (2.1)$$

$$2Al_{(s)} + 2NaOH_{(aq)} + 2H_2O_{(l)} \rightarrow 2NaAlO_{2(aq)} + 3H_{2(g)}. \quad (2.2)$$

2. High temperature corrosion of metals (carbon steel pipes in furnaces, walls of engines):

$$4Fe_{(s)} + 3O_{2(g)} \rightarrow 2Fe_2O_{3(s)}. \quad (2.3)$$

3. Tarnishing of silver in air (corrosion reaction of silver with hydrogen sulphide pollutant in air):

$$Ag_{(s)} + H_2S_{(g)} \rightarrow Ag_2S_{(s)} + H_{2(g)}. \quad (2.4)$$

How may we evaluate the rate of uniform corrosion?

(a) We may measure the thickness of the metallic construction after some period of its contact with the environment and then calculate the change of its thickness. This value is called *corrosion penetration* and is calculated in mm/year or mpy (mils per year, 1 mil = 0.0254 mm). Engineers like this unit because they know the thickness of construction and, therefore, may predict its life period, when to stop its use and to change for new one. Measuring changes of thickness in various places of a metallic construction or equipment may show if corrosion is uniform or localized.

(b) The second way is to measure the weight loss of metal. Certainly it is not really to weigh a large construction or equipment: tank, heat exchanger, reactor, pipe, etc. Therefore, samples of metals (strips, rods, or discs) made of material similar to the construction are installed in various places in industrial streams. You may define the corrosion rate according to formulae (1.8) and (1.9). Scientists like this method very much. They use an analytical balance for this purpose and receive the rate of uniform corrosion in grams/(cm^2·day) units according to (1.8). The engineers like *corrosion penetration* in mm/year and use formula (1.9).

(c) The third way is to measure the volume of hydrogen gas formed in the reaction (2.1). This way is more complicated than measuring of metal thickness or weight loss, but is used both in the laboratory and in industry. In any case, the hydrogen volume can be used to evaluate the weight loss of a metal, and then calculate the *corrosion penetration* in mm/year.

(d) You may measure the uniform corrosion rate of metals in gases (see reaction (2.3)) by weight gain per unit area per unit time. This method can certainly be used only with relatively small samples of metals.

Metals do not always corrode uniformly in the presence of acids. Sometimes pitting corrosion may occur (see Section 2.3). We have observed holes, pits, and cracks formed on metallic equipment and structures in contact with water, soil, atmosphere, and other environments. This means that in practice corrosion of metals is non-uniform, or localized. We shall talk about many appearances of non-uniform (localized) corrosion in the next section.

2.2 Non-Uniform, or Localized, Corrosion

There are some reasons why non-uniform corrosion is more widespread than uniform corrosion. First of all, metals and alloys are polycrystalline materials with many imperfections inside and outside (on the surface). It is impossible to produce a metallic structure without such imperfections as dislocations, inclusions, and voids. Even if we produce the "ideal" metal without imperfections, the environment near the metal changes with time and distance: dissolved oxygen concentrations, pH of media, differences in other variables (temperature, flow rate, concentrations of aggressive species) (Figure 2.2). This means that the heterogeneity of both metals and environment results in non-uniform, or localized, corrosion. Thus, *localized corrosion is a corrosion phenomenon concentrated at some specific sites of a metal surface, which is very small regarding to all the surface of the metal that does not corrode.*

It is explicit that a non-uniform corrosion is more widespread than a uniform corrosion. Here are all kinds of non-uniform corrosion: pitting corrosion, crevice corrosion, galvanic corrosion, microbiologically induced corrosion (MIC), dealloying, intergranular corrosion, exfoliation, filiform corrosion, stress corrosion cracking (SCC), hydrogen damages, corrosion fatigue, erosion, cavitation, impingement attack, and fretting corrosion.

In spite of so many (15!) types of non-uniform corrosion phenomena, they all result in two main forms: pits and cracks. Hydrogen damage may also result in blisters inside metals. Sometimes it is simple to define the corrosion phenomenon type. For example, if two dissimilar metals contact each other in an electrolyte solution, and we observe localized pits on the surface of less noble metal at the place of contact

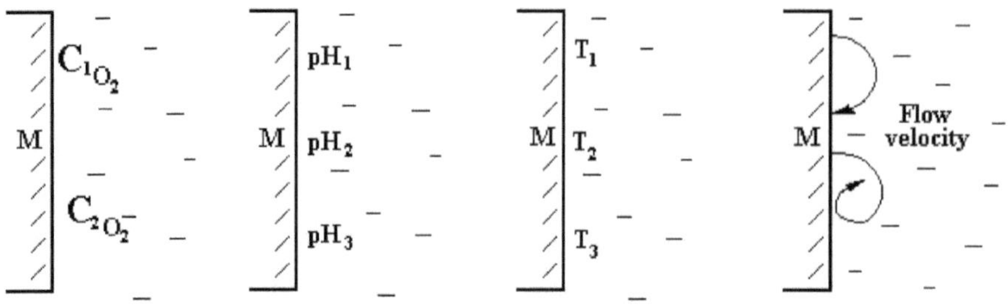

Fig. 2.2 Heterogeneity of the environment on the metal (M) surface. $C_{1(O_2)}$, $C_{2(O2)}$-various concentrations of dissolved oxygen. pH_1, pH_2, pH_3-different values of pH. T_1, T_2, T_3-various temperatures.

with more noble metal after some period, we may say that galvanic corrosion obviously occurred. Sometimes it is not simple to determine the corrosion phenomenon, as pits may be the results of pitting corrosion, crevice corrosion, galvanic corrosion, MIC, erosion, cavitation, impingement attack, and fretting corrosion. Cracks may be the results of intergranular corrosion, SCC, hydrogen damage, exfoliation, and corrosion fatigue. In order to predict and to prevent these corrosion phenomena, we must exactly define them. How to differentiate among such a huge number of corrosion localized phenomena, if there are only two main appearances (pits and cracks) exist on a metallic surface? It is possible if we know what variables, or environmental conditions (chemicals, temperature, flow changes, stresses, etc.) cause a specific phenomenon. Fortunately, we know the reasons resulting in all type of localized corrosion. This will be the theme of the next sections. Only some of the localized corrosion phenomena will be described in this book.

2.3 Pitting Corrosion

Certainly we have heard many times that stainless steel is not recommended for use in sea water or any aqueous solution with high concentrations of chlorides or other halides. This is correct, but not always. We shall recognize this at the end of the chapter. Everyone can observe pits of various dimensions and sizes on carbon steel, stainless steel and copper tubes, devices, utensils, etc. (Figure 2.3). Other metals, titanium, aluminum, magnesium, nickel, and cobalt, are also subject to pitting corrosion because of the presence of chloride anions (Cl^-) in the environment.

Really, pits may appear not only as a result of the presence of chlorides in the environment, but as a result of the existence of differential aeration cells, galvanic corrosion, MIC, erosion, and cavitation. We shall discuss the four last phenomena later. Now let us discuss pitting corrosion that occurs because of the presence of chlorides in the environment. I think that such pitting corrosion is the most mysterious corrosion form. It is not simple to say why pitting is formed here at this point on the metal surface, and not there, several centimeters from the defected point?

a b c

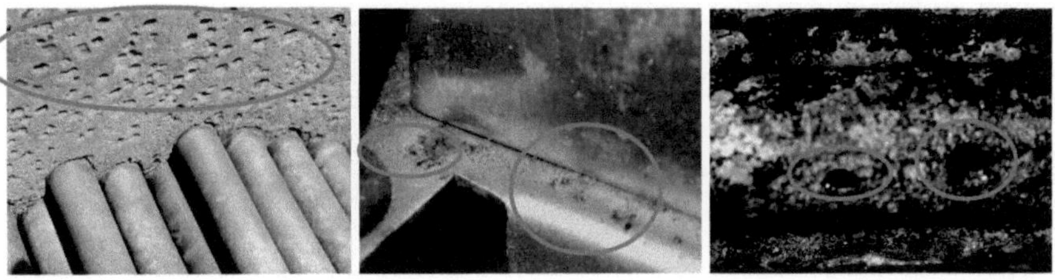

Fig. 2.3 Pits on the carbon steel tube-sheet of a heat exchanger (a), stainless steel 316 bath (b) and Admiralty brass tubes (c). Pits are encircled. (For a full color version of this figure, see the Color Section)

How may we describe "stainless" steel if we know that it is prone to pitting corrosion? Why are chlorides so dangerous for stainless steel, carbon steel, aluminum, and titanium? We shall try to answer these questions.

"Stainless" steel was termed in 1910–1914 immediately after its discovery, because of its high corrosion resistance to nitric acid (see Section 6.1). Really, stainless steel was resistant in many types of environments. After the discovery of a new material, people try to spread its use in other environments, under other conditions. It was very quickly defined that stainless steel was pitted when immersed in water containing chlorides, and even in atmosphere with high chloride content. Scientists soon defined that steel with a chromium content above 12% was "stainless" because of the tenacious protective film consisting of chromium oxide (Cr_2O_3) on the stainless steel surface. This film was called passive and could protect a steel surface under particular conditions. It is *metastable* (not fully stable), and may turn into *active* (see Appendix D). *Passive film* is not homogeneous; defects and impurities exist both in it and on the stainless steel surface under the passive film. Defects in the film may be pores, cracks, any imperfections, inclusions (chemical compounds), and chloride anions may penetrate through them. Chloride anions can destroy this film because of more positive electric potentials at some heterogeneous inclusions, for example, manganese sulphide, than pure alloy. If chlorides enter and penetrate through passive film, they can attract iron or chromium cations from the alloy lattice, and the latter are hydrolyzed with the formation of acidic medium:

$$M^{3+}_{(aq)} + 2H_2O_{(l)} \rightarrow \mathbf{H_3O^+_{(aq)}} + Me(OH)^{2+}_{(aq)}. \tag{2.5}$$

M^{3+} means Cr^{3+} or Fe^{3+}. Ions H_3O^+ and Cl^- make for the presence of hydrochloric acid in the solution and result in low pH (even less than 1!) at the imperfections of the passive film. This acid is formed in very small "reservoirs" on the surface of stainless steel. The solution around these "reservoirs" is neutral. Because of the heterogeneity of stainless steel surface under passive film, pits of various shape, density, and size are formed. Depending on the nature of metals and alloys, different low pH values may be encountered in pits on the surface of stainless steel (see Table 2.1).

Table 2.1 Different cations give various pH on the surface of stainless steel.

Hydrolysis reaction	pH[a] at $a_{Me^{z+}} = 0.01$
$Fe^{2+}_{(aq)} + 4H_2O_{(l)} \leftrightarrow Fe(OH)_{2(s)} + 2H_3O^+_{(aq)}$	7.64
$Fe^{3+}_{(aq)} + 6H_2O_{(l)} \leftrightarrow Fe(OH)_{3(s)} + 3H_3O^+_{(aq)}$	2.27
$Cr^{3+}_{(aq)} + 6H_2O_{(l)} \leftrightarrow Cr(OH)_{3(s)} + 3H_3O^+_{(aq)}$	2.20
$Ni^{2+}_{(aq)} + 4H_2O_{(l)} \leftrightarrow Ni(OH)_{2(s)} + 2H_3O^+_{(aq)}$	7.10
$Mo^{3+}_{(aq)} + 6H_2O_{(l)} \leftrightarrow MoO_{2(s)} + 4H_3O^+_{(aq)} + e^-$	0.97
$Mn^{2+}_{(aq)} + 4H_2O_{(l)} \leftrightarrow Mn(OH)_{2(s)} + 2H_3O^+_{(aq)}$	8.66

[a]pH values were calculated from the Nernst equation (Shreir et al., 1994, p. 1:161). The reader is referred to Appendix B regarding the activity $a_{M^{z+}}$ of metallic cations.

If $a_{Fe3+} = 0.01$, the pH is 2.27. Cations Cr^{3+} and Mo^{3+} cause the greatest pH decrease. Metals and alloys possessing a passive film can be attacked by chloride anions because of the imperfections of such films and because of the heterogeneity of the surface under the films. Iron and carbon steels have films consisting of mixed oxides (FeO, Fe_3O_4, and Fe_2O_3), aluminum – Al_2O_3, titanium – TiO_2, nickel – NiO. Such films are prone to be attacked by chloride anions. Why are we discussing only chlorides? What about other halogenide anions: fluoride (F^-), bromide (Br^-) and iodide (I^-)? Or some other anions, for example, sulphides (S^{2-}), sulphites (SO_3^{2-}) or thiosulphites ($S_2O_3^{2-}$)? We are discussing chloride anions as they are widely spread in the environment and industrial systems: in drinking water, in seas and oceans, in cooling water, etc. All the above-mentioned anions may also result in similar pitting corrosion of many metals and alloys possessing passive films. The shape, density, size, and depth of chloride pitting corrosion depend on chloride concentration, temperature, and flow regime. Unfortunately, different recommendations were given for critical chloride concentrations in water at the proper temperature causing pitting corrosion of stainless steels (Table 2.2).

The higher the temperature, the less chloride concentration is in the water where pitting corrosion may occur. We mentioned that the flow regime influenced the occurrence of pitting corrosion. What does this mean? This means that use of stainless steel is forbidden in sea water under stagnant conditions. Why? Because sea water with a high chloride content will come into contact with the stainless steel surface. In other words, chlorides settle on the stainless steel surface and have enough time to attack the *passive* protective film. If water flows in pipes, in heat exchanger tubes, or in pumps, chlorides are not able to settle and to come into contact with the stainless steel surface. Therefore, stainless steel equipment can be used if you are sure that water flows all time. Do not forget to drain this equipment if the flow of water containing a high chloride concentration is stopped. Not only drain, but also clean the equipment with pure water which do not contains chloride anions.

Table 2.2 Critical chloride (Cl^-) concentrations in water for different stainless steels (without pitting corrosion).

Alloy[a]	Temperature, °C	Critical Cl^- concentration, ppm	Reference
SS 304	25	180	1
		500	2
	40	100	1
	60	60	1
	80	40	1
SS 304L	25	100	3
SS 316	25	500	1
		3,000	2
	40	220	1
	60	150	1
	80	90	1
SS 316L	25	500	3
SS 317L		1,000	3
SS 317LM		2,500	3
SS 317LMN		5,000	3
Type 410		< 5	2
Type 430		200	2
Type 439		100	3
Duplex steel[b]		> 3,000	2
Duplex steel 2205		7,500	3
AL-6XN		18,000	3
SEA-CURE		> 20,000	3

[a]Chemical compositions of stainless steels are shown in Appendix F.
[b]Duplex steel without special type. Duplex steel is a class of stainless steels with a structure approximately 50% ferrite and 50% austenite with better corrosion resistance and mechanical properties than traditional austenitic steels SS 304 and SS 316.

1. R.A. White, E.F. Ehmke, *Materials Selection for Refineries and Associated Facilities*, NACE, USA, 1991, p. 164.
2. C.P. Dillon, *Corrosion Resistance of Stainless Steels*, Marcel Dekker, Inc., New York, 1995, p. 96.
3. John C. Tveberg, P.E. and Daniel S. Janikowski, The performance of superferritic stainless steels in high chloride waters, *Stainless Steel World*, Vol. 17, October 2005, pp. 56–63.

2.3.1 Pitting Corrosion Caused by Factors Different from Chlorides

Sometimes pitting corrosion of carbon steel occurs, but there are no chlorides in water in some industrial solutions or in soil. Why?

(a) Differential aeration cells are responsible for pitting corrosion of steel equipment.
(b) Sometimes inorganic or organic salts forming on the carbon steel surface may cause pitting corrosion. For example, hydrogen chloride (HCl) and hydrogen sulphide (H_2S) gases are present in the distillation columns at oil refineries. In the presence of water, pH may be very acidic (1 to 2) and cause severe corro-

Fig. 2.4 *Under deposit corrosion*: corrosion pits formed under NH_4Cl acidic salt deposits on a carbon steel surface (4 months).

sion of carbon steel equipment. Therefore, ammonia (NH_3) is injected in the overhead of the distillation columns to neutralize acid gases HCl and H_2S. Ammonium chloride (NH_4Cl) salt deposits may be formed under some conditions on the carbon steel surface. This salt is hydrolyzed in the presence of water:

$$\begin{aligned} NH_4Cl_{(s)} + H_2O_{(l)} \rightarrow \mathbf{HCl_{(aq)}} + NH_4OH_{(aq)} \\ NH_4OH_{(aq)} \rightarrow NH_{3(g)} + H_2O_{(g)} \end{aligned} \tag{2.6}$$

Formation of hydrochloric acid causes local decreasing pH to 3 under salt deposits, and then to pitting corrosion (Figure 2.4). Sometimes this corrosion type is called *under deposit corrosion*.

(c) We said earlier that metal immersed in acid was an example of uniform corrosion. We have just described that local formation of acid may result in pitting corrosion of stainless and carbon steels. Another mechanism of pitting corrosion exists in the presence of acid condensation. If acidic gases (HCl, SO_2, or SO_3) with water vapors flow at high temperatures (above 120°C), and then the temperature drops to under 100°C (water condensation temperature, or dew point), acidic drops are suddenly formed on the metal's surface and may result in pitting corrosion (Figure 2.5).

There is another example of pitting corrosion as a result of naphthenic acid condensation during distillation of acidic crude oils. Naphthenic acids are organic acids with saturated ring structures and one or more carboxylic groups –COOH (see Section 3.7). Some crude oils contain such naphthenic acids. They boil above 200°C (216 to 370°C). Therefore, they are distilled together with fractions of kerosene and gas oil in the distillation columns, and pitting corrosion occurs during their boiling and condensation.

Fig. 2.5 Severe pitting corrosion of a carbon steel shell heat exchanger as a result of hydrochloric acid condensation. The surface looks spongy and resembles morphology caused by cavitation.

How can pitting corrosion be prevented? There are several ways:

(a) *Decrease of aggressiveness* of the environment: neutralization of acids, decreasing (or increasing in the case of acid dew point corrosion – see Section 3.3) of temperature, removing of chlorides, washing from acidic deposits.
(b) *Technological changes*: good water circulation (if water contains chlorides), or drainage and washing if circulation is stopped. In this case, stainless steel may be used for heat exchanger pipes and pumps.
(c) *Change of metallurgy*. Addition of molybdenum to stainless steel improves its resistance to pitting corrosion. For example, if naphthenic acids are present in crude oil, stainless steel with molybdenum above 2 wt% (SS 316, or SS 317) must be used. Special stainless steels with molybdenum content about 6 wt% (for example, AL-6XN) may be used in sea water application. The greater the molybdenum content, the greater the resistance against pitting corrosion.
(d) Use of *corrosion inhibitors*. Nitrates (NO_3^-), hydroxides (OH^-) and phosphates (PO_4^{3-}) prevent pitting corrosion in the presence of chloride anions. For example, potable water containing chlorides may be used in hydrostatic tests in reactors, tanks or pipes made of stainless steel if nitrates (about 0.2 wt% – as inhibitors) are injected in water. The use of the above-mentioned anions as the inhibitors of chloride pitting corrosion are based on their preferred adsorption on stainless steel surfaces relative to the adsorption of chlorides.
(e) Use of *coatings*. This means simple isolation of the metal surface from chloride attack.
(f) Use of *cathodic protection*. This is a method of protection of a metal surface from corrosion by making it a cathode by giving a negative charge (see Section 4.2). Remember that anode corrodes and a cathode does not! Chloride anions (Cl^-) cannot come close to metal surfaces in this case.

Fig. 2.6 Heat exchanger: tubes made of titanium and baffles made of carbon steel (cooling water, 4 years). Deposits on the titanium tubes are the salts (scale) formed from the cooling water, and not corrosion products. (For a full color version of this figure, see the Color Section)

Recommended Literature

1. Revie, R.W. (Ed.), *Uhlig's Corrosion Handbook*, Second Edition, Wiley-Interscience, 2006, pp. 173–190, 561–568.
2. Shreir, L.L., Jarman, R.A. and Burstein, G.T. (Eds.), *Corrosion, Vol. 1*, Third Edition, Butterworth Heinemann, UK, 1994, pp. 1:142–1:212.
3. Korb, L.J. and Sprowls, D.O., *Metals Handbook, Vol. 13: Corrosion*, ASM International, USA, 1987, pp. 231–233.

2.4 Galvanic Corrosion

We have seen how workers joined carbon steel (or galvanized steel) water pipes with a tee made of brass (copper–zinc alloy). There is almost no equipment made of only one type of metal or alloy. Look at any tap, a pump, a car, or a heat exchanger, and we will detect the parts made of dissimilar metals or alloys joined together (Figure 2.6). When such equipment is in a dry atmosphere or contacts non-electrolyte liquids (oil or any petroleum distillate possessing low electrical conductivity) nothing occurs. However, if the same equipment contacts electrolyte (aqueous solutions of salts, acids, or alkalis; among them may be organic substances that have high electrical conductivity), corrosion of less noble metal (of a metal with less electrode potential in the electromotive force series – see Table 1.3) will occur. Therefore, this corrosion type is named a dissimilar metal corrosion. We may define that *dissimilar metal corrosion is a corrosion phenomenon occurring when two different metals or alloys (or any conductor, for example, graphite) are in contact in general electrolyte.*

It is difficult to ascertain who was the first to name such a corrosive type *galvanic*, but we have to remember that as in medicine, an illness is often called after the person who first revealed and described it. Many people call this nasty corrosion phenomenon after the name of the Italian anatomist Luigi Galvani: *galvanic corrosion* (see Sections 1.5.6 and 6.1).

One may ask why the different parts of equipment are not made of the same metal or alloy, and what steps must be taken to prevent *galvanic corrosion*. Mechanical, economical, and anti-corrosion requirements define demands to make metallic equipment and structures of dissimilar metals and alloys. In spite of knowing the galvanic corrosion phenomenon for more than 200 years, engineers and architects sometimes ignore it or are not familiar with it, and corrosion failures occur because of *dissimilar metal corrosion (two-metal*, or *galvanic* corrosion). If it is impossible to design and construct the equipment of similar metals, what factors influence *galvanic* corrosion, and how may we prevent or diminish such failures?

2.4.1 Factors Influencing Galvanic Corrosion

Metallurgical, environmental, electrochemical, and geometrical factors influence *galvanic corrosion*. Metallurgical factors include the chemical composition of an alloy, the presence and properties of outer layers on a metal and alloy (oxides, hydroxides, salts, etc.). Environmental factors include the type of electrolyte solution (electrical conductivity, pH, content of aggressive variables) and conditions (temperature, flow rate, volume of electrolyte). Electrochemical factors include the electrode potentials of dissimilar metals (thermodynamic factor) and electrochemical kinetics at the metal–solution boundary (polarization characteristics, rates of the electrochemical reactions on the metals' surfaces; see Appendix C). Geometrical factors include the ratio between anode and cathode area, the distance between them, the geometrical forms of dissimilar metals, and type of joint between anode and cathode (welded, fasteners, or separate by external conductive connection). The large variety and complexity of all these factors point out that it is very difficult to predict the galvanic corrosion occurrence. This is very important for engineers who design new equipment, or upgrade old equipment. Three main factors define the correct choice of dissimilar metal contacts: the difference of electrode potentials of various metals and alloys in equipment, the ratio between anode and cathode areas, and the electrical conductivity of media. The greater the difference of electrode potentials between dissimilar metals, the more severe the galvanic corrosion that will occur. The value of 250 mV is usually defined as a "non-dangerous" one, or galvanic corrosion with a very low insignificant rate. The smaller the ratio of the anode to cathode area and the more the electrical conductance of a medium, the faster galvanic corrosion of the anode will occur. We will now discuss how these factors influence galvanic corrosion.

2.4.2 Galvanic Electromotive Force Series

We may say that it is very simple to predict the occurrence of galvanic corrosion if we know the electrode potential of dissimilar metals, namely, the electromotive force series (EMF) (see Table 1.3). The reversible electrode potentials in Table 1.3 concern the metals in the solutions of their salts with concentrations of 1 mol/L. In practice, equipment is made of dissimilar metals and works in industrial cooling waters, in sea waters, in potable waters, in other solutions, or in soils, and concentrations of metallic cations in these media are usually less than 1 mol/L. That is, we have to take into account the irreversible electrode potential of metals (see Appendix B). If we put aluminum into any aqueous solution and measure its electrode potential, the potential will not be −1.66 V (as written in the EMF series in Table 1.3). We will obtain a value of about −0.7 to −0.9 V. This is significantly more positive than the reversible electrode potential of aluminum! The same concerns such metals as chromium, titanium, niobium, and zirconium. Why? All these metals have oxide films, and the electrode potential of such metals differs from that of pure metals without films. If we remove the oxide films from the surface of these metals, we may receive a reversible electrode potential as shown in the EMF series (see Table 1.3).

The third restriction for use of EMF data in the prediction of galvanic corrosion concerns alloys usually containing two and more metals (with other elements). It is impossible to measure the reversible electrode potential of alloys. The *EMF series* relate only to pure metals.

Thus, we have to know the real electrode potential of metals and alloys in specific media under particular conditions in order to predict their corrosion. The *EMF series* for pure metals does not satisfy our needs, and a *galvanic force series* for metals and alloys in flowing sea water at 25°C has been suggested (Table 2.3). We observe the data both for metals and alloys in this table. Therefore, some alloys (for example, aluminum ones) occupy a wide range of electrode potentials. There are two positions for stainless steel and nickel based alloys: shaded and black rectangles. Shaded rectangles relate to the passive state of alloys, and the black ones to their active state (after destroying the passive layer under some specific conditions: flow rate is low, or a closed volume of water with bad aeration). In such cases, the electrode potential of stainless steel changes by about 0.4 to 0.5 V in the negative direction. Table 2.3 allows the prediction of the occurrence of galvanic corrosion when two or more dissimilar metals (or alloys) contact each other in flowing sea water at 25°C. The greater the difference the electrode potentials of two alloys, the greater is the probability of galvanic corrosion. Here are some notes for using the *galvanic electromotive force series* in Table 2.3:

(a) Ideally, the *galvanic force series* for metals and alloys must be determined separately for every particular solution under specific conditions. It is not easy, sometimes impossible, and in practice we may use the data in Table 2.3.

Table 2.3 Galvanic electromotive force series for metals and alloys in sea water[a] (dark boxes indicate active behavior of active-passive[b] alloys).

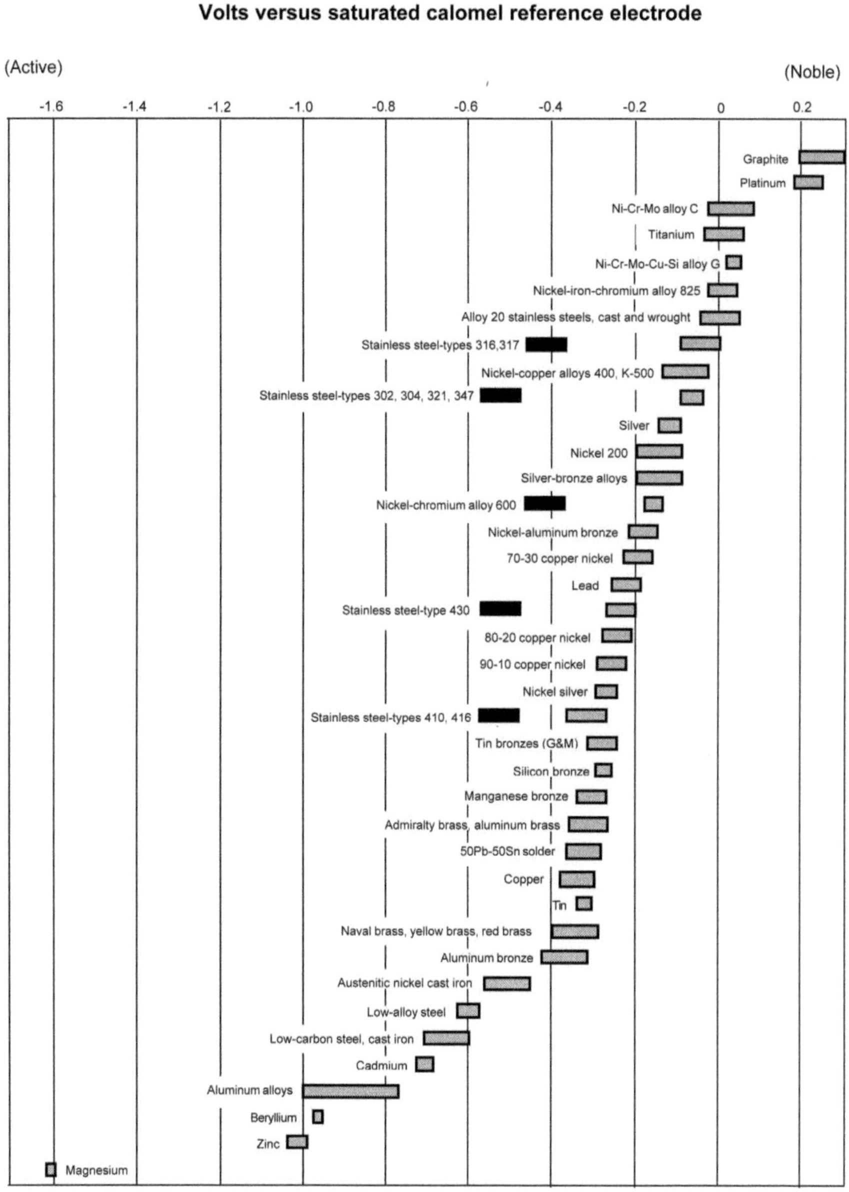

[a]F.L. LaQue, *Marine Corrosion, Causes and Prevention*, John Wiley and Sons, New York, 1975. Chemical composition of alloys is shown in Appendix F.
[b]One can read about passivity in Appendix D.

(b) The order of the metals in the *galvanic force series* in the majority of cases is similar to the order of the metals in the *EMF series* (compare Tables 1.3 and 2.3).
(c) The electrode potentials of dissimilar metals and alloys contacting each other in a solution of electrolytes may change with time because of the formation of corrosion products on anode, cathode, or both surfaces. This situation may result in a decrease in the galvanic corrosion rate, and even changes between anode and cathode (anode changes for cathode). Therefore, it is impossible to predict the occurrence of galvanic corrosion and its rate only on the basis of the *galvanic potential force series* from Table 2.3, or on the basis of the real electrode potential of metals and alloys measured in the solution. These potentials only give the information about thermodynamic possibility of the occurrence of galvanic corrosion, and do not say anything about its kinetics, i.e., the rate of galvanic corrosion.
(d) The electrochemical kinetic properties (polarization curves, see Appendix C) of dissimilar metals and alloys may show the real corrosion situation, which allows the determination of the galvanic corrosion rate and its change with time. We will show how we may predict this rate by means of electrochemical polarization curves.

2.4.3 Prediction of Galvanic Corrosion Rate by Means of Electrochemical Polarization Curves

We have to carry out the polarization curve measurements (see Appendix C) of two metals or alloys (carbon steel and Admiralty brass) that are planned to be in contact in electrolyte (Figure 2.7). The cross-section of the anodic curve for less noble metal (carbon steel – 1) with the cathodic curve for more noble alloy (Admiralty brass – 2′) will give the corrosion rate for dissimilar metal corrosion. As shown in Figure 2.7, the alloy with less electrode potential (carbon steel) will corrode at the rate i_{corr}.

These curves give information about the initial time of contact in general electrolyte. There are many variables influencing the polarization curves of dissimilar metals with time and we can imagine different situations for polarization curves data with time (Figure 2.8). The data shown in Figure 2.8 explain how the corrosion rate of alloy with less electrode potential may change with time. Therefore, it is problematic to predict the occurrence of galvanic corrosion only on the basis of different initial electrode potentials of dissimilar metals. Kinetic factors must be included in such a prognosis.

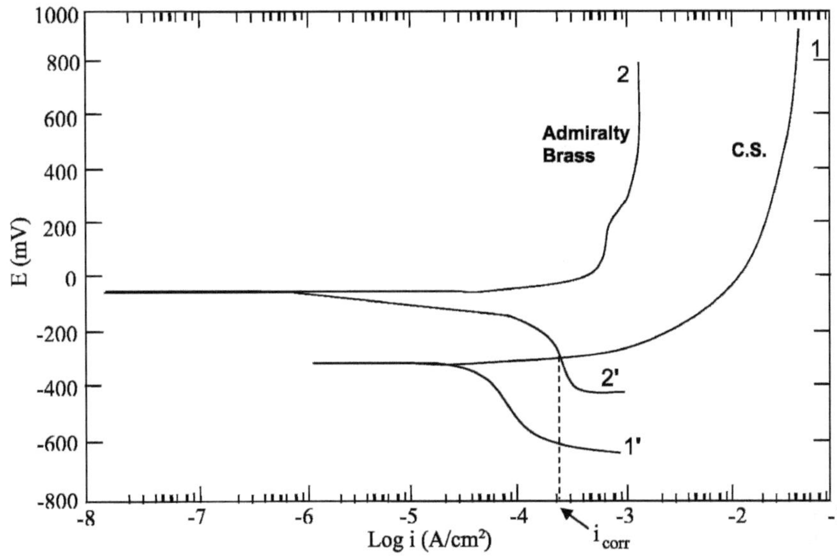

Fig. 2.7 Polarization curves for carbon steel (C.S. – 1, 1′) and for Admiralty brass (2, 2′) in the water of an extinguishing line: 1, 2 – anodic curves; 1′, 2′ – cathodic curves. $\log i_{corr} = -3.85$.

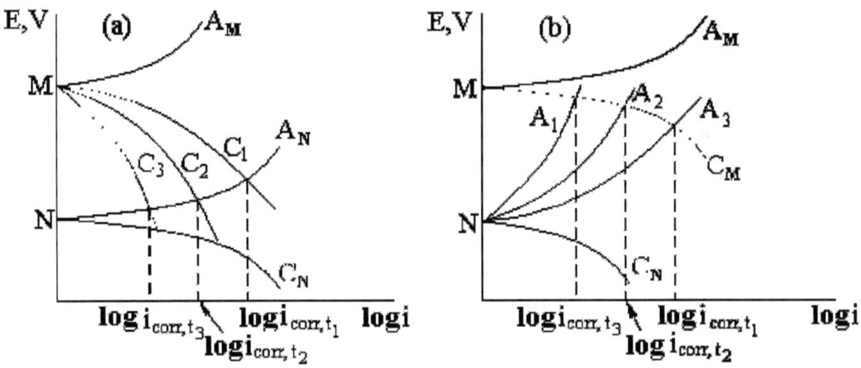

Fig. 2.8 Various situations of polarization curves for dissimilar metals (M and N). (a) Cathodic curves C_1, C_2 and C_3 for the metal M are changed with time; corrosion rates i_{corr,t_1}, i_{corr,t_2}, i_{corr,t_3} are also changed with time. (b) Anodic curves A_1, A_2 and A_3 for the metal N are changed with time; corrosion rates i_{corr,t_1}, i_{corr,t_2}, i_{corr,t_3} are also changed with time.

2.4.4 Interchange of Electrode Potentials between Dissimilar Metals

In order to prove if galvanic corrosion will occur, some electrochemists recommend measuring the electrode potential of dissimilar metals in a particular electrolyte.

This is correct only for the initial time of contact. The conditions on the surfaces of both metals may change with time, for example, corrosion products (salts, oxides, and hydroxides) may be formed on the anode or cathode surfaces. As a result, the electrode potentials may also be changed. Usually the initial electrode potential is measured at 20 to 25°C. Then, dissimilar metals contact in heat exchangers at higher temperatures. As a result of temperature changes, the electrode potential may also change. These changes may be so significant that an anode may interchange with a cathode. For example, zinc is an anode to iron in tap water at 25°C. When the water temperature rises up to 60–70°C, zinc is a cathode to iron, and the latter begins to corrode and protect the zinc. Such "interchange of electrode potentials" depends on the chemical composition of the water, and is known for the following dissimilar pairs: zinc/iron, aluminum/iron, stainless steel (13% Cr)/aluminum, tantalum/high-silicon iron, and tin/iron. The first pair is very important if galvanized steel (steel with a zinc coating) is used. The standard electrode potentials of iron and zinc are −0.44 and −0.763 V respectively (see Table 1.3). These potentials are realized if the concentrations of iron and zinc cations are 1 mol/L in aqueous solutions. We have to analyze the equilibrium

$$\mathrm{Zn_{(s)}} + \mathrm{Fe^{2+}_{(aq)}} \leftrightarrow \mathrm{Zn^{2+}_{(aq)}} + \mathrm{Fe_{(s)}} \tag{2.7}$$

The Nernst equation for this reaction at $T = 298$ K:

$$E = E^\circ + \frac{0.059}{2}\log\left[\frac{\mathrm{Fe^{2+}}}{\mathrm{Zn^{2+}}}\right], \tag{2.8}$$

$$E^\circ = E^\circ_k - E^\circ_a = E^\circ_{\mathrm{Fe^{2+}/Fe}} - E^\circ_{\mathrm{Zn^{2+}/Zn}} = -0.44 - (-0.763) = 0.323\ \mathrm{V}, \tag{2.9}$$

$[Fe^{2+}]$ and $[Zn^{2+}]$ are concentrations of iron and zinc cations in water.

Reaction (2.7) will be at equilibrium if $E = 0$. When $E > 0$, the reaction proceeds with the dissolution of the zinc. When $E < 0$, the reaction proceeds with the dissolution of the iron, from right to left. It is simple to show that reaction (2.7) may occur from right to left if

$$E = E^\circ + \frac{0.059}{2}\log\left[\frac{\mathrm{Fe^{2+}}}{\mathrm{Zn^{2+}}}\right] < 0, \tag{2.10}$$

$$0.323 + \frac{0.059}{2}\log\left[\frac{\mathrm{Fe^{2+}}}{\mathrm{Zn^{2+}}}\right] < 0. \tag{2.11}$$

Iron will dissolve at $[Zn^{2+}]/[Fe^{2+}] > 10^{10.95}$. This is possible if the concentration of Zn^{2+} is much more than Fe^{2+}. For instance, if $[Fe^{2+}] = 10^{-11}$ mol/L and $[Zn^{2+}] = 1$ mol/L, iron will dissolve in the pair iron/zinc existing in galvanized systems. And this is a real situation existing in practice. Another very important factor for galvanic corrosion to occur is the ratio between the anode and cathode areas.

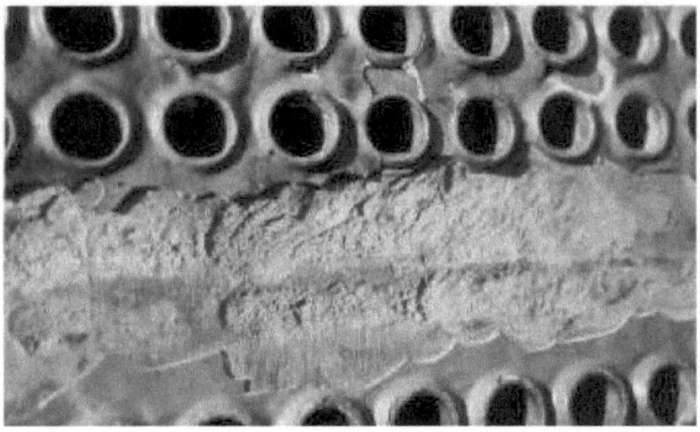

Fig. 2.9 Galvanic corrosion of coated carbon steel tube sheet because of high ratio between cathode (Admiralty brass tubes) and anode area (carbon steel tube sheet coated with epoxy). Gray – failed epoxy coating; yellow – rusted tube sheet. (For a full color version of this figure, see the Color Section)

2.4.5 Ratio between Anode and Cathode Areas

When two dissimilar metals contact in an electrolyte solution, the corrosion current is

$$I_{\text{corr}} = I_{\text{anode}} = -I_{\text{cathode}}. \tag{2.12}$$

If the anodic and cathodic areas ($A_{\text{anode}} = A_{\text{cathode}}$) on dissimilar metals are equal, the current densities ($i = I/A$) are also equal:

$$\frac{I_{\text{anode}}}{A_{\text{anode}}} = \left|\frac{-I_{\text{cathode}}}{A_{\text{cathode}}}\right|, \tag{2.13}$$

or, in the terms of current densities (i),

$$i_{\text{anode}} = |-i_{\text{cathode}}|. \tag{2.14}$$

If the anodic area (A_{anode}) is less than the cathodic area (A_{cathode}), the current density at the anode may be very high and corrosion will be also high. This assertion is very important when we coat an anode in order to protect it from corrosion. Figure 2.9 shows a heat exchanger tube sheet made of carbon steel and tubes made of Admiralty brass. Cooling water flowed inside the tubes. In order to protect the tube sheet anode from galvanic corrosion, epoxy coatings were used for a covering. We may see severe corrosion on the painted tube sheet anode. Why? There is no ideal coating without the defects. The epoxy coating covered the anode, but not ideally, as small porosity remained. Anodes were concentrated in small defective areas under the epoxy coating. If a tube sheet anode was used without an epoxy coating, anode reaction would occur on all large areas of the tube sheet. But if the tube sheet anode was coated, anode reaction would be concentrated in small defective areas. What to do? Paint the inside of the cathodic tubes (about 30 cm length) made of Admiralty brass. It sounds like the paradox "to add sugar to honey", but this is correct. Therefore, the general rule is to design the anode area to be more than the cathodic area, in order to diminish galvanic corrosion of the anode. In any case, we may exclude the galvanic corrosion between tubes and tube sheets, if the tube sheet is clad (type of metal coating – see Section 4.1) with metal/alloy similar to the material of tubes.

2.4.6 The Distance between Anode and Cathode

The greater the distance between anode and cathode, the less galvanic corrosion occurs. This distance depends on the electrical conductivity of a medium. The greater the electrical conductivity of a solution, the greater the distance from the cathode at which corrosion of the anode will occur.

2.4.7 Galvanic Corrosion in the Presence of Corrosion Products on Metals

Corrosion products forming on metals may be *isolators*, *semiconductors*, and *conductors*, and their electrode potentials are more positive than that of pure metals. Therefore, corrosion products are cathodic to pure metals in electrolytic solutions. Reductions of ferrous-ferric oxide Fe_3O_4 and ferric hydroxide $Fe(OH)_3$ are cathodic reactions according to (1.30) and (1.31) with $E^\circ = -0.315$ V and $E^\circ = -0.146$ V, respectively. Thus, these corrosion products on iron surfaces can work in two ways: as cathodes for the reduction of dissolved oxygen in accordance with (1.24), and as particular cathodes, thus becoming involved in the occurrence of galvanic corrosion. In other words, in a galvanic pair "iron/iron oxide or iron/iron hydroxide (rust)", iron oxidizes according to (1.20), and oxygen reduces according to (1.24) on the surface of iron oxide or iron hydroxide which serves as a cathode.

Sometimes engineers are surprised: "We have just changed old corroded underground pipes for new ones, and they have also corroded". Why did this happen? The electrode potential of an old carbon steel pipe coated with corrosion products is higher than that of a new bare pipe (without corrosion products). As a result, the new pipe will be an anode and will corrode. It is better to change all pipes, or to carry out correct corrosion control: to isolate the new and the old pipes, and then to apply cathodic protection (see Section 4.2). Corrosion of bare steel pipes in soil ($E \approx -500$ mV) which are in contact with steel in concrete ($E \approx -200$ mV) also has a galvanic nature.

2.4.8 Galvanic Corrosion Caused by the Presence of Metal Cations in Water

Heat exchangers made of different alloys (carbon steel and Admiralty brass, aluminum and copper) are used in water cooling systems, and these alloys are not in electric contact. Such a combination may cause severe galvanic corrosion and, as a result, pits. If copper corrosion inhibitors are not used in cooling water systems, copper may dissolve in water in the presence of dissolved oxygen, and then deposit on the iron or aluminum surface. Copper cations are now in physical (electric) contact with the atoms of the iron or aluminum surface, and take electrons from these atoms. As a result, severe pits are formed under copper "point" deposits.

If aluminum tubes, fins, or other parts are present in water cooling systems, copper ions give rise to catastrophic failure, as the copper and aluminum are arranged far from each other in galvanic series (see Table 2.3). The difference in electrode potential between copper and aluminum may reach above 1 Volt. It is a pity that it is impossible to use this electric current forming in the galvanic corrosion between copper and aluminum.

If copper ions appear in boiler feed water (this water must be of high purity!), they can result in severe pitting corrosion of carbon steel drums, that is, even water of high electrical resistance does not stop galvanic corrosion when copper ions are present in water.

Galvanic corrosion will not occur in media of high electrical resistance: in oils, in petroleum distillates, in some organic solvents, in very pure water (except in some cases). What about the atmosphere? Now we shall answer this question.

2.4.9 Galvanic Corrosion in the Atmosphere

If relative humidity is very low (not more than 60%), and there are no salts (or other electrolytes) in the air, galvanic corrosion is impossible. Humidity is the amount of water vapor in the air. Relative humidity is the ratio of the content of water vapor in the air to its content at saturation in percent at a given temperature (see Section 3.2). Galvanic corrosion will not occur in a dry and pure atmosphere. But if humidity is high and there are salts in the air (near the sea or ocean), galvanic corrosion can take place. Condensation film must be formed on the metal's surface for galvanic corrosion to occur. Its intensity will depend on period that electrolyte film will be present on the surface and the distance from the source of salt or other electrolytes (sea, ocean, chemical plants, etc.).

2.4.10 Methods to Prevent Galvanic Corrosion

(a) Do not select dissimilar metals, alloys, or other conductive materials (for example, graphite) that have a difference in electrode potentials of more than 0.25 V between them.
(b) Select an anode more than a cathode area.
(c) Use insulators between dissimilar metals.
(d) Coat the cathode and only then may you coat the anode.
(e) Do not use copper and aluminum equipment in one system.
(f) If you use iron and copper equipment in one water cooling system, you have to use corrosion inhibitors not only for carbon steel, but also for copper alloys.
(g) Reduce the electrical conductivity of media (if possible), or inject corrosion inhibitors (see Section 4.3).
(h) Sometimes you may design in such manner so that to use a third metal will work as a sacrificial anode (see Section 4.2) for two dissimilar metals which are in contact.
(i) Design a convenient way for the change of anode parts, and/or produce them thicker.

Recommended Literature

1. Revie, R.W. (Ed.), *Uhlig's Corrosion Handbook*, Second Edition, Wiley-Interscience, 2006, pp. 137–164.
2. Shreir, L.L., Jarman, R.A. and Burstein, G.T. (Eds.), *Corrosion, Vol. 1*, Third Edition, Butterworth Heinemann, UK, 1994, pp. 1:213–1:243.
3. Korb, L.J. and Sprowls, D.O., *Metals Handbook, Vol. 13: Corrosion*, ASM International, USA, 1987, pp. 83–87, 234–238.

2.5 Corrosion That Occurs with the Participation of Microorganisms

When we talk about corrosion, we think about deterioration of a metal because of a reaction with the environment. The word bacteria brings to mind microorganisms giving rise to various infectious diseases in humans. We may use a similar approach to metals. Corrosion which takes place with the participation of some special kinds of bacteria on a metal surface is called microbiologically influenced (induced) corrosion (MIC). Like not all bacteria which are present in the human body can cause illnesses, not all bacteria on a metal surface can give rise to MIC. Only special kinds of bacteria result in MIC. It is not enough that bacteria are present in the environment for MIC to occur. Bacteria must be active and contribute to metallic corrosion. MIC occurs on a metal surface in the presence of a suitable environment and conditions which help special kinds of microorganisms to participate in corrosion. Many

Fig. 2.10 MIC on a metal surface compared with pitting, crevice, differential aeration cell, or galvanic corrosion. The bottom of a crude oil storage tank after a microbial attack. (For a full color version of this figure, see the Color Section)

engineers are surprised how such small organisms can lead to spectacular failures of large industrial systems. But they are not surprised that other smaller particles (molecules) such as oxygen, water, hydrogen sulphide, and hydrogen chloride which we cannot observe with the naked eye also cause severe corrosion failures. Probably, some microorganisms can decrease corrosion. Bacteria which are present in the human body, in animals, in plants, or in food, do not influence corrosion of metals. Otherwise it was impossible to use metallic articles for our needs, as they would corrode under any bacteria activity. Other kinds of bacteria influence corrosion of metals. We shall reply to the following questions in this section: What kinds of microorganisms influence corrosion of metals? Do bacteria "eat" metals? What is the mechanism of corrosion in the presence of bacteria? How may MIC be defined and controlled?

If we take into account that corrosion phenomena have been investigated for about 100 years, people did not immediately understand that bacteria are able to give rise and to accelerate corrosion of metals. Why? Because the results of MIC on metal surfaces were found in the form of pits, which are like the results of pitting corrosion, crevice corrosion, differential aeration cells, or galvanic corrosion (Figure 2.10). Therefore, corrosion specialists gave a humoristic definition for "MIC – a term used to describe pitting when no other corrosion mechanism fits the conditions". This is correct because it is impossible to differentiate MIC appearance from pitting, crevice or galvanic corrosion, as well as from dealloying, stress corrosion cracking, and under deposit corrosion.

MIC is the "youngest" corrosion phenomenon that scientists begun studying in the 1930s. Many questions and many "white points" exist in MIC. We can only be surprised why MIC was not studied earlier as this phenomenon is connected with the invention of the optical microscope by the Dutch scientist Antony van Leeuwenhoek in the 17th century. He observed bacteria in 1674, nearly at the same time when the term "corrosion" started to be used by the English physico-chemist Robert Boyle. The scientists at first did not believe that bacteria existed.

Occurrence of corrosion depends on the type of metal, the environmental conditions, and the boundary conditions between a metal and the environment. This border is a "battle field" between microorganisms and the surface of a metal. The kinetics of the corrosion process depends on the state of this border. Chemical and physical conditions in the biofilm form an aggressive media for metals and alloys. What kinds of inhabitants exist in the biofilm? How do they live and proliferate together?

2.5.1 Existence of Microorganisms

Microorganisms are the "simple" representative of life, and they were the first inhabitants on the Earth. Estimation of bacteria quantity on the Earth gives a value of 5 million trillion of trillion (5 with 30 zeroes). This quantity is more than the quantity of stars in the universe. Sometimes bacteria are called "primitive" organisms. In spite of their "primitivism" (or because of that!), bacteria adapted to difficult conditions and "terrible" life. Today we are familiar with various microorganisms which live and grow in the environment at pH = 0 to 13, at temperatures between -15 to $+150$°C, at pressures up to 1,000 bar, in aqueous solutions with different salt content (from pure water to the Dead Sea – 26% of salts), in oils, in fuels, in the presence of radiation, and even in the presence of biocides (substances intended for killing bacteria).

Certainly microorganisms cannot "eat" metals as we eat bread, or a mouth eats paper and cheese. Microorganisms have no teeth. In order to understand how they take part in the corrosion of metals, we have to be familiar with many kinds of microorganisms called bacteria, fungi, and algae.

Bacteria (singular *bacterium*) are the microscopic organisms that lack a membrane – *bound nucleus* and *organelles*. Bacteria are unicellular and may have spherical, rod-like, or curved bodies. Bacteria inhabit all environments, including the atmosphere, soil, water, organic matter, and the bodies of humans, animals, and plants. Bacteria exist in all spheres around us: from the deep of several kilometers under ocean to dozens of kilometers above the Earth's surface.

Bacteria are surrounded by a lipid membrane that regulates the flow of chemical substances in and out of the cell. A rigid cell wall completely surrounds the bacterium and lies outside the membrane (Figure 2.11). Flagellum is a long filamentous appendage that propels bacteria; the literal meaning is "whip". Pili are thin protein tubes. All bacteria need *nutrients* – the substances that bacteria consume for their existence, growth, and reproduction. These nutrients are carbon, nitrogen, sulphur, phosphorous, inorganic salts, and trace metals. Water, air, dust, microbial byproducts, various materials (metallic components, plastics, even people) may be the sources of nutrients for bacteria.

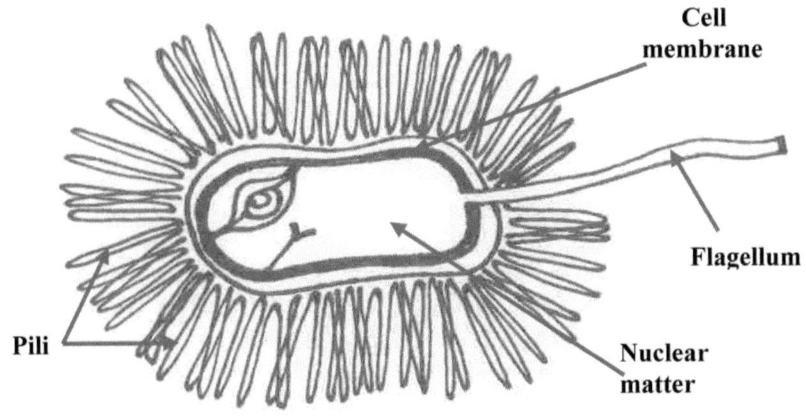

Fig. 2.11 Structure of a bacterium. The picture is by the artist Vladimir Frenkel.

Fungi (singular *fungus*) – organisms (about 50,000 types) including yeasts, rusts, smuts, mildews, molds, and mushrooms. Fungi are among the most widely distributed organisms on the Earth and are of great importance. Many fungi are free-living in soil or water; others form parasitic or symbiotic relationships with plants or animals, respectively. Fungi grow through mycelium and their nutrition is based on absorption of organic matter. Mushrooms form the greatest part of this group of microorganisms and the Latin word for mushroom, *fungus*, has come to stand for the whole group. Fungi need humidity and temperature about 20 to 30°C for their growth.

Algae (singular *alga*) – aquatic, photosynthetic microorganisms, ranged in size from the tiny flagellate of 1 μm in diameter to giant kelp that reach 60 m in length. Algae provide much of the Earth's oxygen, they are the food base for almost all aquatic life.

Microbiological activity of all these microorganisms influence corrosion of metals occurring in water, soil, crude oils, and fuels.

What is the source of microorganisms in the environment of metals? How do they appear on the surfaces of metals?

The dimensions of most microorganisms are about 1 to 5 μm in length and 0.5 μm in diameter. There is information about the existence of *nannobacteria* with dimensions of 0.05 to 0.2 μm, that is one-tenth of the diameter and one thousandth in volume of conventional bacteria. Certainly, Lilliputians exist not only among people but also among microorganisms. There is an assumption that *nannobacteria* play an important role in MIC of iron, copper, and aluminum.

Microorganisms are very light, therefore they move like aerosols (tiny solid or liquid particles suspended in the air) from one place to another, form deposits on metallic equipment and structures in air, water, soil, crude oils, fuels, and can exist for a long time without food (nutrients).

Many species of bacteria swim in liquids by means of *flagellum*, i.e., hairlike structures whose whiplike lashing provides propulsion (see Figure 2.11). Motile bacteria can swim along a chemical concentration gradient towards a higher concentration of a nutrient, and this movement is called *chemotaxis*.

Living organisms are unique in that they can extract energy from their environments and use it to carry out activities such as movement, growth, and reproduction. When applied to bacteria, the term growth is identified with development and reproduction, and refers to an increase in the number of bacteria in a population rather than in the size of an individual microorganism. Bacteria usually reproduce through binary fission, budding, chains of spores, and through the segmentation of elementary units, shortly, in asexual processes, in periods lower than 20 minutes. A single cell and its descendants will grow exponentially to more than 2 million cells in 8 hours. This growth rate is never actually realized because microorganisms are limited by space and available nutrients. They are defined by means of an optical microscope. A large quantity of microorganisms form *biofouling* (*biofilm*) defined with the naked eye and they are slippery to the touch. They choose metal surfaces as a place to live and form *biofilm* with thicknesses from several microns to several centimeters. *Biofilm* consists of microorganisms and substances (mainly products) of their metabolism. The latter is the sum of chemical reactions that take place within each cell of a living organism and that provide energy for vital processes and for synthesizing new organic material. Bacteria in biofilm excrete *extracellular polymeric substances* (EPS), or sticky polymers, which work as glue and hold the biofilm together and cement it to the metal surface. EPS is called *glycocalyx*, which serves for trapping and concentrating nutrients from the water, and acts as a protective coating for the attached cells, and protect microorganisms from biocides and other toxic substances. Because the *glycocalix* matrix holds a lot of water, a biofilm-covered metal surface is gelatinous and slippery. Biofilm is made up of bacteria and a "spiders web" of EPS. We are already familiar with some biofilms: the slippery slime on sea stones and the gel-like film inside a vase of flowers that has been kept for a week. More than 99% of all microorganisms live in biofilm communities. The material of the surface has little or no effect on biofilm development. Microorganisms adhere to carbon steels, stainless steels, and polymers (Teflon, PVC – polyvynilchloride) with almost equal "enthusiasm" within 30 seconds of exposure.

Microorganisms exist and grow on a metallic surface in the presence of liquids (usually water). Air, water, and soil are the source of microorganisms which travel from place to place and stop and congregate at the surface of metallic equipment. They are always present in the environment, but they do not always proliferate and grow. They are often "frozen" in the environment and are waiting for better conditions with good food (nutrients), suitable air, temperature, pH, flow, etc. For example, there are some types of microorganisms which form spores resistant to biocides and high temperatures. When suitable conditions appear for their growth, they begin to propagate and produce substances corrosive towards metals, or food for the growth of other kinds of microorganisms which did not even exist before in the system. There is no explicit and generally established classification of mi-

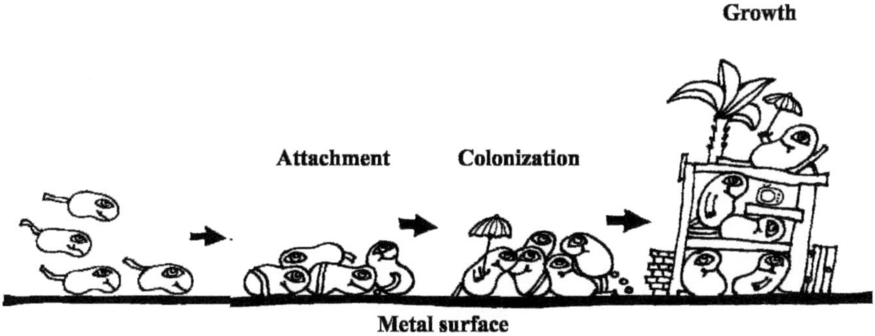

Fig. 2.12 How microorganisms attach to a metallic surface, or biofilm formation. The picture is by the artist Vladimir Frenkel.

croorganisms which are responsible for the corrosion of metals. Usually these are classified into three groups: *aerobic*, *anaerobic*, and *facultative* (*heterotrophic*).

Aerobic microorganisms use oxygen in metabolic processes. *Anaerobic* microorganisms grow in the environment without or almost without oxygen. They were the first representatives of life on Earth. Then with the development of photosynthesis (the process by which green plants and certain organisms transform light energy into chemical energy), oxygen appeared and gave rise to the destruction of anaerobic microorganisms. But not all anaerobic microorganisms were killed. They found their ecological niche under aerobic microorganisms that close and prevent anaerobic ones from obtaining oxygen. Between aerobic and anaerobic microorganisms there are *heterotrophic* ones which can exist and grow both with and without oxygen. They play an important role in the beginning of the microbiological settling on metallic surface (Figure 2.12).

Here is a scenario how three major kinds of microorganisms capture the surface of metal. Probably, the trace of organic compounds and other nutrients are the first substances adsorbing on the metal surface and form "conditioning layer" for the attachment of microorganisms. The heterotrophic microorganisms usually settle on the first. Some of these cells adsorb to the surface for a finite time, and then desorb. This is called reversible adsorption. Some of the adsorbed cells form dense formation and become irreversibly adsorbed. There is no or only a low oxygen concentration underneath. Conditions without oxygen are convenient for the growth of anaerobic microorganisms under heterotrophic ones. Aerobic microorganisms settle above heterotrophic ones and are close to the media (water, crude oil, fuel, etc.). Thus, anaerobic microorganisms are arranged the closest to a metal surface. Therefore, they are the most dangerous. Other scenarios may occur. Certainly, aerobic and heterotrophic microorganisms can separately settle on a metallic surface, form biofilm, and play, in this particular case, an essential role in the corrosion of metals. The development of a mature biofilm may take several hours to several weeks, depending on the system.

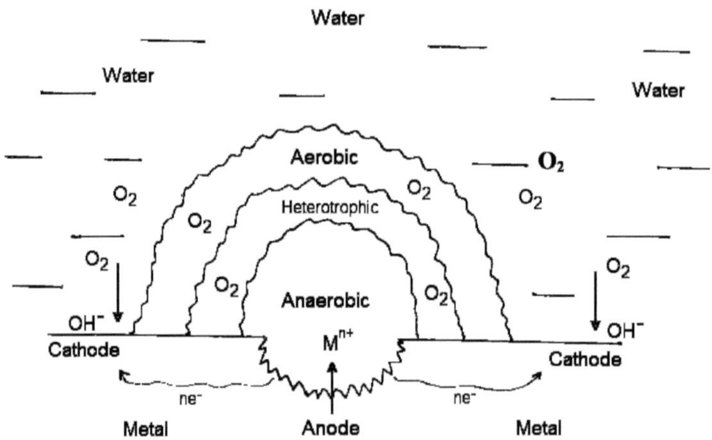

Fig. 2.13 Biofilm contains anaerobic, heterotrophic, and aerobic microorganisms which form the *differential aeration cell.* O_2 – dissolved oxygen; *ne*$^-$ – electrons; M^{n+} – metal cations. Pitting is formed under the biofilm (oxygen depletion area, see Section 1.9).

Microorganisms influence, or take part, in the corrosion of metals in different ways:

(a) They can influence anodic or cathodic reactions (kinetics of corrosion processes) on the metal's surface. Kinetics is defined by physico-chemical conditions formed on the metal–environment boundary: concentrations of dissolved oxygen and various salts, pH, redox potential, temperature, etc.
(b) They can destroy protective films on the surface of metals.
(c) They can form corrosive conditions, for instance, produce acids, sulphides, ammonia, or differential aeration cells (Figure 2.13).

The presence of microorganisms only in the environment or on a metallic surface does not prove MIC occurrence. Microorganisms must create corrosive conditions, grow and produce metabolic substances which are corrosive to metals. First of all, microorganisms need water. They are not able to be reproduced without water, and begin to be active in the presence of suitable humidity. For example, microorganisms exist in dry air, dry soil, dry fuels, and other dry chemicals, but they begin to grow in the presence of water. Besides water, microorganisms need energy source (light or chemical substances), carbon source (organic substances or carbon dioxide CO_2), electron donor (inorganic or organic substances releasing electrons), and electron acceptor (oxygen, nitrates, sulphates, or carbon dioxide – chemical substances that can gain electrons). It is convenient to divide all microorganisms taking part or influence corrosion into five groups:

A. *Sulphate Reducing Bacteria* (SRB). They are anaerobic and play an important role in corrosion of metals.
B. *Microorganisms producing acids.*

C. *Microorganisms which oxidize ferrous (Fe^{2+}) and manganese (Mn^{2+}) cations.*
D. *Slime-forming bacteria.*
E. *Methane (methanogens) and hydrogen producing bacteria.*

Below we shall discuss all these groups.

Sulphate Reducing Bacteria (SRB) are the most distributed in nature and in industrial systems. They exist in water, in soil, in crude oils, in fuels, and in wastes. SRB were historically the first microorganisms which were found to be responsible for corrosion of carbon steel in America (1910, Gains R.H.) and cast iron tubes in soil in Europe (1934, Wolzogen Kühr and Van der Vlugt). The latter scientists were the first to suggest the mechanism how SRB took part in the corrosion of metals. The most widespread kind of SRB, *Desulphovibrio Desulphuricans*, was found in the 1890s. These bacteria contain *hydrogenase enzymes*. An *enzyme* is a *protein* that catalyzes (speeds up) a chemical reaction. A *protein* is a complex, high molecular weight organic compound (bio-macromolecules) that consists of amino acids. Wolzogen Kühr and Van der Vlugt suggested that the *hydrogenase enzyme* allows SRB to consume hydrogen atoms which are formed in the cathodic corrosion process. Thus, SRB accelerate the reducing of sulphates (SO_4^{2-}) contained in soil into sulphides (S^{2-}) which, in their turn, attack metals:

$$SO^{2-}_{4(aq)} + 8H^{+}_{(aq)} + 8e^{-} \rightarrow \mathbf{S}^{2-}_{(aq)} + 4H_2O_{(l)}. \tag{2.15}$$

In the absence of SRB, the reduction of H^+ (cathodic process in neutral media) occurs very slowly. Sulphides forming in this process are corrosive to many metals, especially to iron, copper, and zinc alloys. Up to now, no explicit explanation of the mechanism of how SRB take part in the acceleration of corrosion of metals. Most common strains of SRB grow best at temperatures from 25 to 35°C. It is simple to define if SRB take part in corrosion. One can observe a thick black biofilm reaching 10 cm of thickness on steel surfaces (Figure 2.14a). Iron sulphides as corrosion products forming under SRB biofilm on a steel surface have black color. If several drops of hydrochloric acid (15 wt%) are poured on black corrosion products, it would smell of rotten eggs, a specific smell of hydrogen sulphide evolved as a result of the reaction of iron sulphide with hydrochloric acid:

$$FeS_{(s)} + 2HCl_{(aq)} \rightarrow \mathbf{H_2S}_{(g)} + FeCl_{2(aq)}. \tag{2.16}$$

Usually SRB grow under aerobic or heterotrophic bacteria in the absence of air. Another type, *Desulphotomaculum*, was first described in 1965. Some of them, for example, *Desulphovibrio salixigens* require at least 2.5 wt% NaCl, other SRB, *Desulphovibrio vulgaris* and *Desulphovibrio desulphuricans*, can grow in both fresh and salt water. All SRB require a reducing environment (absence of air) in order to thrive. They are widespread in water cooling pipelines (see Figure 2.14a) and on the bottom of crude oil and fuel storage tanks (Figure 2.14b). SRB have been implicated in the corrosion of cast iron and carbon steels, ferritic and austenitic stainless steels, copper and nickel alloys. A simple agitation or flushing of media

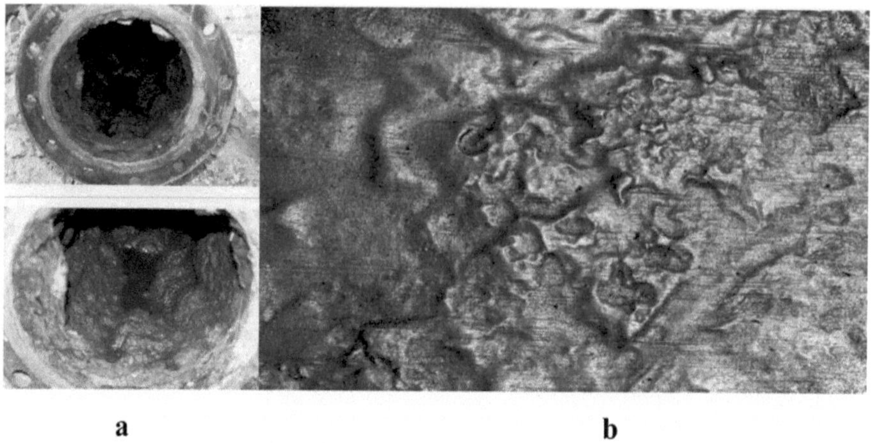

Fig. 2.14 (a) SRB inside cooling water carbon steel pipes (after four years); the thickness of biofilm with SRB was about 6 to 10 cm. (b) Corroded carbon steel bottom of the crude oil aboveground storage tank after 18 years of service as a result of SRB activity. (For a full color version of this figure, see the Color Section)

may kill SRB and prevent their dangerous attack. Mechanical cleaning (scrubbing and scraping) of sludge (biofouling) is also an effective method. One way to restrict the SRB activity is to reduce the concentration of their essential nutrients: phosphorus, nitrogen, and sulphate-containing compounds.

Microorganisms producing acids

Sulphur oxidizing bacteria. In the 1970s, the underground railway system was built in Kiev, the Ukrainian capital. When steel piles were erected under the ground, the workers noticed that they became thinner after several days. It was a big surprise, and when the pH of wet soil was measured, the value was 0.5! A sulphuric acid was found in the underground water. But there was no chemical plant in this region. Microbiological analysis showed the presence of aerobic bacteria with the poetic name *Thiobacillus Thiooxidans* in the soil. These bacteria derive energy from the oxidation of sulphur and its compounds (for example, hydrogen sulphide or other sulphur-containing substances) to sulphates (SO_4^{2-}):

$$S_{8(s)} + 12O_{2(g)} + 8H_2O_{(l)} \rightarrow 8\mathbf{H}_2SO_{4(aq)}, \tag{2.17}$$

$$4H_2S_{(g)} + 7O_{2(g)} \xrightarrow{H_2O_{(l)}} 2\mathbf{H}_2SO_{4(aq)} + 2\mathbf{H}_2SO_{3(aq)}, \tag{2.18}$$

Sulphur is present in nature as molecules containing eight atoms. Aerobic sulphur oxidizing bacteria usually live in soils. They are responsible for acid mine drainage, proliferate inside sewer lines, and when they produce sulphuric acid (up to 10 wt%), are very dangerous to iron, steels, copper, aluminum, zinc, in short, to all materials (concrete, reinforcing steel, some coatings) non-resistant to acids. Sulphur oxidizing

Fig. 2.15 Microorganisms forming biofouling at the water–hydrocarbon boundary (this mixture was taken from a kerosene storage tank).

bacteria are almost always accompanied by SRB. These two different bacteria types, aerobic bacteria in an acid environment, and anaerobic SRB in a neutral medium, may coexist and form their own microenvironment. Sulphur oxidizing bacteria exist on surfaces of stone buildings and statues, and are probably responsible much of the accelerated damage commonly attributed to acid rains.

Heterotrophic bacteria and fungi. Severe pitting corrosion was revealed in the fuel systems of several airplanes in the beginning of the 1950s. The fuel systems in the airplanes were made of aluminum alloy, and kerosene was stored inside. It was difficult to believe that kerosene consisting of hydrocarbons, non-corrosive towards metals, was responsible for such pits. Biological filaments were found on the aluminum surfaces inside the fuel systems. Microbiological analysis showed the presence of fungi *Hormoconis resinae* (formerly known as *Cladosporium resinae*). These filamentous fungi excrete organic acids, not so strong as inorganic acids, but they were strong enough to cause the pitting corrosion of aluminum. The question was how did these fungi appear and proliferate in kerosene? We said that microorganisms might be present, but not be active (not be reproduced) in any environment: air, fuel, water, solid materials, etc. The fungi might enter into the fuel storage system with air through the vents. Kerosene manufactured at oil refineries contains a very small quantity of dissolved water, about 30 to 65 ppm. Fungi cannot grow in kerosene without water, but such small water concentrations are not enough for their proliferation. During the airplanes' flights temperatures changed, air with water vapors ingressed into the fuel system through the vents or broken and unseated gaskets in the caps, and water vapors were condensed. When the quantity of water was enough to be separated from kerosene, a two-phase water–kerosene medium was formed. Now water ("home" for fungi growth) and kerosene (hydrocarbons were food for their growth) were present in separate phases at suitable temperatures and aeration. Growth of microorganisms in fuel storage tanks occurs at the water–hydrocarbon boundary and biofouling is formed (Figure 2.15).

Metabolic by-products of these fungi are organic acids such as oxalic, lactic, and acetic. Microbiological growth produces sludge matter that can clog the stream parts of fuel systems. SRB are able to grow under filamentous fungi. They release hydrogen sulphide into the fuels and thus increase their corrosiveness. Usually *Hormoconis resinae* do not exist alone, but together with various aerobic and anaerobic bacteria. Fungi and heterotrophic bacteria may deteriorate not only kerosene, but also diesel oil and lubricating oils if they have slight water contamination. Therefore, it is very important to keep and store dry fuels and other petroleum distillates free from any water contamination. The microorganisms flourish on the hydrocarbon–water border, living in water and feeding on the hydrocarbons. The fungi can form dense fungal *mats* causing operational problems (filter blockage, etc). Such *biofouling* can also consume rubber gaskets, O-rings, and rubber hoses. The food cycle of the fungi releases more water into the system, which in turn favors an increase in fungal growth. Thus, the corrosive bio-environment is self-perpetuating.

Bacteria oxidizing ammonia (NH_3) to nitric acid (HNO_3). There are some bacteria which use ammonia as food and oxidize it to nitric acid (HNO_3):

$$NH^{+}_{4(aq)} + 2O_{2(g)} + H_2O_{(l)} \rightarrow NO^{-}_{3(aq)} + 2\mathbf{H_3O}^{+}_{(aq)}. \tag{2.19}$$

This acid may attack many metals, alloys, concrete, polymers, and coatings. Ammonia is widely used in fertilizers and agricultural fields are the source of food for bacteria giving rise to severe corrosion of underground pipes and other metallic constructions. Chemical plants and water treatment units may be the source of ammonia in bays. If water from the bays is used for cooling condensers made of the copper, they may be attacked both by nitric acid and ammonia.

Bacteria oxidizing cations Fe^{2+} (iron bacteria) and Mn^{2+}. Tubercles – stalactite-like formation can be observed inside water pipes. This is the place to elucidate how stalactites and stalagmites are formed. There are several kinds of iron bacteria called poetically *Galionella*, *Sphaerotilus*, *Leptothrix*, *Crenothrix*, and *Thiobacillus ferrooxidans* which are able to oxidize cations Fe^{2+} to Fe^{3+}. Therefore, they are called iron or iron-oxidizing bacteria. Bacteria *Galionella* and *Sphaerotilus* may also oxidize Mn^{2+} to Mn^{4+}, but we do not meet manganese cations in all media. Ferrous cations Fe^{2+} are present near a steel surface in many aqueous solutions (in rivers and lakes), especially in water wells. Therefore, drinking water is very sensitive to the growth of iron bacteria. After a short while, iron bacteria oxidize Fe^{2+} to Fe^{3+} and derive energy from this electrochemical process for their growth. Ferric cations (Fe^{3+}) combine with hydroxyl anions (OH^-) and form non-soluble iron hydroxide $Fe(OH)_3$ on the bodies of iron bacteria. Slow-slow, iron bacteria are "coated" with iron hydroxide (rust), and they stop their activity. *Tubercles* are formed with dense outer shells which are porous and fragile inside. These *tubercles* form *differential aeration cells*. Anodic areas are formed under tubercles and severe pits are developed underneath as a result of under deposit corrosion. Therefore, iron bacteria are sometimes called iron depositing bacteria. The deposits are typically brown or red-brown. The iron bacteria tubercles are sometimes called *rusticle* and contain channels to allow water to flow through. They seem to be built in a ring structure similar to the growth rings of a tree. The *rusticles* are very delicate and easily fall apart into fine powder if knocked. *Rusticles* were formed on the wreck of the Titanic and were investigated in 1996 (carbon steel contacted with ocean water 84 years!). There is information that among small *nannobacteria* iron bacteria exist working as a "corrosion catalyst", because they pull Fe^{2+} cations by their negative charged cells, oxidize Fe^{2+} to Fe^{3+}, and form a supersaturated solution with Fe^{3+} cations, which in turn results in the deposition of iron hydroxides (rust).

Why are tubercles forming inside the water tubes dangerous? They clog the inner diameter of tubes and more pump capacity is needed for the supply of water. After removing the tubercles formed by iron bacteria, not only pits, but also holes are revealed in the tube walls. Therefore, iron bacteria form slow-activity mines and are responsible for severe corrosion in water cooling systems, drinking water pipes, water extinguishing lines, and fire sprinkler systems.

Slime-forming bacteria. Fungi and algae belong to facultative microorganisms that can exist and grow both in the absence and in the presence of air. These microorganisms that live and grow well in many media, produce extracellular polymers that make up slime biomassa (biofouling). This polymeric mat is actually a sophisticated network of sticky strands that bind the cells to a metal surface. Slime-forming bacteria use various organic substances (which are contaminants in water, in soil, as well as from fuels) as food. They influence corrosion in two ways: forming *differential aeration cells* (and as a result, cause localized corrosion) and excreting organic acids (oxalic, lactic, acetic, and citric). Many metals and alloys, concrete, and polymeric materials undergo attacks from these acids. Slime-forming bacteria can be efficient "scrubbers" of dissolved oxygen, thus preventing the oxygen from reaching the underlying surface. This creates an ideal site for the growth of SRB and other anaerobic microorganisms. Various types of enzymes exist within a polymeric mass. Some of them are capable of intercepting and breaking down toxic substances (biocides) and converting them into nutrients for other types of microorganisms. Sometimes the use of biocide may even result in a proliferation of microorganisms. We observed such effect when the slime at the bottom of the kerosene storage tank was treated with hypochlorite (efficient biocide in many systems!). After a short period of decrease in the quantity of microorganisms, intensive growth (more severe than previously!) occurred. In these cases, only mechanical cleaning of slime-forming bacteria may help in the prevention of MIC.

Methane (methanogens) and hydrogen producing bacteria. Methanogens are anaerobes and, similar to many SRB strains, consume hydrogen and carbon dioxide to produce methane:

$$CO_{2(g)} + 4H_{2(g)} \rightarrow CH_{4(g)} + 2H_2O_{(l)}. \tag{2.20}$$

Because *methanogens* obtain energy from the synthesis of methane and use hydrogen gas as a product of cathodic corrosion reaction, they accelerate the cathodic process on metal surfaces. The *methanogens* usually live in mud at the bottom of lakes, swamps, sewage, in organic wastes, and wastewater, because they lack oxygen. Some anaerobic microorganisms produce hydrogen gas (H_2) as a product of carbohydrate fermentation from waste materials. Hydrogen gas can diffuse into metals and cause hydrogen-induced cracking. Biological hydrogen production is the most challenging area of biotechnology with respect to environmental problems. Therefore, corrosion problems related to hydrogen attack can appear in bioreactors and other metallic equipment. Sometimes *methanogens* live in symbiosis with hydrogen-generating bacteria.

2.5.2 Control of Microbiological Activity towards Metals

How can we know if microorganisms take part in corrosion in particular systems: cooling water, drinking water, crude oil, fuel, and water storage tanks, heat exchangers, etc.? First of all, we have to examine the water, soil, sludge in tanks, tubes, or heat exchangers for the presence of microorganisms potentially responsible for corrosion of metallic equipment. Determination of *planktonic* microorganisms (free-floating bacteria whose movements are controlled by water or other liquid movement – not attached to surfaces) may not always give a simple and correct answer. Even if the microbiologists give the value of microorganisms in liquid called the total bacteria count (TBC) (for example, 10^2, 10^3, 10^5 or 10^7 microorganisms in 1 ml), is this much or little? What is the allowable value of TBC in liquid to prevent metal equipment from the danger of MIC? There is no clear recommendation regarding such a value. It depends on a specific particular system. Sometimes 10^4 microorganisms in 1 ml cannot be dangerous. In other cases, 100 microorganisms (for example, SRB) in 1 ml may be dangerous. Certainly, if there are no favorable conditions for the proliferation of microorganisms in an industrial system, we may be out of danger using 10^3 to 10^4 microorganisms in 1 ml. But if there are leakages of oil into the water system, or there is a poor water flow, or water appeared in the fuel, even 100 microorganisms in 1 ml may immediately cause severe growth of the microorganisms in the water and then give rise to corrosion over several months and even weeks. Typical microorganisms and the problems they bring with them are shown in Table 2.4. If there are no favorable conditions for the proliferation of microorganisms in industrial water cooling systems and fuel storage tanks, 10^3 to 10^4 microorganisms (not SRB) in 1 ml is an allowable value and it is recommended not to exceed this value. Another approach to the definition of the allowable quantity of microorganisms depends on the goals of the treatment of media. If the aim of the treatment against microorganisms in drinking water, pooling water, or food production, is to reach an absolute lack of microorganisms, it is not economical to kill absolutely all bacteria in industrial systems. We have to control the quantity of microorganisms (and also the conditions!) at a level not favorable for their growth in the industrial systems.

Secondly, but even more important than microbiological control in the volume of media, we have to control microbiological contamination in deposits formed on

Table 2.4 Typical microorganisms and associated problems.

Type of microorganisms	Type of problem
Slime-forming bacteria	Form dense, sticky slime with subsequent fouling
Sulphate reducing bacteria (SRB)	Generate sulphides from sulphates and can cause localized corrosion
Other (than SRB) anaerobic bacteria	Create corrosive localized environments by secreting corrosive wastes
Iron depositing bacteria	Cause the oxidation of Fe^{2+} and subsequent rust deposition
Sulphur oxidizing bacteria	Generate sulphuric acid from substances contained sulphur; can cause severe corrosion
Nitrifying bacteria	Generate nitric acid from ammonia contamination; can cause severe corrosion
Fungi (yeasts and moulds)	Clogging; localized corrosion because of organic acid secretion or excretion, and formation of differential aeration cells
Algae	Clogging; localized corrosion
Methane (methanogens) and hydrogen producing bacteria	Hydrogen induced crackings

a metal's surface, or in other words, to monitor *sessile* bacteria (attached to solid surfaces – opposite to *planktonic*). As many as 30 to 40% of the proteins present in bacterial cell walls differ between sessile and planktonic bacteria. Some of the targets for biocides are not there anymore, so sessile bacteria become difficult to kill.

Thirdly, chemical and physico-chemical analysis of deposits (mixture of corrosion products and other contaminants with microorganisms and their metabolic products) may give important information about microbiological activity. Sometimes the presence of sulphur above 1 wt% in deposits may show SRB activity. The presence of iron sulphides in deposits does not always point out SRB activity. Iron sulphides may be formed as a result of the presence of hydrogen sulphide in the environment: in crude oil and fuels, and their leakage into cooling water. High concentration of chlorides (Cl^-) in aqueous solutions shows the possible activity of iron and manganese bacteria, especially if high concentrations of iron and manganese ions are revealed in a solution. There is no one method of MIC detection. Microbiologists, chemists, metallurgists, and corrosion specialists must work together in order to respond if MIC occurs in a specific system. We have to use chemical analysis, microbiological, electrochemical, and metallurgical methods for the detection of MIC. We shall return to some of these methods in Chapter 5. For now, we shall become familiar with the prevention of MIC.

2.5.3 The Prevention of MIC

If we are sure that microorganisms are responsible for corrosion in a system, we would like to know what measures may be used for their control and MIC prevention:

(a) *Technological measures, or changes of conditions in the system.* This means that you have to create appropriate conditions to prevent the existence and growth of microorganisms. For example, aeration, if anaerobic bacteria are responsible for MIC, or agitation in order to prevent stagnation conditions. Drainage and cleaning of tank bottoms are very useful technological measures against MIC inside tanks. Mechanical scrubbing or scraping, and high-pressure spraying are also used.
(b) *Use of coatings.* Coatings may be conventional and *anti-foulants*. The inner side of storage tank bottoms for crude oil, kerosene, diesel oil, fuel oil, and gasoline must be coated. We shall talk about the types of coatings in Section 3.6. The inside of pipes may also be coated. The carbon steel tube sheet surfaces inside heat exchangers with cooling water must also be coated. Anti-fouling coatings of two types are used. The first one contains biocides which are released during the lifetime of the coating and which discourage the settlement and growth of microorganisms. The second type presents a sufficiently smooth surface to which microorganisms cannot attach. In the past, organotin (tributyltin – TBT) and copper (or copper compounds) were used as biocides in the first type of anti-fouling coatings. The legal use of TBT is now limited to vessels greater than 25 m in length. Copper in the form of copper (I) oxide, copper (I) thiocyanate, powdered or flake copper metal, bronze, copper naphthenate, copper resinate, and copper (I) sulphide are commonly used as biocides in anti-foulants although some microorganisms are tolerant to copper. Organic biocides are also used in anti-foulants, and when they are used in combination with copper, they are termed *booster biocides*. Some anti-fouling coatings can have an impact on non-target organisms. Copper-based anti-foulants cannot be used on aluminum surfaces because of galvanic corrosion of aluminum in contact with copper in an electrolyte solution. Self-polishing anti-foulants contain a cross-linked copolymer in which the cross-linking agent reacts with the sodium ion in water, causing it to slowly dissolve away, leaving a fresh concentration of biocide within the surface micro-layer of water adjacent to the paint. This cycle of dissolution and exposure prevents the settlement and growth of juvenile microorganisms. Another type of self-polishing coatings is the silicones which have very smooth surfaces.
(c) *Change of metallic constructions for non-metallic materials* (polymers, fiberglass, concrete). However, acid-producing bacteria are very dangerous to concrete, some polymers, and fiberglass.
(d) *Boiling of water* is one of the ancient disinfection measures. Many bacteria may be killed at 100°C. But firstly, we are familiar with the *thermofiles* (temperature-resistant bacteria), and secondly, killed bacteria may be food for another kinds of

bacteria. In order to kill the *thermofiles*, water must be boiled at higher pressures (in autoclave), and thus at higher boiling temperatures.

(e) *Filtration*. If we know that the dimensions of microorganisms may vary from 1 to 5 μm, water or fuel deteriorated with them may be filtrated through *biological* filters with a porosity (pore dimension) of 0.2 μm. Because of very small pore dimensions, the flow rate of liquid is very low, and much time is needed to treat media deteriorated by bacteria. Certainly, if large volumes of water or fuels are used, filtration through biological filters is not cost-effective.

(f) *Use of ultraviolet (UV) light*. Sunlight has long been known to kill microorganisms. The rays from the sun contain the UV spectrum used in UV water treatment against microorganisms. In order to kill microorganisms, the UV rays must strike the cell and the energy penetrates the outer cell membrane, passes through the cell body, and disrupts its nucleic acids, thus preventing reproduction. UV treatment is environmentally friendly as it does not add anything to water and creates no disinfective by-products. The treatment has an effective kill rate of nearly 100% of most microorganisms. There are some limitations to the UV process. If water contains iron and organic substances, proper killing may not occur because these substances will absorb the UV energy. If water contains high concentrations of calcium, magnesium, and iron, softeners and filters must be used before the water enters the UV system. The UV units kill microorganisms at one point in a watering system and do not provide any residual germicidal effects downstream. The killed microorganisms in the water are a food source for any bacteria that survive downstream of the UV unit.

(g) *Use of ultrasound. Ultrasound* is based on acoustic energy (sound-wave energy), and is sometimes called *sonication*, i.e., use of ultrasound to disrupt the microorganisms. Ultrasound is sourced with frequencies above the hearing range of human beings and uses two mechanisms. The first mechanism produces conditions of *cavitation* (*hammer of water*), which causes cell walls to break resulting in cell death. Cavitation is the process of formation and collapsing of bubbles (see Section 2.6). Microorganisms subjected to these conditions cannot survive. The second mechanism is based on the *apoptosis*, or "*dance of death*". As soon as the cells are exposed to the ultrasonic waves, they loose their ability to reproduce and will slowly enter into a dying phase. *Apoptotic* cells break into smaller pieces that other cells can eat. In other words, the ultrasonic waves induce microorganisms to commit suicide through a cascade of biochemical, physiological and morphological changes in the cell with membrane rupture. *Apoptosis* is also called *programmed cell death*. Ultrasonic devices based on these principles are manufactured and used in industrial water systems.

(h) *Cathodic protection* (see Section 4.2). In order to protect carbon steel constructions in an aqueous solution of electrolytes, from the thermodynamic point of view, the electric potential must be lower than −0.85 V regarding the copper–copper sulphate reference electrode. The presence of microorganisms in the electrolyte environment (cathodic protection works only in electrolytic solutions) needs diminishing of this protective potential lower than −0.95 V regarding the copper–copper sulphate reference electrode.

(i) *Use of biocides* – toxic substances for microorganisms. A compound which kills microorganisms is a *microbiocide*, though the shortened form *biocide* is widely used. This is the most ancient and most widespread method of microbiological control in different water and fuel systems. Therefore we shall say a little bit more about biocides.

In conclusion, we have to emphasize that if we are determined that a system is severely deteriorated by microorganisms, none of the above-mentioned methods can prevent and decrease MIC. For example, if severe biofouling (several centimeters of thickness) is determined on the bottom of the fuel storage tank, or inside a heat exchanger, or inside a water pipeline, only mechanical and chemical cleaning may help in such cases.

2.5.3.1 Use of Biocides

Not for nothing we just have said that the use of biocides is the most ancient method. We may read in the Bible (Kings, before 1000 B.C.):

> People of the town said to prophet Elisha: ... water is bad, and soil is dead. And he said: Give me a plate and put salt inside. Now put this plate with salt into the water well. I cured water, and there is no death inside water.

Probably this salt used by Elisha was the first biocide on the Earth to be recorded in the literature. Another example, copper strips were nailed to ships' hulls by Phoenicians to inhibit biofouling as copper ions kill microorganisms in water. Use of biocides for microbiological control in drinking water started in the 19th century. It is interesting to emphasize that similar biocides, and some others, are used for microbiological control in industrial water cooling systems (Appendix G). All biocides have different effectiveness, potency and duration of biocidal activity. Suffice it to say, we have to call biocides, according to their anti-microbial activity, *bacteriocide*, *fungicide*, and *algaecide*. All existing biocides are divided into oxidized and non-oxidized types and work by two ways. Some biocides change the penetrating properties of the cell wall of bacteria membranes, and as a result disturb the metabolic processes (interchange between proteins), so important for bacteria's life. Other biocides fully destroy membranes, or prevent the entry of nutrients inside and the outlet of wastes outside the cell. We shall now discuss oxidized and non-oxidized biocides.

Oxidized biocides: chlorine (Cl_2), sodium hypochlorite ($NaOCl$), calcium hypochlorite $\{Ca(ClO)_2\}$, chlorine dioxide (ClO_2), bromine (Br_2), ozone (O_3), peroxy-compounds (hydrogen peroxide, peroxy-acetic acid), bromine-organic compounds containing chlorine (halo-bromine). They are the most effective and widely used in various systems in industry. Chlorine and its compounds are effective in a very narrow pH region (from 6.5 to 7.5) and are probably the least expensive of all biocides. The activity of chlorine against attached biofilms is particularly high; not only planktonic and sessile bacteria are killed, but chlorine also destroys the polysaccharide web and its attachments to the surface. Bromine and its compounds are effective in the wider pH region. Chlorine is a good biocide and algaecide, but does not work against anaerobic *Desulphovibrio* strains (SRB) that are widespread in water systems in industry. When chlorine comes to disinfection of biofilm, a high chlorine concentration for a short duration is more effective than a low concentration for a long duration.

Chlorine dioxide has biocidal activities similar to those of chlorine. Because it is unstable, it must be mixed and prepared on-site. Chlorine, bromine, and their compounds are not effective against fungi. If ammonia is present in the system, chemical reaction with chlorine gives rise to the formation of amine-chlorine compounds. Solubility of chlorine in water decreases if the temperature increases. As to bromine, on the contrary, solubility increases if temperature increases. The synergistic effect of the use of chlorine and bromine compounds was found when they are present together. Synergism is the phenomenon of the increase of action (in this case anti-microbial activity) of two or more substances when they are present together, more than the sum of their activity when they are present separately. Usually bromine is used with sodium hypochlorite, or the latter with sodium bromide or ammonium bromide. Bromine-organic compounds containing chlorine (halo-bromine), for example, 1-bromine, 3-chlorine, 5,5 dimethylhindathoine, are solid at temperatures of about 20°C, slowly dissolve in water and turn into two acids HOBr and HOCl that carry disinfection functions. The first acid (HOBr) is a main toxicant for the microorganisms (especially at high pH, when HOCl decomposes), that gives rise to the formation of bromine anions ($Br^-_{(aq)}$) during disinfection. The latter anions react with HOCl, new HOBr molecules are formed, and so anti-microbial activity increases. This is an example of bromine returning into an active biocide form. Chlorine, bromine, and its compounds are highly toxic and dangerous not only for microorganisms but also for people. Oxidized biocides are highly corrosive. Many of them are not stable and quickly decompose. Monitoring that enables biocide optimization is very important.

Non-oxidized biocides: aldehydes, iso-thiazolones, bis-thiocyanates, quaternary ammonium salts (quats). They work as enzymes' poisons and protein denaturants, namely, change the structure of proteins which cause the loss of its biological activity. They may be more effective as work against a wide spectrum of microorganisms: bacteria, fungi, and algae. Many non-oxidized biocides are not sensitive to the changes of pH, and work over a long period of time, compared with oxidized biocides.

There is no ideal biocide which is suitable and efficient in all media. Therefore, both oxidized and non-oxidized biocides are usually used together in various industrial systems.

When added to fluid, a bacteriocide should be able to reduce the microbial populations (as measured in Colony Forming Units – CFU per ml) by a factor of 100 to 1000 times. Usually biocides in small concentrations (1 to 10 ppm) kill the *planktonic* microorganisms. Sometimes you may hear that larger concentrations of biocides are needed for killing the *sessile* microorganisms. Most biocides using in industrial systems cannot penetrate through thick biofilm. Biofilm associated bacteria (sessile) may be 150 to 3,000 times more resistant to free chlorine and 2 to

100 times more resistant to mono-chloramine than free-floating (planktonic) bacteria. Bacteria in a biofilm can resist biocides because they are shielded in slime, and delivery of the biocides is limited by its rate of diffusion through the biofilm. It takes a higher concentration, over a longer contact time, for the biocide to reach bacteria cells in a biofilm compared to free-floating microorganisms. Therefore, biocides are effective for MIC control only in clean systems. If biofilm was formed, biocides would not be able to accomplish their functions. Some companies produce *biodispersants*, but they are intended for preventing biofilm formation, and not for killing microorganisms in biofilm. Traditional biocidal testing has been done with single-species planktonic cultures. The CT constant for a biocide is the product of C (concentration) $\times$ T (time) required to kill a particular bacteria. However CT values should not be extrapolated to sessile bacteria.

Some biocides are produced by nature. For instance, natural furanones produced by Australian seaweed (algae) *Delisea pulchra* can be used as anti-biofilm compounds (incorporated into coatings).

Conclusions about MIC

(a) MIC does not result in a special phenomenon of non-localized corrosion. The presence and activity of microorganisms may give rise to pitting corrosion, crevice corrosion, selective dealloying, or to increasing galvanic attack.

(b) Microorganisms and metabolic substances produced by them are present in the porous layer, *biofilm*, containing water up to 95 wt%. The arrangement and distribution of microorganisms and metabolic substances in the *biofilm* is not homogeneous. Corrosive conditions (pH changes, different concentrations of oxygen and ions) are formed at the *metal–biofilm* border resulting in localized corrosion.

(c) Monitoring *planktonic* microorganisms is useless in most cases, since there is not often a relationship between the types and numbers of microorganisms on metal surface, where corrosion occurs, and in the bulk liquid. Monitoring the activity of *sessile* microorganisms is needed. Identification of the microorganisms will provide information on biofilm formation and biocide efficiency, but gives no corrosion information.

(d) Changes of corrosive conditions influence the electrochemical parameters that may be used for the estimation of the occurrence of MIC. Currently it is difficult to differentiate between various forms of localized corrosion as many of them have a similar appearance. Therefore, electrochemical methods of MIC monitoring must be used cautiously (see Section 5.4).

(e) Microorganisms exist in any medium: air, water, soil, crude oils, fuels, solvents, and other substances. Therefore, the determination of only the microorganisms does not prove their participation in corrosion. There are particular "corrosive" microorganisms that must be active and result in corrosive conditions at the metal–biofilm border.

(f) There is no particulate method which allows determining MIC occurrence. MIC belongs to complicated phenomenon and its determination is based on a knowledge of chemistry (especially, electrochemistry and surface chemistry), microbiology, biochemistry, materials science, and metallurgy. Only joint action may help to determine the occurrence of MIC.

Recommended Literature

1. Borenstein, S.B., *Microbiologically Influenced Corrosion Handbook*, Industrial Press, New York, 1994.
2. Charaklis, W.G. and Marshall, K.C. (Eds.), *Biofilms*, John Wiley & Sons, New York, 1990.
3. Flemming, H.C. and Geesey, G.G. (Eds.), *Biofouling and Biocorrosion in Industrial Water Systems*, Springer-Verlag, Berlin, 1991, 220 pp.
4. Geesey, G.G., Lewandowski, Z. and Flemming, H.C. (Eds.), *Biofouling and Biocorrosion in Industrial Water Systems*, Lewis Publishers, CRC Press, USA, 1994, 297 pp.
5. Videla, H.A., *Manual of Biocorrosion*, Lewis Publishers, CRC Press, USA, 1996, 273 pp.
6. Korb, L.J. and Sprowls, D.O., *Metals Handbook, Vol. 13: Corrosion*, Ninth Edition, ASM International, USA, 1987, pp. 114–122, 314–315, 492–494.
7. Revie, R.W. (Ed.), *Uhlig's Corrosion Handbook*, Second Edition, Wiley-Interscience, 2006, pp. 915–927.
8. Gaylarde, C.C., Bentom, F.M. and Kelly, J., Microbial Contamination of Stored Hydrocarbon Fuels and Its Control, *Revista de Microbilogia* **30**, 1999, pp. 1–10.

2.6 Erosion – Corrosion and Cavitaion

We presume that the reader is familiar with erosion of soil due to water streams. For instance, beaches are chewed away by erosion and disappear, and wind erosion affects rocks, soil, and sediments. The word erosion is translated from the Latin as *to gnaw* or *to wear away*. Something similar, but more complicated, happens with *erosion* of metals – destruction of metals, alloys, and other materials by the abrasive action of moving fluids and gases, accelerated by the presence of solid particles or liquid drops in gases. The substances flowing against a metal surface may be inert from the chemical point of view (sand or organic liquids) or aggressive-corrosive (steam at high temperature, liquid water or water vapors containing HCl, H_2S, CO_2, O_2, active chlorine, etc.). *Erosion* is a pure mechanical phenomenon resulting in the physical loss of a material. Aggressive species often may influence as corrodents for metals and alloys, namely, take part in corrosive processes. Corrosion is a chemical or electrochemical phenomenon (see Section 1.5). Therefore, the presence of aggressive species in the moving media (fluid or gas) results in a conjoint action involving erosion (mechanical phenomenon) and corrosion (chemical/electrochemical phenomenon) and in this case, the phenomenon is called *erosion-corrosion* (Figure 2.16). It is very difficult to determine separately the contribution of erosion and corrosion in the general erosion-corrosion phenomenon context. Usually, these mutual mechanical-chemical

Fig. 2.16 Erosion-corrosion inside a liquefied petroleum gas (LPG) carbon steel tube. "Grooves" are formed as a result of erosion. (For a full color version of this figure, see the Color Section)

(electrochemical) erosion-corrosion phenomena, occurring with metals and alloys because of the specific flow regime, or the presence of aggressive species in moving media, are differentiated as *impingement attack* (or *impingement corrosion*) and *cavitation. Impingement corrosion* is a form of erosion-corrosion usually associated with the local hits of high velocity streams of fluids, gases, solid particles, or together (for example, suspension) against a metallic surface.

Cavitation is a destructive phenomenon occurring on a metallic surface because of the formation and collapses of gaseous bubbles on it. This term was suggested by the naval architect R.E. Froude to describe the clouds of bubbles produced by propeller blades.

Erosion-corrosion is often called *flow-assisted, flow-accelerated,* or *flow-induced corrosion, cavitation-erosion,* or *liquid-impingement erosion.* This phenomenon relates to localized corrosion because the results are pits (up to holes), "grooves, valleys, and waves" at the surface of metals. Usually erosion-corrosion results in failures within a very short period: from several weeks to several months.

There are almost no metals or alloys which are not subjected to erosion-corrosion. Mild metals and alloys, such as aluminum, copper, and carbon steel, are especially susceptible to erosion-corrosion. We shall explain how erosion-corrosion occurs.

2.6.1 Mechanism of Erosion-Corrosion

If liquids or solids, as well as liquids and gases containing solid materials, and gases containing liquid drops, flow against metal surface, abrasive action will result in the mechanical wear of metal. The metal will gnaw in the form of ions or in the form of corrosion products if these ions can react quickly with the environment.

Mechanism of erosion-corrosion depends on following factors:

- flow velocity and stream character: hydrodynamic regime, namely, flow geometry, turbulence or laminar; the presence of obstacles to flowing liquid and angle of flow to a metal surface;
- environmental conditions: chemical and physical properties, for instance, pH, presence of water drops in steam, oxygen in water, acid gases, one-phase water or two-phase water-air or water-steam medium;
- the nature of metals and alloys: chemical composition, hardness, metallurgical factors (presence of various phases such as ferrite, bainite, martensite, and austenite in steels); presence, type, and structure of passive films at the surfaces of metals; roughness and geometry of metallic equipment.

There are two main reasons for the occurrence of erosion-corrosion. The first reason is wear because of the impact of particulate matter or collisions of liquid drops on the surfaces of metals. People have known from ancient times that such "weak" materials as wood and skin might be cut and treated with such "hard" materials as stone and metal. The same phenomenon occurs on a metal's surface. If two solid contacting materials (metal and eroded particles) have different hardnesses, and move against one another, the material with a higher hardness would scratch (erode) the less hard material. Kinetic energy of the particulate matter (particles) and liquid drops moving with high velocity transfers the energy needed to cut, clip, or break the outer layer of metals. Erosion is a mechanical action of metal wear. Corrosion is a chemical (electrochemical) action of the dissolution of metal. The flow velocity increases the transport rate of the aggressive species to the surface of the metal, and corrosive products in the opposite direction from the metal surface to the environment. The flow rate also reduces the electric double layer on the surface of the metal and, as a result, the diffusion rate of species taking part in anodic and cathodic corrosion reactions increases. The application of the electrochemical methods for monitoring and controlling erosion-corrosion points out the appearance of the electrochemical mechanism aspect in this complicated phenomenon.

If a particulate matter is present in a flowing liquid or gas, erosion of metallic surface may occur. In many cases, the threshold of the content of a particulate matter for the occurrence of erosion is not yet known. In some cases there are no solids in liquids at all, but erosion may occur because of flow turbulence. In these cases, erosion occurs because of the severe mechanical action of two-phase mixture liquid-gas. If the theory of erosion under particulate matter abrasion is understood well, the theory of erosion under high-velocity turbulent flow of liquids is completely unclear. Probably, severe kinetic energy of turbulent flow may result in the same effect as the hits of solid particles on the surface of metals.

The second reason for the erosion-corrosion occurrence, namely, the cavitation, is the formation and collapse of the bubbles in the liquid (the first type of cavitation), or condensation of molecules of vapors (the second type of cavitation) on a metallic surface during flow. In the first type of cavitation phenomena, turbulent flow of liquids (violent flow under agitation) may result in changes of pressure in the flowing liquid near a metallic surface. This may occur on the surface of a propeller or in the centrifugal pumps. During the passage of liquid into a centrifugal pump, the pressure sometimes drops under specific conditions. If this pressure diminishes, for instance, to a vacuum of 0.037 atm, water can boil at 38°C. Certainly such organic liquids as gasoline, acetone, or benzene which boil at temperatures of under 100°C at 1 atm, can boil at room temperature if the pressure becomes too low. It is simple to detect cavitation in centrifugal pumps if specific noises and vibration with "sounds" like stone motion against metallic surfaces are heard. The second type of cavitation relates to the sudden collapse of vapor bubbles during their condensation on a metallic surface. For instance, such type of cavitation may occur in the condensing lines if vapor water molecules become liquid drops. While vapor molecules collapse, local stresses on the surface of metal are formed because of shock wave release. These stresses possess greater energy and forces than the chemical electrostatic forces between the atoms in a lattice of metals. Such hits on a metal surface are not uniform, and shallow pits like cavities are formed. These cavities gave the name cavitation, which was formed from the Latin word *cavitas – hollow, cavity*. Cavities on the surface of a metal resemble a honeycomb, or the peculiar form when somebody ate an ice-cream with a small spoon. Cavitation of the first type is spread in turbines, centrifugal pumps, propellers, blades; the cavitation of the second type – in steam lines and other equipment with a temperature drop under dew point (the temperature of a liquid formation from vapors). Chemical factors (if aggressive chemicals are present in liquids or in the vapor phase) can accelerate cavitation.

Are there any ways to differentiate between the contribution of *erosion* and *corrosion* when two phenomena are present together? Visual examination of a metallic surface can sometimes answer the question, if *erosion* occurs, but it is impossible to determine the *erosion* and *corrosion* components in a general *erosion-corrosion* phenomenon. It was suggested identifying these two phenomena according to an investigation into how the corrosion rate depends on flow velocity. Corrosion can take place at zero flow velocity – stagnant conditions. Erosion can only occur if the flow velocity prevails at some critical marginal value. There is a common factor between erosion and corrosion: the existence of an incubation period. Certainly this period is different for erosion and corrosion. Many metals and alloys have oxide films on their surface: Al_2O_3, Cr_2O_3, TiO_2, FeO, Fe_2O_3, Fe_3O_4, CuO, NiO, etc. Erosion results in the removal of these films. If a flow velocity is not high, oxide protective films may reappear on the metal surface every time after removing. If a flow velocity rises above some critical value, there is not enough time for protective films to reappear, and the removal of the metal ions occurs from the lattice without an outer protective film. In other words, the rate of film removal is larger than the rate of film healing. It is possible to find various critical (limit or allowable) values for the flow velocity of water for different metals and alloys in the literature but these values have to be used carefully because small changes in flow regime, temperature, chemical content of liquids, and alloys may result in a shift of the critical flow velocity.

The following metallic equipment and structures may fail because of erosion-corrosion: tubes and pipes, especially bends, elbows, tees, pumps, impellers, bellows, taps, orifices, valves, turbine blades, propellers, channels, baffles in heat exchanges, etc. It is very difficult to foresee the failures and predict life longevity of metallic equipment if erosion-corrosion occurs because small changes in the flow regime, unexpected air ingestion in a circulating water, addition and operation of new pumps, appearance of new obstacles resulting in a velocity change and turbulence, the presence of aggressive gases (H_2S, HCl and CO_2), temperature drop under dew point, and construction changes of equipment, may result in the sudden development of erosion-corrosion. Therefore, failure may occur for a short period if hydrodynamics system has worked well for a long period before any changes in flow.

It is impossible to reach a conclusion about erosion-corrosion based only on visual examination of a failed surface, because similar shallow pits may be formed, for instance, as the result of the condensation of acids on the metal surface. Metallurgical examination may give the answer if erosion-corrosion has occurred. The following *preventive measures* are recommended:

(a) Strictness of engineering design, namely, prevention of places (if possible) with hydrodynamic changes, use of impingement plates or impingement sleeves on the outer surface of the tube bundles.
(b) Cathodic protection (see Section 4.2) in some cases may prevent or diminish erosion-corrosion, because hydrogen gas which may be formed on a metal surface from water during negative electric potentials can act as a protective layer against erosion.
(c) The use of alloys less sensitive to high velocity: the cobalt-base stellite alloys, stainless steel and titanium.
(d) The use of elastomeric coatings: polyurethane and rubber.
(e) The use of ceramic coatings.

We should not forget to use monitoring methods such as periodical inspection of equipment, use of ultrasonic and radiographical devices, and recording the physical and chemical content of flowing media (see Chapter 5).

Recommended Literature

1. Fontana, M.G., *Corrosion Engineering*, Third Edition, McGraw-Hill, 1986, pp. 91–95.
2. Powell, G.W. and Mahmoud, S.E., *Metals Handbook, Vol. 11: Failure Analysis and Prevention*, Ninth Edition, ASM International, USA, 1986, pp. 163–171.
3. Korb, L.J. and Sprowls, D.O., *Metals Handbook, Vol. 13: Corrosion*, Ninth Edition, ASM International, USA, 1987, pp. 7, 311–313.
4. Shreir, L.L., Jarman, R.A. and Burstein, G.T. (Eds.), *Corrosion, Vol. 1*, Third Edition, Butterworth Heinemann, UK, 1994, pp. 8:197–8:207.

5. Revie, R.W. (Ed.), *Uhlig's Corrosion Handbook*, Second Edition, Wiley-Interscience, 2006, pp. 233–294.
6. Levy, A., *Solid Particle Erosion and Erosion-Corrosion of Materials*, ASM International, USA, 1995, 220 pp.
7. Berge, P., Modelling Corrosion for Life Prediction of Nuclear Reactors Components, In: *Life Prediction of Corrodible Structures, Vol. 1*, R.N. Parkins (Ed.), NACE, 1994, pp. 14–27.
8. Coulon, P.A., Wear Prediction for Power Generating Corrodible Structures, In: *Life Prediction of Corrodible Structures, Vol. 2*, R.N. Parkins (Ed.), NACE, 1994, pp. 879-891.
9. Lichtenstein, J., Condenser Tube Erosion – Corrosion Repairs Using Coatings, *Materials Performance* **39**(2), 2000, pp. 52–53.
10. Groysman, A., Kaufman, A., Feldman, B. and Man, Y., Cavitation Corrosion in a Continuous Catalytic Reformer Unit, *Materials Performance* **39**(6), 2000, pp. 62–66.

2.7 Other Corrosion Phenomena

In this section we will describe *fretting corrosion*, *intergranular corrosion*, *dealloying*, *stress corrosion cracking*, *hydrogen damages*, and *corrosion fatigue*.

2.7.1 Fretting Corrosion

Fretting is *wear* that occurs between tight-fitting surfaces subjected to a cyclic relative motion of extremely small amplitude. The results of *fretting* are pits and transfer of metal particles from one surface to another. Small particles of metal which are broken off oxidize, forming oxides which are harder than the mated metallic parts. These particles become trapped between the mating surfaces and cause abrasive damage and scoring. *Fretting* is a mechanical failure and appears to be aggressive in cases of gears, pulleys, wheels, flywheels, ball, roller or bridge bearings, riveted and pin joints, riveted lap joints, bolted flanges, couplings, hubs, orthopedic implants, and electrical contacts. It is worst under perfectly dry conditions, increases with contact load, slip amplitude, and the number of oscillations. Soft metals are more susceptible to *fretting* than hard metals and alloys. *Fretting* may cause serious dimensional loss of accuracy of closely fitted components and reduce their fatigue strength.

Fretting corrosion is a corrosion phenomenon that occurs at the contact area between two materials under load and subject to minute relative motion by vibration. In other words, it is the *fretting* that occurs if the environment participates in metal destruction. Thus, it is the combining of *corrosion* and *wear*, or the *abrasive* effect of corrosion debris often occurring in equipment with moving or vibrating elements. During vibrating, the protective film on the metal surfaces is removed by a rubbing action and exposes fresh, active metal to the corrosive action of the atmosphere. Other terms for *fretting corrosion* are *brinelling*, *false brinelling*, *chafing fatigue*, *cocoa*, and *friction oxidation*. Reddish brown oxide debris and spots usually appear as a result of this corrosion on ferrous metal contacts. *Fretting corrosion* causes pits, grooves, seizing, and *galling* of mating surfaces. The latter occurs as a result of excessive friction between high spots resulting in localized welding with subsequent spalling and a further localized roughening of rubbing surfaces and the creation of protrusion above the metal's original surface. *Fretting corrosion* is most detrimental to electrical contacts and causes the formation of hard oxide particles between soft metal contacts. These ceramic particles are pressed between two soft electrical contacting layers which cause an increase in electrical resistance.

Prevention and control of *fretting corrosion*:

- Use of lubricants, surface treatment (phosphating) and coatings; ensure that the film of lubricant is constantly maintained between the surfaces. Lubrication has a positive effect against fretting corrosion by reducing friction and wear, but can promote slip rather than prevent it.
- Increase the hardness of surfaces.
- Decrease bearing loads on mating surfaces.
- Use of barriers to limit ingress of a corrosive environment to mating surfaces.
- The best and most effective method of preventing fretting corrosion is to avoid the relative motion of mated parts by a *special contact design*: selecting materials and combinations that are less susceptible to fretting corrosion, prevention of slip, namely eliminating the source of vibration, increasing the friction between the surfaces, or restricting the movement.

2.7.2 Intergranular Corrosion

The German physicist Max von Laue (1879–1960) discovered in 1912 the diffraction of X-rays by crystals, namely that metals and alloys consist of grains. Various grains differentiate from each other with boundaries. The latter differ from the grains. *Intergranular (intercrystalline) corrosion* is the selective dissolution of grain boundaries. This dissolution is caused by electrochemical heterogeneity between precipitates at the grain boundaries and the grains themselves. The mechanism differs with each alloy, but most often intergranular corrosion occurs with stainless steels. What can happen with the latter (for instance, SS 304) containing iron, nickel, chromium, carbon, and nitrogen if it is heated to between 425 and 875°C? Such temperatures may occur during production, fabrication, and welding (heat-affected zone). The carbon content of the early (1920–1930) austenitic stainless steels was ~0.15 wt%. Chromium and carbon previously dissolved in an iron lattice forming the solid alloy solution, react at temperatures of 425 to 875°C with the formation of carbides $Cr_{23}C_6$ or Cr_7C_3. Carbon diffuses in austenite faster than chromium and, therefore, comes from the volume of alloy and meets the chromium. Time and temperature defines the amount of carbide precipitation. There is not enough space for these carbides in the grains, and they are precipitated preferentially at the grain boundaries where nucleation is favored. As a result, depletion of chromium occurs in the alloy adjacent to the precipitate. This state of steels is called *sensitization*, and they become susceptible to intergranular attack in a corrosive environment. Austenitic stainless steels with 0.03% C and above are at risk at 425 to 875°C. Carbon diffuses too slowly at $T < 425$°C. Chromium diffuses fast enough at $T > 875$°C to replenish local depletion. A decrease of the chromium content in grains causes the loss of corrosion resistance and can lead to an increase in corrosion rate if such steel comes into contact with specific chemicals, for instance, a ferric sulphate–sulphuric acid solution. *Intergranular corrosion* in the heat-affected zones close to welds is called *weld decay*. We can imagine *sensitization* similar to the particular state of when a building was made of bricks but cement was removed. Contact with a specific environment may cause such a building to fall.

Not only austenitic stainless steels, but also aluminum alloys, ferritic stainless steels (SS 430, SS 446), nickel alloys (600, 601, 800, 800H, Nickel 200, Hastelloy B and C) are susceptible to intergranular corrosion. The latter with aluminum alloys in an industrial or marine environment is called *exfoliation*. Susceptibility to intergranular corrosion depends on the environment and on the extent of intergranular precipitation, which is a function of alloy composition, fabrication, and heat treatment parameters.

What are the remedies against intergranular corrosion?

(a) *Heat treatment*. If we heat austenitic stainless steels to temperatures >1050°C (annealing treatment), carbides $Cr_{23}C_6$ (Cr_7C_3) dissolve, and carbon is in a solid solution. Then rapid cooling by water (quenching) through the precipitation range of 875 to 425°C will not allow precipitating carbides.
(b) *Low carbon content*. If we know that carbon can react with chromium and form carbides, another way is to diminish the carbon content to such low concentrations that no precipitates are formed during welding or other thermal processes. Such a border carbon concentration is 0.03 wt% was defined and extra-low carbon grades 304L, 316L and 317L were introduced in practice. We should take into account that reduction of carbon content lowers the strength of stainless steels, and as a result heavier equipment must be designed. Decreasing the carbon content is not effective in the cases of long-term exposure at 425 to 875°C.
(c) *Use of stabilizers*. Titanium (Ti) and niobium (Nb) form more stable carbides (TiC and NbC) than chromium. If Ti and Nb are added in low concentrations to stainless steel, they combine with (*stabilize*) carbon and prevent the formation of chromium carbides. Stabilized steels SS 321 (with Ti) and SS347 (with Nb) were introduced in the 1930s. Niobium is a better carbon scavenger than titanium, and SS 321 can be attacked. Intergranular corrosion of stabilized steels near the welds is called *knife-line attack*.

2.7.3 Dealloying

Metal alloy is the mixture of several elements, and at least one of them must be a metal. This mixture creates thermodynamic heterogeneity, because the chemical potentials of elements contained in metal alloys are different. If a metal alloy is immersed in the electrolyte environment under specific conditions, different electrode potentials of elements in a metal alloy may cause *selective leaching*, or *dealloying*, of one of the elements. Elements are added to metal alloys, first of all, for improving their mechanical properties. If one of the elements is leached from the alloy, the mechanical properties deteriorate. Many alloys are subjected to *dealloying*, but *brass* and *gray cast iron* are most susceptible this attack under particular conditions. This phenomenon has many names: *dealloying attack*, *parting corrosion*, *parting*, *selective leaching*, *selective dissolution*, *selective corrosion*, *preferential leaching*, or *leaching*. When zinc selectively dissolves from solid brass, the phenomenon is called *dezincification*; if carbon is leached from the alloy, it is called *decarburization*; if cobalt dissolves from its alloy, it is called *decobaltification*, nickel – *denickelification*, aluminum – *dealuminification* (*dealuminization*), tin – *destannification*, silicon – *desiliconification*, etc. In other words, this phenomenon can be called *grain boundary depletion*. Only selective leaching of iron from gray cast iron is wrongly called *graphitic corrosion*. It would be more correct to call it *deironification* but historically specialists continue to use the term *graphitic corrosion*. *Dezincification* is the most widespread *dealloying* phenomenon, because brass is widely used in industry and in daily life (many elements of valves and tees for drinking water are made of brass). Brass consists of copper and zinc which are both amphoteric metals and can dissolve in acidic (copper only in the presence of dissolved oxygen) and alkaline aqueous solutions. Brass is resistant in contact with neutral aqueous solutions, but zinc as a component of brass can dissolve selectively under particular conditions: slightly acidic or alkaline solutions, in the presence of chlorides, high concentration of dissolved O_2 and CO_2, stagnation or low rates of liquid flow, relatively high temperatures (above 60°C), formation of deposits on a brass surface, in crevices with low degrees of aeration, and exposed in particular atmosphere.

The mechanism of *dezincification* is not clear, and two explanations were suggested. In the first explanation, zinc as the metal with less electrode potential than copper, selectively dissolves from the brass lattice structure, and porous copper with weaker mechanical properties than the alloy remains on the surface of the previous brass. In the second explanation, the brass alloy Cu-Zn dissolves as one compound, but copper is immediately reduced at these places, and porous spongy copper is formed. In both cases, mechanical properties (strength, hardness and ductility) of brass deteriorate. After the removal of zinc from the brass surface, its color changes from yellow to red. There are several preventive measures against *dezincification*. First of all, only copper alloys containing above 15 wt% Zn are subject to *dezincification*, and the first way is to decrease the content of zinc to less than 15 wt%. This is not always possible because the mechanical properties decrease in this case. Adding tin (Sn, 1 wt%) and three elements (arsenic As, antimony Sb, and phosphorus P at 0.02 to 0.06 wt%) separately or together with Sn, improves the resistance of brass to *dezincification*. When these elements are added to brass, it is called *Admiralty inhibited brass* (Cu and Zn with Sn and one of the three elements As, Sb, or P), or *Naval brass* (Cu and Zn with Sn only). The inhibited brasses are recommended for use when aqueous solutions are expected to cause *dezincification*. Cathodic protection can also be used for the prevention of *dealloying*. One important measure against *dezincification* is to retain the conditions of aqueous solutions which cannot cause it, that is neutral or near-neutral aqueous solutions must be used, the temperature should be low (not more than 60°C), stagnant conditions and formation deposits on brass surface should be eliminated. The environmental conditions are not clearly specified for preventing *dezincification*, and it may occur even with inhibited brass which was supposed to be immune. Lead is added to brasses to enhance their machinability. Valves and tees for drinking water may be produced from these brasses. The problem is that lead may selectively dissolve and deteriorate drinking water, and thus be dangerous for people.

Another example of *dealloying* is *graphitic corrosion*, which occurs in *gray cast iron*. The latter consists of iron (>95 wt%), carbon (2.1 to 4 wt%) and silicon (1 to 3 wt%). Silicon, during manufacture of *gray cast iron*, causes the carbon to rapidly come out of the solid solution as graphite (inside the alloy), leaving a matrix of iron. The latter is an anode and graphite is a cathode in the presence of electrolyte (see Table 1.3). Iron dissolves, leaving a brittle and weak graphite. The form of the cast iron object maintains, but mechanical strength decreases. Slight loading can cause fracture. *Graphitic corrosion* is often not readily apparent from visual examination. It occurs in buried *gray cast iron* pipes after many years of exposure to soil in the presence of SRB, and in *gray cast iron* cannons in ships that have been sunk at sea. There are several methods of graphitic corrosion control: replacing the cast iron, suitable heat treatment, applying a coating to cast iron, use of cathodic protection, altering the environment (increasing pH to neutral or slightly alkaline levels, decreasing Cl^- and SO_4^{2-}, use of inhibitors), and avoidance of stagnant conditions.

Selective leaching is sometimes intentionally used to prepare catalysts with large surface areas, for instance, Raney nickel (see Section 6.2).

2.7.4 Stress Corrosion Cracking

Two events related to the death of people and the phenomenon of the unusual failure of bullets and the explosion of riveted boilers occurred more than 100 years ago. The first event occurred in the British Army in India when soldiers noticed that cartridge cases made of brass cracked after storage during the rainy season. This failure of the cartridge cases was called *season cracking*, and occurred because of the presence of ammonia resulting from the decomposition of organic matter. Even 1 ppm of ammonia can cause the *cracking* of brass. The second event, the explosion of riveted boilers made of carbon steel, was attributed to *caustic embrittlement* because of brittle failures at the rivet holes and whitish deposits of caustic sodium hydroxide (boiler water was treated with alkalies to minimize corrosion). Both terms are now obsolete. J.C. Hodge and J.L. Miller distinguished failures by cracking in austenitic stainless steel under tensile stresses in the presence of chlorides from intergranular corrosion in the mid-1930s and called it *chloride stress corrosion cracking* (Figure 2.17).

Fig. 2.17 Chloride stress corrosion cracking of an SS 316 tube in the atmosphere containing HCl gas.

Then it was shown that dissolved oxygen or other oxidizing species accelerate stress corrosion cracking in the presence of chlorides. Cracking of solid metals induced by liquid metals and fused substances is called *liquid metal embrittlement.*

What is common between all these corrosion phenomena? Metals and alloys subjected to constant (static) tensile stresses and exposed to certain environmental conditions (type, concentration of aggressive species and temperature) within certain electrode potential range may develop cracks, and this phenomenon is called *stress corrosion cracking* (SCC) or *environment-induced cracking*. The result is fracture caused by a combination of mechanical loading and chemical attack. If the load varies, the phenomenon is called *corrosion fatigue*. Not all environments and concentrations of aggressive species can cause SCC, but sometimes even pure water may induce SCC of some alloys. Tensile stresses may be residual and operating (applied). Residual stresses are formed as a result of welding, thermal treatment or other fabrication methods (cold deformation, such as bending and drawing), and they involve inhomogeneous deformation in materials. Such stresses must be relieved by proper heat treatment, named stress-relief annealing. Applied stresses are formed during the operation of equipment. Compressive stresses do not cause SCC, and this fact is used as one of the preventive methods. Most likely locations for SCC are welded, bolted, and riveted joints, notches and crevices. Usually cracks start at surface discontinuities (preexisting or corrosion-induced surface features), or at pits. Time of failure may last from minutes to several years. The result of SCC is cracks (namely, brittle fracture which means sudden rapid fracture under stress), similar to that formed in *intergranular corrosion*. The two phenomena, SCC and intergranular corrosion, are different as the latter occurs in a specific environment without stresses. Usually, during SCC most of an alloy surface is not attacked, while fine cracks appear and progress through it. SCC is associated with intergranular or transgranular paths through the metal. Intergranular cracking proceeds along grain boundaries, while transgranular cracking advances without preference for boundaries. Cracking proceeds generally perpendicular to the applied stress, and its morphology can vary from a single crack to branching similar to a "river delta". No unifying mechanism of SCC exists, and it is not clearly understood. Most suggested mechanisms invoke either a process of localized embrittlement of the metal in the vicinity of the crack tip or of localized electrochemical dissolution in that region. Four main factors of SCC exist:

Table 2.5 Alloy-environment systems exhibiting SCC[1–4].[a]

Alloy	Environment
Carbon steels	Hot nitrate, hydroxide, carbonate/bicarbonate solutions, anhydrous liquid ammonia, aqueous amines, molten Na-Pb alloy, liquid Li
Ferritic steels	Hydroxide, nitrate, phosphate solutions, anhydrous liquid ammonia, high temperature water, CO and CO_2 in H_2O, CS_2 in H_2O
Martensitic steels	Aqueous solutions of hydroxides, nitrates, sulphides, and chlorides
High-strength steels	Aqueous electrolytes (containing H_2S)
Austenitic stainless steels	Chlorides (in liquid water and steam), bromides, hydroxides, polythionic acids ($H_2S_xO_y$), sulphurous acid, Cl^- + H_2S (aqueous)
Duplex steels	Chlorides, hydroxides, Cl^- + H_2S (aqueous)
Nickel alloys	High-purity steam, caustic soda solutions, polythionic acids, chlorides, fluorides, HF
Copper and copper alloys	Ammoniacal solutions (ammonia and amines), nitrites, nitrates, steam, liquid Li, Hg, Bi
Aluminum alloys	Chlorides, bromides and iodides in water, high-purity hot water, CCl_4; liquid Hg, Ga, Zn
Titanium alloys	Chlorides, bromides and iodides, red fuming nitric acid, anhydrous N_2O_4, methanol-halides, HCl, CCl_4; liquid Hg
Magnesium alloys	Chlorides, NaCl-K_2CrO_4 solutions, distilled water
Zirconium alloys	Chlorides, aqueous solutions of $FeCl_3$ and $CuCl_2$, halogens in water, halogen vapors, methanol + halides, HNO_3 (concentrated), CCl_4; fused salts
Gold alloys	$FeCl_3$ and acidic sulphate solutions, acetic acid-salt solution, aqua regia[b]
Monel	Fused caustic soda, HF, H_2SiF_6
Nickel	Fused caustic soda

[a] This table is not intended to be an exhaustive list.
[b] Aqua regia is an aqueous solution of HCl (3 volume) and HNO_3 (1 volume).

(a) *Metallurgical*: type of alloy (metal composition, strength), its structure, presence of passive film on surface.
(b) *Environmental*: type of substances and their concentration, pH, temperature, physical state (one phase or alternate wetting and drying conditions).
(c) *Electrochemical*: electrode potential of metal surface.
(d) *Mechanical*: tensile stresses and stress intensity factor.

Environmental conditions for SCC of particular metals and alloys are summarized in Table 2.5. This table shows that we know which alloy–environment combinations may cause SCC, and by excluding their use, we can apply preventive measures especially at the first stage of a project. Environment may be aqueous and non-aqueous media, fused substances, liquid metals, and alloys. It is important to note that stainless steels which have protective films require the presence of aggressive chlorides to promote SCC, but carbon steels require the presence of passive solutions of anodic inhibitors (nitrates, hydroxides, carbonates, and bicarbonates). The tensile stress value, or stress intensity factor, is a very important feature, and there is a threshold value for SCC occurrence. Usually the susceptibility to SCC increases with the strength of alloys. The electrochemical factor is very important at the first stage, and the mechanical factor is very important during further stages of SCC development. If we know the factors causing SCC, we can prevent or control it by using them.

Prevention and Control of SCC

1. *Metallurgical measures* include selecting a resistant alloy in a particular environment, or changing the alloy composition and structure. For instance, use of stainless steels with a maximum content of carbon 0.02% and suitable heat treatment can avoid sensitization or redissolve carbides and eliminate chromium depletion, and thus prevent SCC. Ferritic stainless steels are much more resistant to chloride SCC than austenitic stainless steels. Pure metals are more resistant to SCC than alloys. Addition of aluminum, titanium, or niobium with titanium, which react preferentially with carbon and nitrogen in steels, improve resistance (but are not immune) to SCC.
2. *Metallic* and *organic* coatings isolate the surface of alloys from the environment and thus prevent SCC. But if coatings using for thermal insulation on stainless steels release chlorides, SCC may occur.
3. *Environmental measures* include removing aggressive chemical species (for instance, chlorides or dissolved oxygen), addition of inhibitors (nitrates and phosphates to boiler waters; sometimes inhibitors are added to priming paints), injection of oxidizing substances increasing electrode potential above the range where SCC can occur, and keeping the temperature under the danger level. The presence of small concentrations (0.02 ppm) of oxygen in anhydrous liquid ammonia can cause SCC of carbon steel. Addition of 0.2% water can prevent SCC in this case.
4. *Mechanical measures* include decreasing of residual and operating stresses. The first may be decreased by thermal stress relief. The operating stresses may be decreased by diminution of load and avoiding stress concentrators. Shot peening (named also shot blasting or grit blasting) results in formation of surface compressive stresses which are harmless. If the surface was cracked, peening will not help against SCC.
5. *Cathodic or anodic protection.*

In some cases, SCC occurs during startup and shutdown periods. Special measures for removing aggressive substances are recommended in these cases. For instance, in order to protect refinery equipment made of austenitic stainless steel from polythionic acid SCC while idle, nitrogen purging, washing by alkaline aqueous solutions (2 wt% Na_2CO_3), and dry air purging is used.

The presence of hydrogen in the environment can cause SCC, called hydrogen induced cracking (HIC), or stress-oriented hydrogen-induced cracking (SOHIC), but is simple to distinguish such failure from other species. Cathodic protection is an effective preventive measure of SCC, whereas it rapidly accelerates HIC and SOHIC. Now we shall discuss hydrogen damage.

2.7.5 Hydrogen Damages

Hydrogen is the lightest, the smallest, and the most abundant element in the universe, the third most abundant element on the Earth's surface, and can cause very dangerous corrosion phenomena under the general term of cracking which relates to its penetration into metals and alloys. This hydrogen damage is close to SCC, but is caused only by hydrogen atoms and molecules and tensile stresses, and has four types (Figure 2.18): *hydrogen blistering*, *hydrogen-induced cracking* (HIC), *stress-oriented hydrogen-induced cracking* (SOHIC), and *sulphide stress cracking* (SSC). Cracking occurring because of hydrogen is also called *hydrogen embrittlement* or *hydrogen cracking*. What is the source of hydrogen? Hydrogen gas is formed and used in many industrial and technological processes: during chemical cleaning by acids, thermal dissociation of water in metallurgical processes (casting and welding), decomposition of gases, cathodic protection, galvanic plating, and some corrosion reactions. The role of hydrogen increases all the time in the context of its use as an energy carrier. Hydrogen can enter metals and alloys either from a gas or a liquid phase. The mechanism of entering from the gas phase includes adsorption of H_2 (or other gas containing hydrogen, for instance, H_2S), its dissociation into adsorbed atoms H_{ads} on the surface and the diffusion into the metal lattice. In the liquid phase, hydrogen is formed as a result of an electrochemical reaction in acid

$$H_3O^+_{(aq)} + e^- \rightarrow H_{ads} + H_2O_{(l)}, \tag{2.21}$$

or in neutral and alkaline solutions:

$$H_2O_{(l)} + e^- \rightarrow H_{ads} + OH^-_{(aq)}. \tag{2.22}$$

Cathodic protection of steel can result in hydrogen forming on the surface of a bare or under-coated construction according to reaction (2.22) of the decomposition of water under proper electrode potentials. Hydrogen atoms forming in these cathodic processes are called *nascent* and are very active, which can diffuse into metals. Corrosion inhibitors are used for reducing this process. Oil and gas, oil refining and petrochemical industries particularly suffer from *hydrogen damage*, because most media in these industries contain hydrogen sulphide which reacts with metals with hydrogen formation:

$$Fe_{(s)} + H_2S_{(aq)} \rightarrow FeS_{(s)} + \mathbf{H_{2(g)}}. \tag{2.23}$$

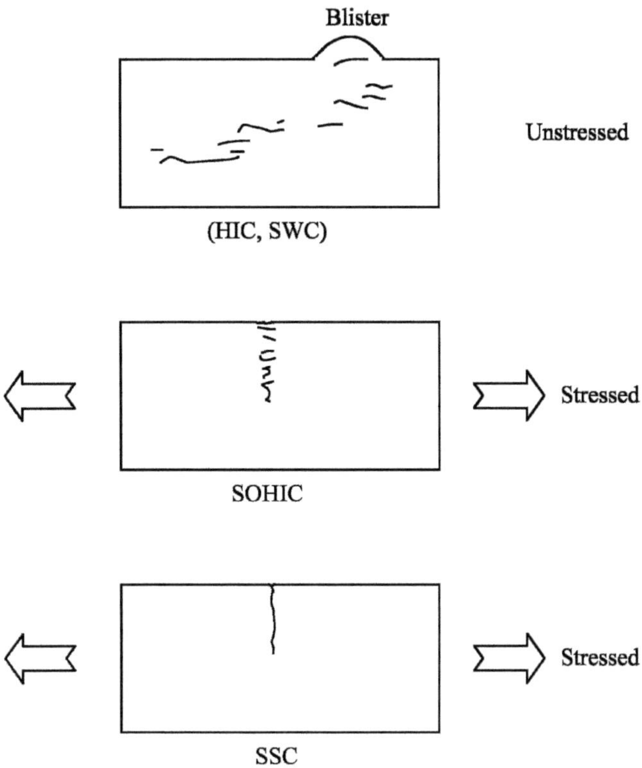

Fig. 2.18 Typical hydrogen damage in metals and alloys. HIC – hydrogen-induced cracking, SWC – stepwise cracking, SOHIC – stress-oriented hydrogen-induced cracking, SSC – sulphide stress cracking.

Sour (H_2S-containing) environments (gas, crude oil, hydrocarbon condensate, and other media) can also result in hydrogen damage. All types of hydrogen damage occur in three stages:

1. Formation of hydrogen atoms and their adsorption on metal surface.
2. Diffusion (penetration) of adsorbed hydrogen atoms into metallic lattice.
3. Accumulation of hydrogen atoms inside metals, leading to increased internal pressure, and thus to blistering or cracks.

Similar to SCC, hydrogen damages occur in the presence of tensile stresses which may be residual or applied (as in SOHIC and SSC), or internal stresses resulting from the internal pressure of hydrogen gas (as in hydrogen blistering and HIC).

Hydrogen atoms (H) can combine inside metallic lattice forming molecular hydrogen (H_2), which causes the internal pressure to increase to a level at which cracks initiate and propagate. Several ppm H_2 is enough for cracking to occur.

Hydrogen blistering is the phenomenon of accumulation of hydrogen molecules H_2 inside metals and the formation of blisters because of large hydrogen pressure.

HIC, also known as *stepwise cracking* (SWC), is a form of *blistering* in which laminating-type fissures parallel to the metal surface link in the through-surface direction (see Figure 2.18). HIC can occur without external stress and mainly in low-strength steels. Environmental and metallurgical factors influence hydrogen damage.

SOHIC is a variation of HIC, where the laminations are arranged in parallel arrays perpendicular to the surface.

SSC is a form of *hydrogen embrittlement* that occurs in the presence of hydrogen sulphide (H_2S) and water in high-strength steels (quenched-and-tempered and precipitation-hardened steels) and on localized hard zones in welds of susceptible alloys. Some oil-well brines and natural condensates in Canada, France, and Russia contain up to 25% H_2S. SSC is related to the amount of atomic hydrogen dissolved in the metal lattice and usually occurs at <90°C. Unlike SCC, cracks caused by SSC usually do not branch, and the fracture mode is often a brittle intergranular fracture.

Prevention and Control of Hydrogen Damage

There are two main ways to prevent hydrogen damage. The first one consists in isolating the metal surface from hydrogen penetration, and the second one consists in reducing the concentration of hydrogen traps inside metals.

(a) Use of *organic, inorganic, and metallic coatings*. Sometimes carbon steel may be clad with nickel. It is not recommended to use stainless steels (SS) for cladding because hydrogen can diffuse through them.
(b) Injection of *inhibitors* of hydrogen penetration. They are adsorbed on metal surface forming a protective film.
(c) Change of *environmental conditions*: increasing pH and removing aggressive species from solutions, such as sulphides, cyanides, and arsenic compounds.
(d) *Metallurgical* measures. Addition of only copper above 0.2 wt% to steel, or together with molybdenum, cobalt and tungsten, or cobalt and nickel differentially is effective. Reducing the sulphur content in metals and alloys to 0.002 wt% may also inhibit hydrogen damage as the number of inclusions (MnS) is reduced. Nickel alloys have low hydrogen penetration. But nickel is detrimental to SSC resistance, and steels with >1 wt% Ni are not recommended for use in sour environments. Addition of calcium and rare earth elements to steels can also inhibit hydrogen damage. Carbide-forming elements (chromium and molybdenum) increase the resistance of steels to hydrogen damage. Stainless steels (especially, austenitic SS) are resistant to hydrogen damage.
(e) Suitable *heat treatment*.
(f) Proper *welding*, namely, maintaining dry conditions during welding.

2.7.6 Corrosion Fatigue

We remember that in order to break a metallic wire, you have to bend it several (sometimes many) times before the wire will crack. This is an example of the phenomenon called *fatigue*. Usually, metals cannot bear such repeated reverse bending. *Fatigue* is a phenomenon (and certainly a type of damage), of multiple fluctuating stresses leading to fracture. The stresses are usually significantly lower than the ultimate strength of the material. All metals and alloys are susceptible to *fatigue* and have *endurance limit, fatigue limit* or *fatigue life*. The *fatigue limit* is measured in pressure units, and *fatigue life* is the number of cycles to failure. The latter is the number of cycles of stress that can be sustained prior to failure. If environment influences fatigue, the phenomenon is called *corrosion fatigue*. Thus, *corrosion fatigue* occurs as the result of the combined action of a fluctuating (cyclic) stress and a corrosive environment. *Corrosion fatigue*, or *corrosion fatigue cracking*, is closed to SCC, but fluctuating stresses exist instead of constant stresses, and combinations of environmental conditions and metal types are not specific for corrosion fatigue. Similar to SCC, corrosion fatigue depends on mechanical (loading), metallurgical, and environmental factors. All environments influence the fatigue life, and its value decreases drastically when air is replaced by an aqueous media. The mechanism includes crack initiation, its propagation, and overload fracture. The first stage begins from local corrosion defects which act as stress concentration sites, namely, pitting, rupture of passive film, or localized embrittlement by hydrogen. Prevention measures of *corrosion fatigue*:

- Reducing either stress or the amplitude of the cyclic stress: altering the design, stress-relieving heat treatment, or shot peening.
- Correct material choice.
- Use of corrosion inhibitors.
- Use of coatings (metal and ceramic).
- Cathodic and anodic protection.

Recommended Literature

1. Fontana, M.G., *Corrosion Engineering*, Third Edition, McGraw-Hill, Inc., USA, 1986, pp. 51–59, 143–152, 181–184, 529–534.
2. Revie, R.W. (Ed.), *Uhlig's Corrosion Handbook*, Second Edition, Wiley-Interscience, USA, 2006, pp. 205–220, 1129–1137.
3. Uhlig, H.H. and Revie, W.R., *Corrosion and Corrosion Control: An Introduction to Corrosion Science and Engineering*. Third Edition, John Willey & Sons, USA, 1985, pp. 142–147.
4. Korb, L.J. and Sprowls, D.O., *Metals Handbook, Vol. 13: Corrosion*, ASM International, USA, 1987, pp. 108–113, 163–171, 303–310, 1277–1280.
5. Shreir, L.L., Jarman, R.A. and Burstein, G.T. (Eds.), *Corrosion, Vol. 1*, Third Edition, Butterworth Heinemann, UK, 1994, pp. 8:84–8:114.
6. Engel, L. and Klingele, H., *An Atlas of Metal Damage*, Wolfe Science Books, 1981, pp. 121–132.

7. Davis, J.R. (Ed.), *Corrosion. Understanding the Basics*, ASM International, USA, 2000, pp. 180–189.
8. Powell, G.W. and Mahmoud, S.E., *Metals Handbook, Vol. 11: Failure Analysis and Prevention*, Ninth Edition, ASM International, USA, 1986, pp. 245–251.

Chapter 3
Corrosion in Natural and Industrial Environments

The ancient Greeks differentiated four main elements: water, air, earth, and fire. These four environments are very important for the corrosion of metals. We shall discuss separately the corrosion of metals and alloys in water and in air (atmosphere). The corrosion of metals in fire (oxidation at high temperatures) and in soil is beyond the scope of this book.

3.1 Corrosion in Water

If carbon steels take first place in industry and other aspects of life among metals and alloys, water takes the main place among media on the Earth and in industry. If astronauts from other worlds would approach our planet for the first time, they would see water and not earth. The average temperature of water in oceans and seas is +17.4°C: from −4°C in the Sea of Okhotsk to +39°C in the Red Sea. Ships, platforms, piers, pipes, and other metallic constructions and structures are exploited in oceans, seas, lakes and rivers; pumps, heat exchangers, condensers, reactors, and tanks come into contact with water and aqueous solutions of electrolytes (Figure 3.1).

With the word *water* we mean many natural and artificial aqueous media. Natural waters are oceans, seas, lakes, ponds, rivers, streams, waterfalls, groundwater, springs, and wells. Artificial waters are potable (tap) water (or any water treated for drinking), industrial cooling waters, extinguishing waters, boiler feed water, demineralized water, and condensate. In general, *water* is a mixture of pure water, H_2O, with dissolved gases, salts, and other electrolytes, suspended solids, microorganisms, and various organic substances including aqueous fauna and flora. We deal with pure water H_2O only in some particular cases: distilled water, demineralized water, and boiler feed water. It is very difficult to obtain absolutely pure water as "something" is dissolved in the pure water. Distilled water is the water obtained by condensation of water vapors. A criterion for water purity is a constancy of its properties, for example, electrical conductivity. Only after 35 to 40 distillation pro-

A. Groysman, *Corrosion for Everybody*, DOI 10.1007/978-90-481-3477-9_3,
© Springer Science+Business Media B.V. 2010

Fig. 3.1 Carbon steel tubes used in contact with water. (For a full color version of this figure, see the Color Section)

cesses repeated in a vacuum in devices made of quartz, the properties of water do not change. Pure water is corrosive towards many metals and alloys, and aggressive to some organic coatings. Corrosiveness of pure water (even in the absence of dissolved oxygen) is explained by the thermodynamic tendency to react with metals and alloys (see Section 1.3 and Appendix A). Usually we do not deal with pure water H_2O, but with aqueous solutions of dissolved gases, inorganic salts, organic substances, and microorganisms. The chemical composition of waters and environmental conditions (temperature and flow rate) are the main factors influencing electrochemical corrosion of metals in waters. Here we will analyze each of them.

1. The values of pH of water can differ from 6 (for rivers) to 8.5 (for oceans and seas), from acidic (pH = 1) to alkaline (pH = 13) for springs. Some springs may contain high concentrations of dissolved hydrogen sulphide (H_2S) gas. We will not analyze such critical cases. The pH values of most natural and artificial waters are about neutral (pH = 6 to 8.5), and alkaline for boiler feed water (pH $\approx$ 9 to 9.5). In the last case, special alkaline species (for example, ammonia) are intentionally injected into water. Corrosion of metals in neutral and slightly alkaline waters occurs according to the electrochemical mechanism with the participation of dissolved oxygen in cathodic reactions (see Section 1.6).

2. *Gases* contained in the atmosphere and soil can be dissolved in water. Their role in the corrosion of metals can be different. *Dissolved gases* in waters are *oxygen* (O_2), *carbon dioxide* (CO_2), *ammonia* (NH_3), *sulphur oxides* (SO_2 and SO_3), *nitrogen oxides* (NO_x), *hydrogen sulphide* (H_2S), *hydrogen* (H_2), and noble gases (He, Ne, Ar, Kr, Xe, and Rn). Nitrogen (N_2), the main component of the atmosphere (about 78%), being dissolved in water does not influence the corrosion of metals. In spite of the small solubility of oxygen (non-polar gas presents about 21% in the air) in polar water, 8 ppm (8 mg O_2 in 1 kg H_2O) at 25°C, this concentration is enough for the existence of 10,000 plants and more than 150,000 animals living in water and for the corrosion of metals to occur in water. Algae are small microorganisms

that live in water and, like plants, produce a large amount of oxygen which goes into the atmosphere. The ratio of oxygen to nitrogen in air is 21:78 (volume). This ratio in surface water changes to 35:65, as oxygen dissolves in water better than nitrogen. This does not mean that it is "easier to breathe" in water. The lower the oxygen concentration in water, the lower the rate of cathodic processes on metallic surfaces and, consequently, a lower corrosion rate. Therefore, all factors causing a decrease in solubility of oxygen in water, such as an increase of electrolyte concentrations (see Appendix E) and temperature, will slow down the corrosion of metals, but this influence is complicated and the corrosion rate, as a function of the concentration of electrolyte or temperature, is described by a curve with maximum (see Section 1.8). Concentration of dissolved oxygen in the upper layers of oceans, seas, and lakes is greater than the inner layers because of photosynthesis and efficient mixing of the upper water layers with air. The water–air interface is the most dangerous place for the corrosion of metallic structures. Just as conflicts between countries take place at their borders, so differential aeration cells are responsible for severe corrosion of metals at the water–air interface (see Section 1.9).

Carbon dioxide (CO_2) is a non-polar gas but dissolves relatively well in polar water. As a result, carbonic acid H_2CO_3 is formed,

$$CO_{2(g)} + H_2O_{(l)} \leftrightarrow H_2CO_{3(aq)}, \tag{3.1}$$

$$H_2CO_{3(aq)} + H_2O_{(l)} \leftrightarrow \mathbf{H_3O^+_{(aq)}} + HCO^-_{3(aq)}, \tag{3.2}$$

and the pH of pure water can decrease to 5. This is the reason why water distilled in air does not have pH = 7, but 5 to 5.5. The concentration of CO_2 in air is 0.03%, and this value increases because of the anthropogenic activity of people (development of industries and vehicles). Concentration of CO_2 in water is 25 times greater than that in air.

Ammonia (NH_3) is a polar gas and dissolves well in water. The source of NH_3 in nature is the destruction of organic substances, and man-made fertilizers (ammonium sulphate, ammonium phosphate and ammonium nitrate), sewage, and wastes. Ammonia being dissolved in water, results in a slightly alkaline solution because of the formation of OH^- ions:

$$NH_{3(g)} + H_2O_{(l)} \leftrightarrow NH_4OH_{(aq)} \leftrightarrow NH^+_{4(aq)} + \mathbf{OH^-_{(aq)}}. \tag{3.3}$$

This case is dangerous for equipment made of copper and its alloys when water from bays is used. Very small concentrations of ammonia in water (several ppm) can cause corrosion of pumps and heat exchangers made of copper alloys:

$$Cu_{(s)} + 4NH_{3(aq)} \rightarrow [Cu(NH_3)_4]^{2+}_{(aq)} + 2e^-. \tag{3.4}$$

Silver artifacts can also react with ammonia:

$$Ag_{(s)} + 4NH_{3(aq)} \rightarrow [Ag(NH_3)_4]^{2+}_{(aq)} + 2e^-. \tag{3.5}$$

Ammonia is also dangerous to zinc:

$$Zn_{(s)} + 4NH_{3(aq)} \rightarrow [Zn(NH_3)_4]^{2+}_{(aq)} + 2e^-. \quad (3.6)$$

However, ammonia dissolved in water is a corrosion inhibitor of iron and carbon steels.

The concentration of ammonia in water is restricted by various rules and requirements of environmental services to 0.2 ppm. This concentration is not dangerous for copper and not beneficial for carbon steels. Aluminum is an amphoteric metal like zinc, but ammonia in water is less dangerous to aluminum than to zinc.

Hydrogen sulphide (H_2S) is one of the dangerous polar gases dissolving well in water. This gas, with a smell of rotten eggs formed as a result of bacterial activity, is particularly corrosive to iron, copper, zinc, and their alloys. Iron sulphides and copper sulphides, forming on surfaces of iron and copper respectively, are cathodic to base metals and even after purification of water from hydrogen sulphide, corrosion of iron and copper will occur in purified water under cathodic films of sulphides. Some springs contain high concentrations of H_2S, therefore, care is needed not to use tubes and structures made of iron, copper, zinc, or their alloys in contact with such water. Aluminum is resistant to H_2S.

Sulphur oxides (SO_2 and SO_3) being dissolved in water result in the formation of sulphurous (H_2SO_3) and sulphuric (H_2SO_4) acids. Their concentrations are not high and do not significantly influence corrosion of metals in waters.

Nitrogen oxides with the general formulae N_xO_y can dissolve in water, but not all of them are dangerous for metals. The most dangerous is NO_2, because it dissolves in water with the formation of two acids HNO_3 and HNO_2:

$$2NO_{2(g)} + H_2O_{(l)} \rightarrow HNO_{3(aq)} + HNO_{2(aq)}, \quad (3.7)$$

$$3NO_{2(g)} + H_2O_{(l)} \rightarrow 2HNO_{3(aq)} + NO_{(g)}, \quad (3.8)$$

$$4NO_{2(g)} + 2H_2O_{(l)} + O_{2(g)} \rightarrow 4HNO_{3(aq)}. \quad (3.9)$$

Nitric acid (HNO_3) is aggressive to most metals and alloys and this aggressiveness depends on its concentration. Anion NO_2^- is an inhibitor of carbon steels.

3. *Mineral components* (*salts*) in the form of various ions are present in water. The most abundant cations and anions contained in water are Na^+, Ca^{2+}, Mg^{2+}, Cl^-, SO_4^{2-}, NO_3^-, HCO_3^-, SiO_3^{2-}, and sand SiO_2. Water may contain cations Fe^{3+}, Fe^{2+}, Li^+, and Sr^{2+}, anions F^-, PO_4^{3-}, and CO_3^{2-}, and manganese compounds (MnO_2 and $MnCO_3$). The salt content may vary from 0.05 wt% for freshwater to 31–36 wt% for brackish water (the Dead Sea and the Kara-Bogas-Gol lagoon of the Caspian Sea). Oceans contain on average 3.5 wt% dissolved salts, river and freshwater contain about 0.05 wt% dissolved salts. Seawater consists mainly of sodium chlorides (88.7 wt% of the total amount of salts) and sodium sulphates (10.8 wt%), magnesium, calcium, and potassium ions. River water consists mainly of bicarbonates, sulphates and chlorides of calcium, sodium, magnesium, and silicates. The latter is present in river water, because they flow through soil and rocks containing silicates. The influence of salts on the corrosion of metals is complicated. The influence of NaCl concentrations in water on the corrosion of iron was discussed in

Fig. 3.2 Scale (calcium and magnesium carbonates) formed inside of heat exchanger with cooling water after 4 years. (For a full color version of this figure, see the Color Section)

Section 1.8. We can summarize that the maximum (see Figure 1.22) is typical for the dependence of the corrosion rate of metals from concentrations of salts and other electrolytes in water. Of course, the maximum corrosion rate value depends on the type of salts and metals, and temperature. Cations Ca^{2+} and Mg^{2+} dissolved in water can result in a deposition of non-soluble carbonates of calcium and magnesium (named *scale*) on a metal surface when increasing temperature or pH:

$$Ca^{2+}_{(aq)} + 2HCO^{-}_{3(aq)} \rightarrow CaCO_{3(s)} + H_2O_{(l)} + CO_{2(g)}, \quad (3.10)$$

$$2Mg^{2+}_{(aq)} + 2HCO^{-}_{3(aq)} + 2OH^{-}_{(aq)} \rightarrow (MgOH)_2CO_{3(s)} + H_2O_{(l)} + CO_{2(g)}. \quad (3.11)$$

The concentrations of these cations dissolved in water show the *hardness* of water. If their concentration is high, water is called *hard*; if their concentration is small, water is called *soft*. Total hardness is measured in the units "ppm $CaCO_3$". *Hardness* is a very important parameter of water. Inhabitants in Israel suffer from high *hardness* of drinking water with a total *hardness* of 255 to 465 ppm $CaCO_3$. You could observe the results of the two reactions (3.10) and (3.11), the formation of *scale*, white-gray deposits, inside your kettle if you used hard water, namely water containing more than 250 ppm $CaCO_3$. Water in oceans and seas has a total hardness of 6,400 ppm $CaCO_3$. People dealing with heat exchangers and condensers can also observe white-gray scale inside tubes and tube sheets with cooling water (Figure 3.2).

Such scale protects metal surfaces from corrosion but prevents heat conductance, and can cause inefficiency in boilers and heat exchanger tubes, and even lead to explosions of boilers. In addition, thick scale can decrease the flow. *Soft* waters are more corrosive than *hard* waters, because a protective scale is not formed on metal surfaces.

Silicates (SiO_3^{2-}) are able to combine with ferrous (Fe^{2+}) and magnesium (Mg^{2+}) cations to form $FeSiO_3$ and $MgSiO_3$ deposits. Sulphates (SO_4^{2-}) are able to combine with calcium (Ca^{2+}) to form $CaSO_4$ deposits.

4. *Organic substances* in the form of fats, fuels, oils, and active surface agents (surfactants) may be present in water as suspended particles. Their source may be natural (aqueous flora and fauna) and artificial (fertilizers used in agriculture, wastes, and sewage). When entering water they can be in the form of suspended substances, then deposit on metal surfaces and result in the formation of differential aeration cells which intensify corrosion of metallic equipment. Organic deposits can cause plugging, overheating, and even an explosion of tubes.

5. *Microorganisms* (*bacteria*), *fungi* and *algae* in water are responsible for biofouling and MIC (see Section 2.5).

6. Water *flow rate* accelerates corrosion in various ways. First of all, *flow rate* accelerates corrosion by supplying dissolved oxygen to metal surfaces and the formation of differential aeration cells, which cause severe corrosion (see Section 1.9). Then, if the *flow rate* is low, suspended substances may deposit on metal surfaces and cause under deposit corrosion; even holes may be formed after several months. Severe *flow rate* can result in erosion and impingement attacks on metallic equipment (see Section 2.6).

7. *Temperature* of water also influences corrosion of metals (see Section 1.8.5).

8. *Electrical conductivity* of water influences the corrosion rate of metals, especially when two metals contact each other (*galvanic corrosion*). The more electrical conductivity of water, the larger is the galvanic corrosion rate (see Section 2.4). Electrolytes and dissolved oxygen in water accelerate corrosion according to the electrochemical mechanism (see Sections 1.5 and 1.6).

3.1.1 Prediction of Corrosiveness of Water on the Basis of Chemical Content

Because of many types of water differing on chemical content, their corrosiveness can also be different and depends on many factors. Many scientists tried to build different models which could allow the determination of the rate of corrosion of metals as a function of the chemical composition of water, temperature, and flow rate. There are some general rules. If the ratio of concentration of chlorides to carbonate ions in water is above three, *pitting* corrosion of iron can occur. If the ratio of concentration of carbonates to chloride ions in water is above three, *dezincification* of brass can occur. The chemical composition of water gives a rough estimation of its corrosiveness. In any case, the actual corrosiveness of water can only be determined experimentally. Methods are determining the weight loss of coupons, measuring the concentrations of cations of the metal or alloy in water that is needed for the determination of the corrosiveness, or by means of any other corrosion monitoring method, including electrochemical ones (see Chapter 5).

Fig. 3.3 Metallic structures and equipment in the atmosphere. (For a full color version of this figure, see the Color Section)

3.1.2 Methods of Prevention of Metallic Corrosion in Water

The following methods are used for corrosion control in water: use of coatings (Section 4.1), cathodic protection (Section 4.2), change of chemistry of water, for instance, injection of corrosion inhibitors, or deaeration (Section 4.3), and use of corrosion-resistant materials (specific alloys, and polymeric materials).

Recommended Literature

1. Korb, L.J. and Sprowls, D.O., *Metals Handbook, Vol. 13: Corrosion*, ASM International, USA, 1987, pp. 487–494, 893–906.
2. Shreir, L.L., Jarman, R.A. and Burstein, G.T. (Eds.), *Corrosion, Vol. 1*, Third Edition, Butterworth Heinemann, UK, 1994, pp. 2:43–2:72.
3. Revie, R.W. (Ed.), *Uhlig's Corrosion Handbook*, Second Edition, Wiley-Interscience, 2006, pp. 529–554.

3.2 Corrosion in the Atmosphere

A lot of metallic structures, equipment, and devices are exploited, transported, or stored in the atmosphere (Figure 3.3).

Corrosion in the atmosphere is the corrosion of metals occurring under atmospheric conditions. It is wrong to call this phenomenon *atmospheric corrosion* as the atmosphere of the Earth includes several layers: troposphere, stratosphere, mesosphere, and thermosphere. The atmosphere comes from the Greek *vapor ball*. The nearest to the Earth layer is called the *troposphere*, therefore it is more correctly to call it *tropospheric corrosion*. This layer extends to an altitude of 10 km (over the poles) to 18 km over tropical regions, and contains approximately 80% of the total air mass. The word *troposphere* stems from the Greek *tropos* and *sphere* that means "*change, turning or mixing ball*". In other words, *troposphere* is a *sphere of change*. The *troposphere* is the place where all weather takes place. This is very important for corrosion. Nitrogen N_2 (78%) and oxygen O_2 (20.95%) are the main gases within *troposphere*. Therefore we live and all metallic structures are exploited in oxidizing (because of the high content of oxygen) corrosive atmosphere. In spite of using metallic equipment in the troposphere, we shall use the term *atmospheric corrosion* as it is so described in the literature. In addition to nitrogen and oxygen gases, pure atmosphere contains small quantities of inert (noble) gases, carbon dioxide (CO_2), and water vapors (H_2O). Not all gases influence the corrosion of metals. Nitrogen (N_2) and inert gases (He, Ne, Ar, Kr, Xe, and Rn) do not have any influence on metals at all. Water vapors are always present in the air because of evaporative processes occurring on the surfaces of rivers, lakes, seas, and oceans. The amount of water vapors in the air at any given temperature is usually less than that required to saturate the air. The ratio of actual water vapor concentration to saturated water vapor concentration in percent at a given temperature is called *relative humidity* (RH, %). For instance, if the actual water vapor concentration is 10 g/m^3 at 20°C and the saturation water vapor concentration is 17.3 g/m^3 at 20°C, the relative humidity is

$$\text{RH, \%} = \frac{10\ \text{g/m}^3}{17.3\ \text{g/m}^3} \cdot 100\% = 57.8\% \text{ at } 20^\circ\text{C}. \tag{3.12}$$

Water vapor concentration is proportional to water vapor pressure (suggesting that they behave as ideal gases), therefore relative humidity of air can be defined as the ratio of the pressure P of water vapors in the air to the pressure P_s of saturated water vapors at the same temperature:

$$\text{RH, \%} = \frac{P}{P_s} \cdot 100\%. \tag{3.13}$$

As water vapor pressure (moisture content) in the air at any given temperature is less than that required to saturate the air, relative humidity (RH) is less than 100%. If the air is gradually cooled while maintaining the moisture content constant, the relative humidity will rise until it reaches 100%. The temperature at which the moisture content in the air will saturate the air is called the *dewpoint*. If the air is cooled further, some of the moisture will condense. This phenomenon is very important for plants in geographical areas with small quantities or even a lack of rain. It is surprising how the plants in Lower Galilee in Israel receive water from April until November when there is no rain at all. Flora in such areas receives water by the condensation of water vapors contained in the air at dewpoint. If this phenomenon is useful for plants, it is detrimental for metals. The corrosion rate and lifetime of metallic constructions depend on the aggressiveness of the atmosphere, which is defined by *climate* and *contaminations* in the atmosphere. The climate is defined by nature and its factors are relative humidity, changes of temperature during the day-night cycle, the content and type of sediments, direction, and the strength of winds. The sediments can be rain, dew, fog, snow, hail, and smog. Atmospheres are differentiated into industrial, urban, rural, marine, arctic, and tropic. Different concentrations of gases, salts, moisture, dust, and other contaminants are responsible for corrosiveness in various atmospheres. There are no comments that the corrosion rate of carbon steel in the vicinity of oceans or seas is larger than that in the desert, and aggressiveness of the industrial atmosphere is larger than that of the rural one. *Contamination* may be *natural* and *anthropogenic*. The first are salts (NaCl, $MgSO_4$) coming from the oceans, sand (SiO_2) from the deserts, mountains, and hills, various gases and dust from volcanoes. *Anthropogenic* contamination is defined by the industrial activity of people: manufacturing energy, chemicals, movement of different vehicles, etc. The result of industrial activity is the emissions of acid gases, dust, water vapors, and other aggressive contaminants for metals. The burning of fuels, gases, and coal at power stations, chemical plants, and in vehicles contribute much towards *anthropogenic* contamination in the atmosphere. How do contaminants influence the corrosion of metallic structures in the atmosphere?

3.2.1 Factors Influencing the Corrosiveness of the Atmosphere

Our basic point is that *corrosion is the reaction between a metal and the environment*. Therefore, we have to analyze two sides of the corrosion process. From the side of metal: its nature and surface conditions, the type of metallic structure (closed or open to atmosphere), the presence of stresses, and heat treatment. From the side of environment: content of moisture (relative humidity), gases (SO_2, SO_3, NO_x, H_2S, CO_2, NH_3, etc.), salts (NaCl, $MgSO_4$), dust (particles of coal, sand, metals and their oxides), presence of microorganisms, temperature and its changes, the value and period of the presence of a water layer on a metallic surface (time of wetness). For instance, this period is equal to 4,000 hours (half a year!) in northern Europe, but about 750 hours (one month in a year!) in Central Asia. The two most dry places in the world are Khartoum (Sudan) in Africa and Kushka (Turkmenistan) in Asia. As a result of the dry climate, the corrosion rate of carbon steels in these places is very low, about 0.003 mm/year. The influence of some important factors in the environment on the corrosion of metals is as follows:

1. The influence of *relative humidity* on the corrosion rate of iron in air is shown in Figure 3.4.

When air is pure there is no significant influence of relative humidity on the corrosion rate of iron. If acid gas SO_2 is present in the air (~0.01% volume), the relative humidity increases to 60% which scarcely influences the corrosion rate of iron. The corrosion rate drastically increases at a relative humidity above 60% in the presence of SO_2. This value is called the *critical relative humidity*. Similar dependence between corrosion rate and relative humidity exists not only for iron, but also for aluminum, zinc, copper, and nickel, and the *critical relative humidity* varies

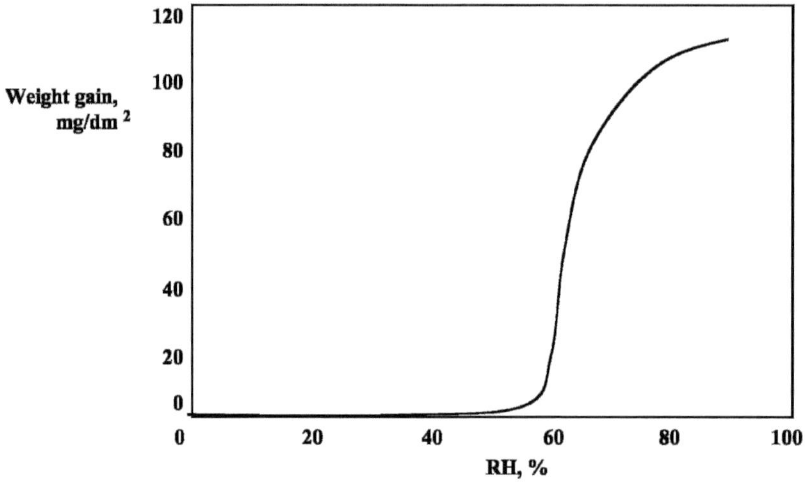

Fig. 3.4 The influence of relative humidity (RH, %) on the corrosion rate (weight gain) of iron in air with 0.01% SO_2.

from 60 to 80%. The main cause of the formation of water layers on a metallic construction is the condensation of water in cracks and crevices on metal surfaces, in oxide films and other corrosion products, and on hard foreign particles (dust, salts, soldering fluxes, and even sweat residues) on metal surfaces. Chloride and sulphate salts are the most aggressive corrosive agents. The number of layers of water on a metallic surface increases with an increase in relative humidity. Therefore, salt mist from the ocean, sea or chemical enterprises can significantly induce atmospheric corrosion. It is interesting to emphasize that such a critical value of water content was defined not only in the atmosphere, but also in gasoline, naphtha, and kerosene (see Section 3.5). This fact points out similar corrosion (electrochemical) mechanisms in the atmosphere and in these fuels with participation of water and dissolved oxygen. The existence of a critical relative humidity of about 60% explains interesting case of resistance to corrosion of the Delhi iron pillar which is attributed to low relative humidity, from 21 to 52%, during the year in that particular region of India where the pillar has been stood since 400 AD. The Indian metallurgists explain high resistance to corrosion by a high content of phosphorus in steel which causes the formation of passive protective layers of iron phosphates on the surface of pillar.

2. Various gases have different influences on the corrosion of metals in the atmosphere.

2.1. *Carbon dioxide* (CO_2) is an acidic natural gas. Its concentration is 0.03% in pure atmosphere. Intensive development of industry and the use of motor vehicles, and as a result a constant increase in the burning of fuels gives rise to large quantities of CO_2 in the air and the so-called *hothouse effect* (increase of temperature of the Earth). CO_2 dissolves in water droplets in the atmosphere, weak carbonic acid H_2CO_3 is formed, and the pH of the water condensate may decrease to 4. This is

Fig. 3.5 Natural patina formed on a bronze statue. (For a full color version of this figure, see the Color Section)

one of the causes of *natural* acidic rains. These rains not only cause corrosion of metallic constructions, but also the destruction of stones and plants.

2.2. *Sulphur dioxide* SO_2 and *sulphur trioxide* SO_3 in the atmosphere are referred to as SO_x. They are formed as a result of anthropogenic processes of the combustion of fuels containing sulphur compounds, and natural processes of decomposition of vegetation and emission from volcanoes. When sulphur oxide acidic gases meet water droplets, two acids, sulphurous H_2SO_3 and sulphuric H_2SO_4, are formed. These acids are very aggressive to many metals and alloys: Fe, Zn, Ni, Cu, Al, and others, as well as to concrete. The results of acid attack of bronze statues, domes, and roofs – a green-blue corrosion product called patina (Figure 3.5) can be observed many times.

This consists of mixed salts-hydroxides of copper and is formed according to the following reactions:

$$2Cu_{(s)} + O_{2(g)} + H_2O_{(l)} + CO_{2(g)} \rightarrow Cu(OH)_2 \cdot CuCO_{3(s)}, \tag{3.14}$$

$$8Cu_{(s)} + 5O_{2(g)} + 6H_2O_{(l)} + 2SO_{2(g)} \rightarrow 2[3Cu(OH)_2 \cdot CuSO_4]_{(s)}. \tag{3.15}$$

The first corrosion product of copper is green *malachite*. Artists make artificial patina on their products and obtain a good color effect. This is an example of corrosion not always being a detrimental phenomenon, and is even beneficial (see Section 6.2).

2.3. *Nitrogen oxides* N_xO_y (also called NO_x) gases are formed as a result of natural (microorganisms) and anthropogenic (automobile emissions, jet engines, and nitrogen-based fertilizers) processes, and consist of four oxides N_2O, NO, NO_2, and N_2O_4. Only one of them (NO_2) takes part in atmospheric corrosion. Nitrous oxide N_2O (known as *laughing gas* discovered by the "father" of cathodic protection, the great English chemist Sir Humphry Davy) is the natural product of the action of bacteria in the soil on nitrogen gas N_2. It does not take part in corrosion processes and can be oxidized to nitrogen dioxide NO_2 which is a toxic gas. Nitric oxide NO and nitrogen dioxide NO_2 are produced by vehicles and by lightning. The latter gas is yellow-orange-brown and may contribute to the color of smog. Sometimes ginger *fox tail* containing NO_2 from the stacks of chemical plants can be observed. Nitrogen dioxide NO_2 dissolves in water drops forming very corrosive nitric acid HNO_3 (see reactions (3.7–3.9)).

Nitric oxide NO may catalyze the oxidizing of SO_2 to SO_3, and may catalyze the destruction of ozone. The first catalyzing reaction is very important, because SO_3 gas in contact with H_2O gives sulphuric acid (H_2SO_4) which is stronger than the sulphurous one (H_2SO_3).

2.4. *Ammonia* (NH_3) gas can be formed by reducing NO_x. Fertilizers, plastics, and explosives may be also the source of ammonia in the atmosphere. Everyone has smelled the sharp, penetrating odor of ammonia which is a toxic and reactive gas. It dissolves well in water droplets giving rise to an alkaline solution (see (3.3)). Ammonia gas and its solutions are corrosion inhibitors of iron and carbon steels, but are very dangerous for copper, zinc, and their alloys, and may cause stress corrosion cracking of copper alloys.

2.5. *Hydrogen sulphide* (H_2S) is found in crude oils, natural condensates, petroleum distillates, and sometimes is present in groundwater. For instance, deposits of natural condensate near Astrakhan (southern Russia) contain up to 25% H_2S. Small concentrations in the air cause tarnishing of silver and copper: black sulphides of silver (Ag_2S) and copper (CuS and Cu_2S) are formed. Therefore, the jewelry from these metals and alloys sometimes blacken. This is the reason why silver and copper strips are used for qualitative determination of traces of hydrogen sulphide in fuels. Hydrogen sulphide may cause severe corrosion of silver and copper relay electric contacts. How do metals corrode in the atmosphere?

3.2.2 Mechanism of Atmospheric Corrosion

The *time of wetness* is the main factor of atmospheric corrosion; therefore it influences the corrosion mechanism and corrosion rate of metals in the atmosphere. Four types of atmospheric corrosion mechanisms are differentiated according to the time of wetness and, as a result, to the thickness of the water layer on a metallic surface.

1. *Dry oxidation* occurs in the absence of a water layer on the metal surface. This is a typical chemical mechanism, when oxidation of metals occurs by oxygen gas.

2. *Damp corrosion* occurs in the presence of a thin water layer which is impossible to observe on a metal surface with the naked eye (RH $< 100\%$).
3. *Wet corrosion* occurs in the presence of a water layer which is possible to observe on a metal surface with the naked eye (RH $= 100\%$). Wet corrosion takes place through rains, artificial water spraying and water vapor condensation under dewpoint.
4. *Sheltered corrosion* occurs inside of structures and equipment closed from the atmosphere when water vapors containing corrosive gases and salts are condensed on surfaces inside metallic structures which do not dry for a long period. Inside surfaces of monuments, sculptures, tanks, and profiles are subjected to sheltered corrosion.

The presence of condensed water saturated with dissolved oxygen on metallic surfaces in the atmosphere is responsible for the electrochemical mechanism of corrosion (see Section 1.6). Water is a weak electrolyte but if salts and acidic gases are present in the atmosphere, water after condensation on a metallic surface becomes a strong electrolyte. Therefore, the laws of electrochemical corrosion are applicable for atmospheric corrosion but there are some specific features. First of all, atmospheric corrosion takes place not in the volume of electrolyte but in its thin layers. Such layers of electrolyte provide very intensive diffusion (penetration) of oxygen from the atmosphere to a metallic surface compared with that in the volume of an electrolyte. Therefore, the rate of cathodic processes under a thin layer of an electrolyte is higher than in its volume. At the same time, thin layers of electrolytes are well saturated with corrosion products and if they have protective properties, the anodic process of metallic dissolution decreases in thin layers compared with that in the volume of electrolyte. In addition, corrosion products prevent the diffusion of oxygen gas to metallic surfaces, and the cathodic processes with oxygen participation also slow down. The corrosion rate of metals in the atmosphere is less than in water and aqueous solutions of electrolytes. For instance, the corrosion rate of carbon steel is 0.11 mm/year in the industrial atmosphere in Israel (2 km from the Mediterranean Sea), 0.17 mm/year in an industrial atmosphere in England, and 0.6 mm/year in cooling water (untreated with anti-corrosion chemicals). We can mention the mutual influence of the corrosion of metals and the environment.

3.2.3 Methods of Prevention and Control of the Atmospheric Corrosion of Metals

(a) First of all, we have to *study the aggressiveness of an atmosphere* where metallic constructions will be used: industrial, rural, ocean, how it is close to the sea, quantity of rains, etc. Table 3.1 compares the corrosion rates of uniform corrosion of carbon steel and zinc in various atmospheric environments. The data shown in this table relate to corrosion of carbon steel and zinc after a year of exposure (short term, or non-steady state) and after 20 years of exposure (long

Table 3.1 Corrosion rates of carbon steel and zinc (μm/year) in atmospheric environments.[a]

Corrosivity	Carbon steel		Zinc		Typical environment
	First year	Steady state	First year	Steady state	
Very low	<1.3	<0.1	<0.1	<0.05	Dry indoors
Low	1.3–25	0.1–1.5	0.1–0.7	0.05–0.5	Desert or non-polluted urban
Medium	25–50	1.5–6	0.7–2.1	0.5–2	Mild marine or mild industrial
High	50–80	6–20	2.1–4.2	2–4	Marine (calm sea)
Very high	80–200	20–90	4.2–8.4	4–10	Marine (surf beach)

[a]ISO 9223, Corrosion of metals and alloys – Classification of corrosivity of atmospheres, 1992, 13 pp.; ISO 9224, Corrosion of metals and alloys – Corrosivity of atmospheres - Guiding values for the corrosivity categories, 1992, 5 pp.; ISO 12944-2:1998(E), Paints and varnishes – Corrosion protection of steel structures by protective paint systems, Part 2: Classification of environments, 5 pp.

term, or steady state). The corrosion rate is high during the first year of exposure and then it diminishes reaching the steady state. We have to mention that the corrosion rate 0.11 mm/year of carbon steel in the industrial atmosphere in the vicinity of the Mediterranean Sea relates to a *very high* corrosion rate (compared with the very high corrosiveness of 0.08–0.2 mm/year in Table 3.1). Corrosion rates of zinc in atmospheres is less than that of carbon steel and this fact supports the use of zinc coatings for the protection of carbon steel surfaces in the atmosphere (see Section 4.2).

(b) Then we should inspect the territory where a metallic construction will be used and its industrial neighbors. Electric power stations, chemical plant manufactured acids, fertilizers, cement, and other chemicals, furnace activities, etc., result in the appearance of corrosive species in the atmosphere. It is very important to be watchful regarding cooling towers and equipment capable of spraying water on metallic structures in the atmosphere. Periodical cleaning of metallic surfaces, protection from dust sediments and purifying the atmosphere may be effective, but is not always possible, especially if there are large areas of surfaces and volume of the atmosphere. This may be effective for corrosion control of electronic and electrical devices (contacts, etc.) exploiting in the atmosphere.

(c) We should select correct measures for corrosion prevention and control in atmosphere.

The use of greases, waxes, slushes, and pastes based on petroleum products was historically probably the first method of corrosion control of metallic equipment in the atmosphere. They are used for temporary protection from atmospheric corrosion. Water and oxygen may dissolve in grease and cause corrosion of a metallic surface protected with greases. Corrosion inhibitors may be added to modern greases (see Section 4.3). The disadvantage of using grease is that it must be removed before using protected equipment or structures. Organic solvents must be used for removing temporary greases, and this usage is dangerous to people and the environment.

Use of coatings (organic, inorganic, and metallic) resistant under specific conditions. It is important to paint the surface of the construction with primer before its transportation. *Electrolytic, hot-dip coatings, metal spraying, chemical plating, vapor- deposited coatings, and diffusion treatment* are used both for protective and decorative purposes (see Section 4.1). In practice, electrolytic coatings (gold, silver, nickel, and chromium) are invariably porous, and liable to mechanical damage. Thick metal coatings produced by high-temperature processes are comparatively robust. Zinc and aluminum coatings have experienced good results, and are widely used for corrosion control in various atmospheres. We should remember that zinc is not resistant to H_2S, and in areas with a possible presence of this gas in the atmosphere, aluminum coatings are preferred.

The existence of critical relative humidity of about 60% points towards the *drying of the atmosphere* (*control of RH*) by heating or by use of substances such as silica gel, zeolites, activated alumina, calcium oxide (CaO), phosphorous oxide (P_2O_5), and sulphuric acid (H_2SO_4) which are capable of absorbing water vapors from the atmosphere. These substances are called *dessicants*. The drawback to using dessicants is that it is not simple to calculate the amount needed because of the unpredictable nature (moisture content) of the environment. Heating means to electrically heat the air flowing through metallic equipment or structures. Stoves and gas heaters may be used too. This method is called *dehumidification*. The target is to reach and to keep the relative humidity of the atmosphere at not more than 50%. *Dehumidification* means to use electricity which may be relatively expensive. Another way to "keep a dry atmosphere" is to fill the inner volume of metallic constructions (tanks, pipes, reactors, boilers, etc.) with inert gas: dry nitrogen or dry air under pressure above atmospheric one.

Use of *vapor phase inhibitors* (VPI), also called vapor corrosion inhibitors (VCI), volatile corrosion inhibitors (VCI), or inhibitors of atmospheric corrosion. We can mention that solid iodine or naphthalene may be changed from a solid to vapor state. This process is called *sublimation*. The same process occurs during performances in theatres using solid "dry ice" CO_2. When it is put on the stage for a performance, CO_2 changes its aggregate condition from the solid to the vapor phase. Some organic substances possess such a property at ambient temperature and atmospheric pressure. The first VPI was discovered in 1941–1943 by the American chemist Aaron Wachter. Inner surfaces of the cannons in the US Navy were protected against atmospheric corrosion with paraffinic slushing compounds. The first, test firing, in order to clean the inner surface of the cannon before the main firing was fruitless. Wachter suggested using organic compound, *dicyclohexylamine nitrite*, which was put inside of cannon in the form of powder and was able to sublimate and to protect the inner surface of the cannon against atmospheric corrosion. When sublimating, dicyclohexylamine nitrite molecules are transported to the metal surface and form a monomolecular adsorbed layer on it. This layer is unstable in the presence of water. VPI are used only for protecting inner surfaces of metallic constructions hermetically sealed from the atmosphere: tanks (see Section 3.6), vessels, various metallic parts, devices, and articles, packed in polymeric materials. If construction is open to the atmosphere (for instance, inner surfaces of tanks and tankers – see Section 3.6), air may ingress from the atmosphere and then escape from the construction, and thus the organic molecules of VPI which are introduced inside the construction can disappear, and new portions of a volatile inhibitor must be added inside to protect the construction. Various organic compounds, mainly amines, were synthesized to protect ferrous and non-ferrous metals and alloys from general, pitting, and galvanic corrosion in various atmospheres. VCI materials are used as powders, tablets, plastic pellets, vapor capsules, diffusers, emitters, and liquids. It is unnecessary to prepare the metal surface prior to using VPI, and their vapors can reach remote areas of an enclosed space. This method has found wide application in military, electronics, telecommunication, and radar equipment, especially during transportation, storage, and exploitation.

Combining VCI with grease, paints, polymeric materials, and paper. When VCI additives are used in grease and paints, temporary soft coatings and oil-based coatings are formed. Examples of thin flexible materials are VCI-treated paper or impregnated plastic films, sheets, and bags. Surface preparation similar to that before painting should be carried out. These anti-corrosion materials are used for packaging, shipping, and storage of different metallic constructions and devices.

Use of alloys resistant to various atmosphere conditions. When metallurgists added various quantities of chromium to iron at the beginning of the 20th century, and evaluated the corrosion rates of these alloys in pure atmosphere, they defined a significant decrease in the corrosion rate of steel with a chromium content above 12 wt%. These steels did not corrode in pure atmosphere, and were called *stainless steels*. Then it was revealed that stainless steel is not stainless in atmosphere polluted with chlorides and metallic or metallic oxide particles, particularly in the vicinity of sites where welding processes are carried out.

People who worked on ships at the Caspian Sea at the end of 1970s pointed out that constructions on ships built in the 1950s corroded quicker in the marine atmosphere than similar constructions built in the 1920s. Chemical analysis showed that the carbon steel of ships built in the 1920s contained about 0.4 wt% copper, and carbon steel built in the 1950s did not contain copper at all. The addition of 0.4 wt% copper to carbon steel diminishes twice its corrosion rate in the atmosphere (Figure 3.6). This is one of the methods to increase the resistance of carbon steels in the atmosphere. The addition of small quantities of copper, phosphorous, chromium, and nickel (totaling a few percent maximum) to carbon steels also significantly improves their resistance in the atmosphere. These steels are called *weathering steels* (or CORTEN) and were developed in the 1930s, especially for use in the atmosphere. *Weathering steels* slightly corrode and dense, tightly adherent reddish or orange-brown rust with a rough texture is formed which has protective properties in the atmosphere. In contrast to this, rust formed on a carbon steel surface is porous, spongy, coarse, and flaky and has no protective properties in the atmosphere. Sometimes protective rust formed on a *weathering steel* is called *artistic rust* because sculptors and architects like to create various buildings and statues of *weathering steel* (Figure 3.7).

Bridges, bolts, and stacks are also made of *weathering steel* and are left uncoated in the atmosphere during use.

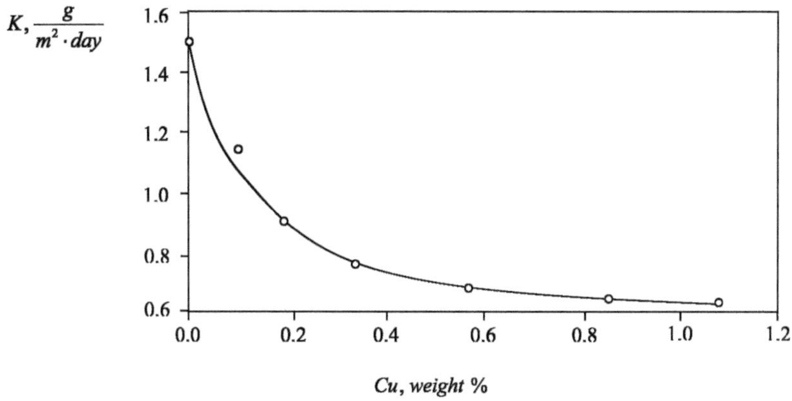

Fig. 3.6 Effect of copper content in carbon steel on its corrosion in the atmosphere (4 years).

Fig. 3.7 Sculpture "Sun – Man" (1998) made of *weathering steel* (CORTEN) by Jorge Vieira (Lisbon, Portugal). (For a full color version of this figure, see the Color Section)

The *correct design* may be a very important control method at the design stage and during use of metallic structures and equipment. Parts should be designed to avoid trapping moisture. Suitable drainholes should be provided for hollow sections. Segments should be designed to prevent insulating materials from becoming soaked with water.

Control of the atmosphere for aggressive corrosive items must be carried out, especially during the use of electronic devices (see Section 5.5).

Note. In spite of the corrosion rate of metals in the atmosphere being less than that in aqueous solutions, one factor is very important: the *aesthetic* sight of a metallic construction. Therefore, even small corrosion rates of metals in the atmosphere may result in the loss of *aesthetic view*, and their protection is very important.

Recommended Literature

1. Revie, R.W. (Ed.), *Uhlig's Corrosion Handbook*, Second Edition, Wiley-Interscience, 2006, pp. 305–328, 515–528, 569–580.
2. Rosenfeld I.L., *Atmospheric Corrosion of Metals*, Academy of Science of the USSR, Moscow, 1960, 372 pp. [in Russian].

3.3 Dewpoint Corrosion

I live in a small settlement in the heart of Lower Galilee at the top of the Schekhaniya hill at a height of 448 m above sea level. I go out my house every morning and observe the beautiful view of the Mediterranean Sea, Mount Carmel and Upper Galilee. Suddenly I look at the green leaves of plants and trees, then at the brilliant surface of the blue, white, green and gray cars of my neighbors. I discover early dew on all these objects. In the evening after sunset I find nightdew on the surface of the same shining cars. What is this? When the air is gradually cooled while maintaining the moisture content constant, the relative humidity will rise until it reaches 100%. The temperature at which the moisture saturates the air to 100% is called the *dewpoint*. In other words, the *dewpoint temperature* of vapor species is the temperature at which the rate of evaporation of the condensate from a clean plane surface is equal to the rate of condensation of the vapors. If the air is cooled further, some of the moisture will condense and we observe the resultant drops of water on leaves, grass, and cars. Does this phenomenon relate to corrosion of metals? I am sure that you have observed many rusted dumping exhaust systems on cars (Figure 3.8).

Fig. 3.8 The exhaust system of a vehicle after dewpoint corrosion. (For a full color version of this figure, see the Color Section)

What is common between this corrosion failure and the natural phenomenon of *dew-point*? A metal surface may be under a film of water condensate, exactly as green leaves of plants. What will happen if acid gases (for instance, sulphur oxides SO_2 and SO_3) are present in the vapor phase? As a result of burning gasoline, exhaust gases contain water vapors H_2O, carbon dioxide CO_2, and sulphur oxides SO_2 and SO_3 at 500 to 600°C. In any car, at the end of the exhaust, the temperature can decrease to *dewpoint* (100°C) and less. Acidic water is formed in this case, *dew-point corrosion* occurs, and the result is holes in exhaust systems. This case is very important for industrial structures and equipment. Let us analyse what happens in the upper part of the distillation column at an oil refinery (see Figure 5.7) where crude oil is distilled. Crude oil is a mixture of hydrocarbons and may contain water, some salts (NaCl, $CaCl_2$, $MgCl_2$), and hydrogen sulphide H_2S. Light hydrocarbons rise together with water vapors, hydrogen chloride HCl (forming as a result of hydrolysis of calcium and magnesium chlorides) and hydrogen sulphide gases. The *dewpoint* for pure water vapors is 100°C at 1 atm. When the temperature is above the dewpoint, there is no danger for water condensation. Therefore the temperature in the upper part of the distillation column is kept at about 120 to 130°C. The volatile hydrocarbons, water vapors, hydrogen chloride and hydrogen sulphide gases are then cooled in the air coolers and the temperature goes down to 100°C. Water vapors condense, acid gases HCl and H_2S are dissolved in the water condensate, and the pH of the acidic solution is about 1. If it is taken into consideration that pipes and air cooler tubes are made of carbon steel, severe acidic corrosion will take place at about 100°C. This is typical *acid dewpoint corrosion.*

Another place favorable for this specific type of corrosion is the column with tail gases from Claus Units at oil refineries. The tail gases contain H_2O, H_2S, SO_2, and SO_3 emitting from the upper part of the column at ~200°C. In order to prevent condensation of water vapors and formation of acidic solution in the overhead of this column, all parts, usually including manholes, are well insulated from the environment by wool, the temperature of the environment ranging from −30 to +30°C. However, if the insulation is disturbed, the metal surface will have the environmental temperature. Condensation of water vapors and dissolution of acidic gases will take place on a bare (non-isolated) metal surface. As a result, severe *acid dew-point corrosion* takes place. Therefore, one of the important methods of control of *acid dewpoint corrosion* is keeping insulation in a good, non-disturbed condition.

The most widespread places in industry vulnerable to *acid dewpoint corrosion* are equipment and structures that come into contact with flue gases. Fuel oil contains sulphur compounds. In spite of constantly decreasing the sulphur content in fuel oils producing at oil refineries, it is impossible to completely remove sulphur from the fuels. Flue gases contain the burning products, water vapors (H_2O), carbon dioxide (CO_2), and sulphur oxide (SO_2, SO_3) gases. The temperature is about 500°C in furnaces. When flue gases flow through ducts and expansion joints to the stack, they are cooled to 200°C. All metallic constructions in contact with fuel gases are insulated from the environment with bricks, concrete, or wool, in order to prevent the cooling of flue gases. Sometimes bricks, concrete, and wool are disturbed and then the upper part of the stack is without insulation. As a result, temperature can decrease to 160, 150°C, and even less. What happens with flue gases in this case? If water vapors condensate alone, this happens at 100°C and 1 atm. But if sulphur oxides are present in the vapor phase the *acid dewpoint* is significantly higher than 100°C and depends on the concentration (the partial pressure) of water vapors and sulphur oxides. If the concentration of water vapors in flue gases is high enough (about 10%), and the content of the sulphur oxides is also high enough (100 ppm), sulphurous H_2SO_3 and sulphuric H_2SO_4 acids with pH = 2.7 are formed at ~150°C. These hot acids will attack carbon steel at a high corrosion rate according to the mechanism described in Section 1.7. Condensation of sulphuric acid in a typical combustion atmosphere frequently occurs at 120 to 150°C. The concentration of the sulphuric acid condensate formed on a steel surface depends on the concentration of the water vapors and sulphur oxides in the flue gases, and temperature.

Acid dewpoint corrosion is very dangerous during shutdown periods. Then there is a decrease of temperature. At relatively high temperatures of about 100 to 150°C, the high concentration (70 to 85%) of sulphuric acid condenses (if 8% H_2O are present in the flue gases). Usually maximum corrosion rate of carbon steel occurs in 50 to 60% H_2SO_4 (see Figure 1.21). When temperature decreases during shutdown of a plant, the sulphuric acid, being hygroscopic, will absorb water leading to dilution of acid solutions. When sulphuric acid condenses on a steel surface, typical shallow pits are formed. The main corrosion products, ferrous and ferric sulphates, are formed on the steel surface. These salts are acidic because being in contact with water, they hydrolize and pH ~ 2 as a result of sulphurous and sulphuric acid formation:

$$FeSO_{4(s)} \xleftrightarrow{H_2O_{(l)}} Fe^{2+}_{(aq)} + SO^{2-}_{4(aq)}, \tag{3.16}$$

$$Fe^{2+}_{(aq)} + 2H_2O_{(l)} \rightarrow \mathbf{H_3O}^+_{(aq)} + FeOH^+_{(aq)}, \tag{3.17}$$

$$Fe_2(SO_4)_{3(s)} \xleftrightarrow{H_2O_{(l)}} 2Fe^{3+}_{(aq)} + 3SO^{2-}_{4(aq)}, \tag{3.18}$$

$$Fe^{3+}_{(aq)} + 2H_2O_{(l)} \rightarrow \mathbf{H_3O}^+_{(aq)} + (FeOH)^{2+}_{(aq)}. \tag{3.19}$$

Therefore, the surface of carbon steel and low-alloy steel pipes covered by deposits of acidic salts should be immediately neutralized with a dilute aqueous soda (Na_2CO_3) solution and washed with pure water. How to control and to mitigate the *acid dewpoint corrosion*?

1. *Control and inspection.* Special probes and devices for measuring the dewpoint may be used and, of course, temperatures must be kept at 20 to 30°C above the dewpoint.
2. The *selection of materials resistant to acids* forming during acid dewpoint corrosion. High nickel alloys with a high content of molybdenum, such as Hastelloy C-276 (57% Ni, 16% Cr, 16% Mo, 5.5% Fe, 4% W) and B-2 (70% Ni, 28% Mo, 2% Fe) are resistant to acids. Special concrete stack linings and fluoroelastomers are also resistant.

Fig. 3.9 Corrosion of a heat exchanger under thermal insulation. (For a full color version of this figure, see the Color Section)

3. The following *coating systems* resistant to acid dewpoint corrosion were developed: *glass flake polyesters, isocyanate-cured epoxy, enamel coatings, arc-sprayed chromia*, and *alumina materials*.
4. *Reducing the sulphur content in the fuels.*
5. The *use of insulation, and keeping it in good condition.*
6. The *use of neutralizing additives*: calcium and magnesium oxides (CaO and MgO), calcium and magnesium hydroxides [$Ca(OH)_2$ and $Mg(OH)_2$], or ammonia NH_3. These oxides and hydroxides are injected into the flue gases flowing through the ducts. The result is neutralization of acid gases, but a drawback is the formation of fouling.

Recommended Literature

1. Holmes, D.R., *Dewpoint Corrosion*, Ellis Horwood Limited, England, 1985, 292 pp.
2. Korb, L.J. and Sprowls, D.O., *Metals Handbook, Vol. 13: Corrosion*, ASM International, USA, 1986, pp. 1001–1004.

3.4 Corrosion under Thermal Insulation

One of the most important properties of metals and alloys is that they have good heat conductance. This property is used in heat exchangers, air coolers, and condensers for heating or cooling substances in various technological processes in chemical, petrochemical, oil and gas, power generation, pulp and paper, pharmaceutical, food and beverage, and other industries. However, in many cases heat conductance of metals plays a negative role, and pipes, heat exchangers, reactors, and other metallic equipment must be isolated from the environment: from cooling, if hot steam (at 250°C) or fuel oil (at 300°C) flow inside the pipes, or from heating, if liquid ammonia, carbon dioxide, or nitrogen at low temperatures flow inside the pipes or are stored in metal vessels. Thus, thermal insulation is used for maintaing temperature, and is intended to reduce the energy loss, controling surface temperatures of equipment for personal protection, facilitating temperature control of a chemical process, and preventing vapor condensation on metallic surfaces having a temperature below the dewpoint of the surrounding environment. In spite of this positive duty, thermal insulation creates conditions that cause corrosion problems on the outside surface of heat exchangers, pipes, tanks, vessels and reactors (Figure 3.9) and we can say that thermal insulation accelerates corrosion of metals. Corrosion can occur under any type of insulating material. Why does thermal insulation not protect metal equipment from corrosion? What occurs with a metallic surface under thermal insulation? What is the mechanism and which factors influence corrosion under thermal insulation? Generally, carbon steel, low-alloy steel, and austenitic stainless steel equipment undergo *corrosion under thermal insulation*. Temperature changes, type of insulation material, type of metal and protective coatings, equipment design, weather barriers, climate, and maintenance practices are the main factors of *corrosion under thermal insulation*.

Imagine that somebody sweats under a warm, thick coat or was wet because of rainfall. Even if the sun shines after this, and the coat dries, the body will be humid for a long time at a body temperature of 36.6°C. A person does not feel well in such a situation and can become ill. Something similar happens with the outer surfaces of pipes, vessels, heat exchangers, tanks, and reactors under thermal insulation. Water once penetrating under insulation, remains there for a long time, and cannot escape. Water is trapped on the metal surface under insulation.

3.4.1 Types of Insulation

Generally, insulations are divided into low temperature (under ambient temperatures of 0 to 25°C) and high temperature (from ambient till 650°C). Low-temperature insulations are organic foams, such as polyurethanes, polyisocyanurates, polystyrene, flexible elastomerics, and phenolics, cotton, wood, and cork. High-temperature insulations are mineral wool, fibrous glass, cellular glass (foamglass), perlite (siliceous rock, amorphous glass mineral of volcanic origin), vermiculite (natural mineral), calcium silicate, and ceramic materials. Usually both low- and high-temperature insulations are porous which facilitates the entry and retention of water with dissolved oxygen. Some insulation materials which have been in contact with water may decrease in pH and increase the chloride content. Phenolic foam and polyurethane foam can diminish water pH from 7 to 4.3 and even to 2.2. Polyurethane is a “good” source of chlorides and bromide ions which are able to accelerate corrosion under insulation.

3.4.2 Corrosion Mechanism

Rust formed on steel surfaces under thermal insulation (see Figure 3.9), showed that water and dissolved oxygen took part in the spontaneous corrosion process according to the electrochemical mechanism. If water is in the vapor phase, there is no large corrosion problem because there is no electrode potential on the metal surface. The corrosion problem begins if water condenses and is present in the liquid phase on the metal surface. Now electrode potential is formed on the metal surface and there are conditions for the occurrence of electrochemical corrosion. What are the sources of water under insulation? Rain, leakage, deluge systems, or wash water. Sweating occurs when operating temperatures fall (for instance, during shutdowns), or cycle below the dewpoint, or at low-temperature operations such as refrigeration units. It is impossible to prevent the insulation becoming wet with moisture. For instance, polyurethane can absorb about 100% of water, or 100 gram of polyurethane can be wet to 200 gram. Calcium silicate can absorb up to 400% of its own weight when immersed in water. Generally, this water is saturated with oxygen. Water does not remain in the insulation material and penetrates to the metallic surface. Even if water does not penetrate from the outside inside of the insulation to the metallic surface, water can condense inside insulation during a temperature drop of under 100°C. Insulation that has become wet might never completely dry out. Water which has been condensed, or entered, becomes trapped under insulation, which then becomes like a sponge. About 5% moisture in insulation material results in double its thermal conductivity and results in energy loss. Now insulation plays the role of a barrier which does not allow the removal of condensed water from the outside of the metallic surface. Hot condensed water with dissolved oxygen is trapped on the metal surface under insulation: "two lovers, iron and water, can frivolously fall in corrosion sin". Iron reacts with water with the participation of dissolved oxygen, and thus an electrochemical mechanism is realized. The result in the form of rust is shown in Figure 3.9.

Carbon and low-alloy steel equipment undergo general (sometimes pitting) corrosion at a rate of ~4 mm/year, causing the formation of loose thick rust and its exfoliation. Austenitic stainless steels suffer from *pitting corrosion* and *chloride stress corrosion cracking*. Chlorides and bromides can reach a metal surface by way of rains or may be leached from insulation material (for example, polyurethane foam).

Corrosion becomes significant for carbon steels at 0 to 150°C, and for stainless steels at 60 to 205°C. General corrosion is most severe at temperatures close to dewpoint (about 95°C) but rarely takes place when operating temperatures are constantly above 150°C. Vessel and column heads are particularly vulnerable since they often have many nozzles which sometimes have no suitable insulation.

3.4.3 Prevention of Corrosion under Thermal Insulation

1. Use of appropriate *coatings* before equipment is insulated: organic (high solids amine epoxy, epoxy-phenolic, epoxy novolac, glass flake epoxy phenolic, high solids epoxy mastic, fusion-bonded epoxy) and aluminum metalizing (thermally sprayed) coatings. Use of organic coatings is limited by operating temperature. For instance, high solids amine epoxy is used up to 120°C, epoxy-phenolic and epoxy novolac until 180°C, glass flake epoxy-phenolic up to 200°C. Thicknesses of these coatings usually vary from 250 to 400 μm. Silicone aluminum coatings of 100 to 125 μm thickness may be used up to 550°C. All organic coatings are used on both carbon and stainless steels. Zinc coatings are not suitable because they are not resistant in a closed, wet environment under insulation. Insulation cannot prevent the ingress of water, air (oxygen), and contaminants from outside sources, therefore use of coatings resistant to water at high temperatures is of critical importance.
2. Careful and correct *design. The most effective measure is to keep the insulation in dry form.* This is simple to say but is usually impossible. The only non-absorptive insulation is a cellular glass. Insulated systems must be designed in such a manner that corrosives are minimized, that is to lessen the intrusion of water. Generally, thermal insulation has lagging, or jacketing, providing mechanical and weather protection for the insulation. The materials that are used for jacketing are aluminum, aluminized steel, and galvanized steel. It has to be taken into account that galvanized steel, or zinc coatings are not resistant to an industrial atmosphere containing hydrogen sulphide and other sulphur-containing gases. Mastics, sealants, and caulks must not contain polyvinyl chloride, brominated compounds, and acetic acid derivatives because they can cause stress corrosion cracking of stainless steel.
3. Careful, regular *inspection*: visual examination and removal of insulation, ultrasonic thickness measurement, radiography, acoustic emission, eddy current, X-ray transmission, magnetic flux leakage, infrared examination, and neutron backscatter.
4. *Correct maintenance*. Insulation systems are disturbed for repairs and are not properly reinstalled and sealed, allowing water ingress under insulation. Expansion joints have also to be given special attention because they are susceptible to uncontrolled movement and failure of insulations. Lagging replacement is sometimes also needed.

It is obvious that there are many ways to prevent corrosion under thermal insulation.

Recommended Literature

1. Pollock, W.I. and Barnhart, J.M., *Corrosion of Metals under Thermal Insulation*, ASTM Special Technical Publication 880, ASTM, USA, 1985, 242 pp.

2. Korb, L.J. and Sprowls, D.O., *Metals Handbook, Vol. 13: Corrosion*, ASM International, USA, 1987, pp. 1144–1147.
3. NACE Standard RP0198-98, The Control of Corrosion under Thermal Insulation and Fireproofing Materials – A System Approach.
4. A State-of-the-Art Report on Protective Coatings for Carbon Steel and Austenitic Stainless Steel Surfaces under Thermal Insulation and Cementitious Fireproofing, Publication 6H189, NACE, Houston, USA, 1989.
5. Posteraro, K., Thwart, Corrosion under Industrial Insulation, *Chemical Engineering Progress*, October, 1999, pp. 43–47.
6. Goldberg, L., Take a Systems Approach to Corrosion under Insulation, *Chemical Engineering Progress*, March, 2000, pp. 63–67.
7. Twomey, M., Detect Corrosion under Insulation, *Hydrocarbon Processing*, January, 1997, pp. 61–64.
8. Lasarte, C., de Rincon, O.T. and Montiel, A., Predicting Performance of Coatings under Thermal Insulation at High Temperatures, *Materials Performance*, October, 1994, pp. 36–39.

3.5 Corrosion in Fuels

In this section we shall answer the question as to why metals may corrode in fuels. Crude oil is a source of petroleum distillates, or fuels, and consists of a variety of chemical substances, the majority of which are organic compounds (hydrocarbons). Some of these compounds contain oxygen, sulphur, and nitrogen (Appendix H, Tables H.1 and H.2). One can often find water, salts (NaCl, $CaCl_2$, $MgCl_2$), hydrogen sulphide (H_2S), and other inorganic compounds in crude oils. All these corrosive species are removed during various processes such as desalting, distillation of crude oil, and then treatment of petroleum distillates (desulphurization, caustic treatment, etc.). First of all, regarding terminology. The word "*petroleum*" means "*rock oil*" from the Latin "*petra*" (*rock* or *stone*), and "*oleum*" (*oil*), and was first used in 1556 by the German mineralogist Georg Bauer (known as Georgius Agricola). Various petroleum fractions are manufactured during distillation of crude oils by increasing the boiling temperature: naphtha, gasoline, kerosene, gas oil (diesel), and fuel oil. All of them contain tens and even hundreds of hydrocarbons which are differed by chemical structure and as a result by molecular weight (Appendix H, Tables H.3 and H.4). We have to differentiate between naphtha and gasoline. When the mixture of hydrocarbons leaves the overhead distillation column at about 120 to 130°C, it is called naphtha. This naphtha is not yet suitable for use in automobile engines. If this mixture of hydrocarbons is sent for treatment for purification and increasing octane number, and is intended for use in automobile engines, it is called gasoline. If the same mixture of hydrocarbons is sent for treatment at a petrochemical plant, it is called naphtha. Therefore, there are storage tanks for naphtha (intermediate product containing H_2S), and storage tanks for gasoline – final product (also a mixture of many hydrocarbons) which is then used in automobiles. All organic mixtures received during distillation of crude oil are called petroleum distillates but not all *petroleum distillates* are *fuels*. Naphtha is not a fuel. Only distillates such as *gasoline*, *kerosene*, *diesel oil*, and *fuel oil* are called fuels. Each fuel is not a pure chemical compound; it is a mixture of different organic compounds called hydrocarbons. Petroleum distillates consist mainly of non-polar hydrocarbons that are not aggressive towards metals and alloys under normal conditions.

The paradox is that practice shows that equipment (storage tanks, pipelines, and pumps) contacting with petroleum distillates/fuels is being damaged. The presence of water, dissolved oxygen, salts, and microorganisms in fuels is the main cause of corrosion during storage and transportation. The water content in fuels manufactured at oil refineries is usually very low (30 to 80 ppm) and is not sufficient to make them corrosive. If gasoline is stored in dry conditions in carbon steel equipment, the latter does not corrode. Water vapors may ingress from the atmosphere into fuels during their storage and transportation. Then water vapors may condense as a result of temperature changes, and the water content in fuels may reach more than 80 ppm. An increase of water in fuels results in a drastic increase in the corrosion rate of carbon steel. For instance, the corrosion rate in "pure" gasoline (80 ppm water) is 0.001 mm/year, and when 200 ppm water is added, 0.4 mm/year. This means that there is a "critical" concentration of water and on the minimum thickness of the water layer on carbon steel in fuels for corrosion occurrence. For example, the "critical" water concentration in gasoline is 200 ppm and the minimum thickness of the water layer on steel in the presence of gasoline is 10 μm. This means that a water layer formed on a steel surface is responsible for the electrochemical corrosion mechanism in fuels (Figure 3.10). This mechanism is similar in any hydrocarbon–water environment. In this environment, a metal is partially wetted by the water that creates a thin water layer between the metal and the organic phase. The average thickness of the water layer on the metal surface is 3 to 10 μm. The oxygen solubility in hydrocarbons (60 to 70 ppm) is higher than in the aqueous phase (8 ppm). We may think of this two-phase system as a *differential aeration cell*. Hence, the corrosion processes proceed at the interface between the two phases, fuel and water. An anodic area is formed in the water phase (low oxygen concentration) and a cathodic area (high oxygen concentration) in the hydrocarbon phase, or fuel.

Factors affecting the corrosion process in the hydrocarbon–water system are water concentration in the mixture, electrical conductivity, type and concentrations of electrolytes in water, temperature, and fluid velocity. Salts and organic acids in the hydrocarbon phase undergo extraction from the petroleum distillates into the aqueous phase and cause an increase in the corrosion rate of carbon steel. Organic acids (formic and acetic) can appear in petroleum distillates as a result of oxidation of hydrocarbons during their treatment or storage. Corrosion products formed on carbon steel in contact with gasoline and kerosene at small concentrations of water have the form of "balls and cylinders" (Figure 3.11).

These "balls, cylinders, and jars" are empty inside and their walls are fragile. A scanning electron microscope (SEM) photo of one of such "cylinders" with a length of 7.4 mm is shown in Figure 3.12a. SEM photo of inner surface of this "cylinder" magnified 3,500 times is shown in Figure 3.12b. We can observe a wonderful "ball" (it resembles a human brain?) consisting of iron corrosion products formed inside of carbon steel pipeline with gasoline.

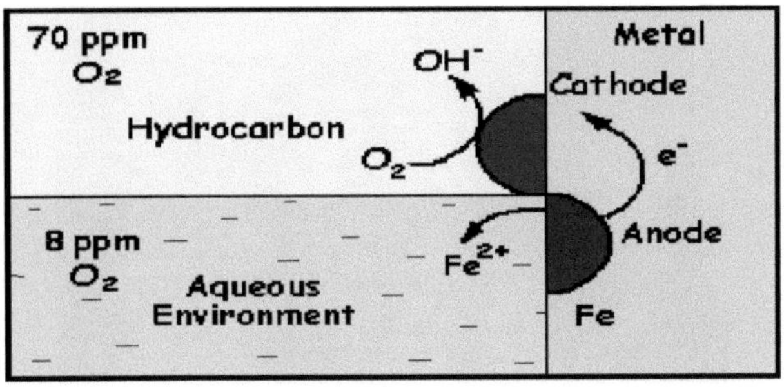

Fig. 3.10 Corrosion of carbon steel (cathodic and anodic areas) in a hydrocarbon-electrolyte environment.

Fig. 3.11 Corrosion "balls" formed on a carbon steel surface (strip coupons) after 3 days' immersion in kerosene with water (1 wt %) added. (For a full color version of this figure, see the Color Section)

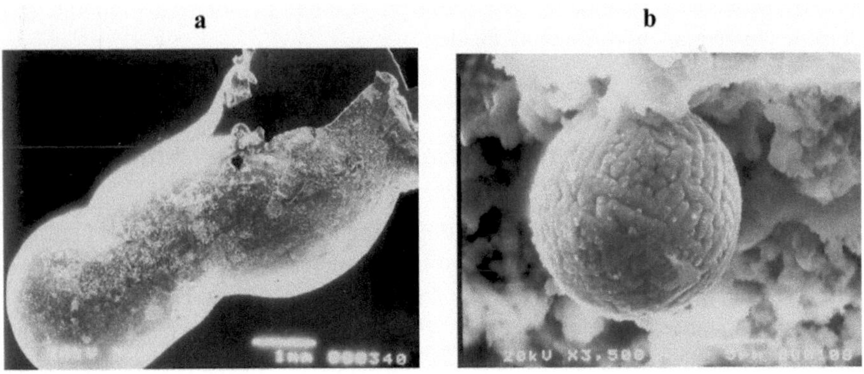

Fig. 3.12 SEM photo of a corrosion "cylinder" (a) with length 7.4 mm and magnification (×3,500) of the inner surface of the "cylinder" (b). Diameter of the "ball" 17 μm.

Similar corrosion products made of "balls, cylinders, and jars" are formed on the inner surfaces of pontoons in gasoline and naphtha storage tanks, as well as inside heat exchangers at the top of crude oil distillation columns, that is when carbon steel comes into contact with light hydrocarbons of gasoline and naphtha in the presence of small quantities of water (Figure 3.13).

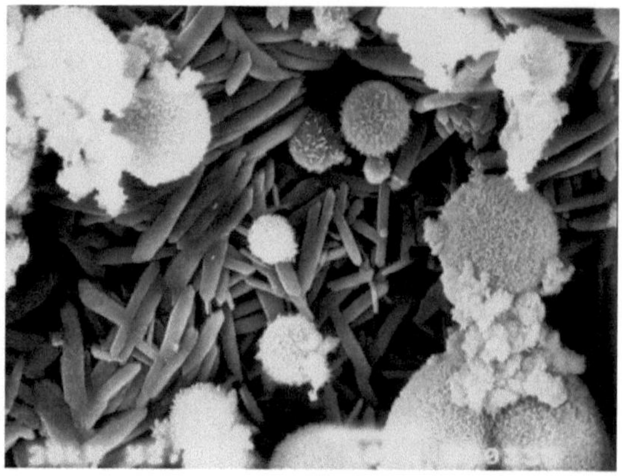

Fig. 3.13 SEM photo (×2,500) of corrosion products formed on carbon steel in contact with light hydrocarbons of gasoline and naphtha in the presence of small quantities of water: corrosion balls – "hedgehogs", platelets – iron sulphide.

An explanation is needed for such an appearance and structure of corrosion products formed on carbon steel contacting with gasoline, naphtha, and kerosene. We may imagine the mechanism of the formation of "corrosion cylinders, jars, and balls" as follows. Certainly, corrosion products of iron begin to grow at the border anode (water)–cathode (gasoline). Gasoline is captured inside such a growing "jar". The circular dense rust consisting of different hydroxides and oxides of iron (II) and (III) probably grow around the gasoline phase. Then gasoline can evaporate and thus the "rusted jar" will be empty inside.

3.5.1 Synthetic Chemicals in Gasoline, Polymers, and Ecology

Some chemicals are added to fuels for improving their properties. Tetra-ethyl-lead [TEL-$(C_2H_5)_4Pb$] has been used in gasoline to increase its *octane number* since 1926. The *octane number* is an indication of gasoline quality and is used for indicating the resistance of motor fuel to knock. Octane numbers are based on a scale on which iso-octane is 100 (minimal knock) and heptane is 0 (bad knock). Typical octane values for gasoline used in passenger cars are between 80 and 100. A gasoline with an octane number of 96 has the same knock as a mixture of 96% iso-octane and 4% heptane. A fuel with a high octane number exhibits better resistance to autoignition. The TEL is not corrosive to metals but has ecological problems. Various chemicals (oxygenates including ethers and alcohols, aromatic solvents toluene and xylene) instead of TEL appeared in order to increase the octane number of gasoline (Table 3.2). They are not corrosive to metals, but are aggressive to some polymers and organic coatings. Therefore, all polymers and organic coatings which were examined for resistance in contact with gasoline before use of oxygenates, and aro-

Table 3.2 Synthetic chemicals adding to gasoline for increasing its octane number.

Chemical Type	*Name*	*Formula*
Ether (oxygenate)	Methyl Tertiary Butyl Ether (MTBE)	$(CH_3)_3C\text{-}O\text{-}CH_3$
	Ethyl Tertiary Butyl Ether (ETBE)	$(CH_3)_3C\text{-}O\text{-}C_2H_5$
	Tertiary Amyl Methyl Ether (TAME)	$C_2H_5C(CH_3)_2\text{-}O\text{-}CH_3$
Alcohol (oxygenate)	Methanol	CH_3OH
	Ethanol	C_2H_5OH
Aromatic	Benzene	C_6H_6
	Toluene	$C_6H_5-CH_3$
	Xylene	$C_6H_4-(CH_3)_2$

Note: Oxygenates – chemical compounds which contain oxygen and are added to gasoline to boost its octane number. They are: MTBE (Methyl Tertiary Butyl Ether), ethanol, TAME (Tertiary Amyl Methyl Ether), ETBE (Ethyl Tertiary Butyl Ether), TBA (Tertiary Butyl Alcohol), DIPE (Di-Isopropyl Ether).

matic solvents added to gasoline, must be examined in gasoline with these additives once more.

Many types of polymer materials, such as Viton, Nitrile Butadiene Rubbers (NBR), and epoxy coatings are resistant to pure gasoline, but some of them fail in gasoline when new synthetic chemicals are added. Viton is resistant to aromatic solvents, but is not resistant to MTBE. NBR is resistant to MTBE, but is not resistant to aromatic solvents. Fluorocarbon polymers (Teflon) are resistant to both aromatic solvents and to MTBE which helps to keep clean air by replacing TEL as a gasoline additive, but gasoline leaks containing MTBE have prompted officials to close municipal water wells in some areas in the USA. People can readily detect the presence of MTBE in water because of its strong odor and taste. In some countries (for example, in Latin America) ethanol is used instead of MTBE. But the gasoline industry has many problems with ethanol. It is usually more expensive than MTBE and it has to be blended with the gasoline at the distribution terminal, not at the refinery, because ethanol tends to separate.

3.5.2 Change of Chemical and Physical Properties of Fuels Because of Metallic Corrosion

Corrosion products may cause a change of quality (chemical and physical properties) of fuels, namely, can accelerate oxidation of hydrocarbons and, as a result, they increase the acidity of fuels and concentrations of resins and slime in fuels. Therefore, long (more than 5 years) storage of gasoline in tanks made of carbon steel can diminish the chemical stability of gasoline (because of an increase of oxidative products). There are many examples of increasing contamination in gasoline, kerosene (jet fuel), and diesel oil after 5 years of storage in tanks. Usually contaminants in gasoline and kerosene consist of corrosion products and organic suspended solids (slime), or sludge. Some of them settle at the bottom and at the wall surfaces in the tanks, but a large quantity of contaminants remain in fuels and then appear in engine systems, including the combustion chamber. These contaminants are not fully burned and, as a result, the mass of soot increases and engine efficiency diminishes. Slime is the source of MIC in storage tanks. Besides corrosiveness and aggressiveness, some petroleum components and contaminants (especially solid corrosion products such as iron oxides and sulphides) influence the tribological properties of fuels, namely, they may increase wear on metallic surfaces of important engine parts.

We conclude that corrosion control of storage tanks for fuels is important, in spite of the fact that the pure fuels are organic non-corrosive mixtures. We shall discuss corrosion and corrosion control of fuel storage tanks in the next section.

3.6 Corrosion of Storage Tanks for Fuels and Their Corrosion Control

Usually fuels are stored in carbon steel tanks of different types and sizes. These are subject to corrosion, depending on many factors. According to tradition, we begin every new section with queries. What is the uniqueness of corrosion of the tanks? What are the causes of corrosion and its mechanism in tanks? Which parts of storage tanks suffer from corrosion? Does this depend on the fuel type? What anti-corrosion techniques are used for preventing corrosion and its control in tanks? How may we select effective coating systems for the protection of the inner and outer surfaces of tanks? We shall discuss the corrosion of aboveground storage tanks (ASTs) containing fuels and crude oil and provide an analysis of the causes of AST corrosion. Crude oil and fuels are stored in carbon steel tanks at ambient temperatures, except fuel oil. The latter is stored usually at 80 to 90°C because of its high viscosity. Steam coils and electrical heaters are used for heating fuel oil in the tanks.

The corrosion mechanism in fuels was described in Section 3.5. The corrosiveness of crude oil is explained by the presence of hydrogen sulphide, water, salts, and microorganisms.

Dissolved water and oxygen may be present in fuels after various types of treatment of petroleum distillates. Not all substances containing sulphur (H_2S, mercaptans) may be removed. Light organic acids (formic and acetic) and chlorides may be present in fuels and, in the presence of water, may be highly corrosive. Fuels are transported to storage tanks through pipes and pumps. During storage, water vapor and oxygen from the air may dissolve in fuels. Water vapors and air can

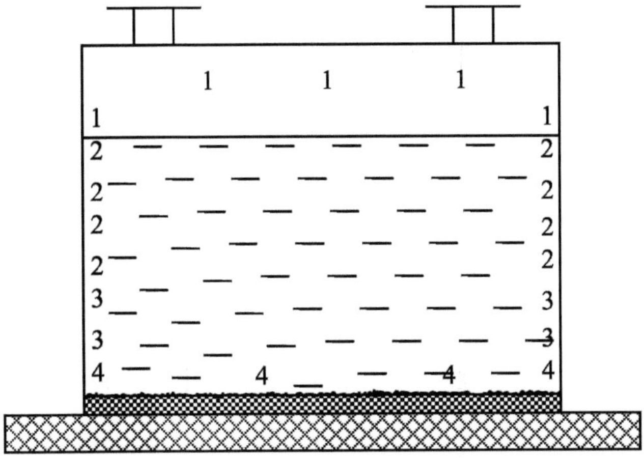

Fig. 3.14 Four corrosion zones forming in the inner surfaces of an AST.

ingress into tanks during emptying operations as well as during the tanks "breathing" through vents. During temperature changes in ASTs at day-night cycles, water with dissolved aggressive variables (oxygen, chlorides, H_2S, light organic acids) may separate from fuels. A thin layer of an aqueous solution of electrolytes may form on the wall and then flow to the bottom of the tanks. Thus, the conditions exist for the electrochemical mechanism to occur.

Four corrosion zones are formed in the inner surfaces of an AST (Figure 3.14):

1. Upper part – inner surface of a roof and shell in contact with the gaseous phase.
2. Splash zone – the border between the liquid and gaseous phase. This boundary is not constant and its dimensions depend on the operations of filling and emptying.
3. Liquid zone – continually in contact with liquid fuels. The dimensions of this zone also depend on filling and emptying factors.
4. The bottom and sometimes the first strip of the shell. These parts of the AST are in contact with water, dissolved salts, organic and inorganic deposits (sludge). Organic deposits consist of hydrocarbons and biofouling; the inorganic ones consist of corrosion products, sand, and some salts.

Some authors describe three zones: gaseous phase, splash zone, and bottom (with the first strip). We differentiated the zone of liquid fuel because we found that severe corrosion of this zone occurred in the AST containing gasoline and naphtha.

The corrosion mechanism occurring on the inner surfaces of the AST is more complicated than on the outer surfaces because of the existence of these four corrosion zones. Corrosion intensity and its forms in the AST depend on the petroleum distillate (fuel) type; the solubility of water and oxygen in the fuel; the volume of the tank; the technology of tank exploitation (the frequency and rate of filling and emptying); temperature and its fluctuation; constructive features of the tanks (the roof type – fixed or floating; the presence of a pontoon; the operation of the "breathing" vents; the condition and form of the inside tank surface – presence of rust, coating, etc.); climate (geographical location of tanks); the geographical direction of tanks (north, south, east, or west). Storage tanks for crude oil, naphtha and gasoline are usually equipped with a pontoon or floating roof to minimize vaporization and subsequent loss of fuels through the vents. In order to diminish the fuel loss from the tanks during storage under sunrays, the outer surfaces of the tanks are coated with white paint colors. We shall analyze corrosion of tanks with different fuels.

3.6.1 Gasoline ASTs

The corrosion rate (CR) of the inner surface of the gasoline AST depends on the strip number and the direction (south, north, west, and east). The average corrosion rates usually are 0.04 to 0.13 mm/year. Maximum corrosion occurs on the southern part of the AST, probably because of considerable fluctuations of temperature during the day-night cycle. Because of the direct exposure to the sunlight, the temperatures are higher on the southern part of the AST and the solubility of water in gasoline increases. When the temperature decreases in the evening, the solubility of water in gasoline lessens. Polar water molecules separate from the mixture "gasoline–water" on the steel surface, and an electrochemical mechanism takes place in the presence of dissolved oxygen (see Section 3.5).

The inner side of the shell and the floating roof is subject most to corrosion. Usually the bottoms in gasoline ASTs are in good condition. Therefore, the corrosion mechanism is probably related to the temperature fluctuations, the presence of dissolved water and oxygen in gasoline, and periodical water separation on the steel shell surface of the tanks. To sum up, the inner side of the shell must be protected against corrosion.

3.6.2 Gas Oil ASTs

The inner side of the shell usually exhibits no corrosion. The inner side of the roof is attacked by H_2S that evolves from the gas oil (Figure 3.15). Pitting corrosion usually occurs at a rate of $\sim$1 mm/year. The bottoms are attacked by microorganisms. Shallow pits are formed on the bottom after 20–25 years of service. The localized corrosion rate inside the bottoms reaches up to 0.5 mm/year. Heterotrophic bacteria and SRB are found in the sludge that contacts the bottoms. To sum up, the inner sides of roofs and bottoms must be protected against corrosion.

Fig. 3.15 Corroded roof of the gas oil AST after 5 years of service. (For a full color version of this figure, see the Color Section)

3.6.3 *Heavy Fuel Oil Storage Tanks*

The inner side of the shell usually shows little corrosion. The corrosion rate varies from 0.006 to 0.014 mm/year. There is no influence of geographic direction. The main corrosion problem in these tanks happens on the inner side of the roofs and to a lesser extent on the bottoms. Usually the roofs are changed every 10 years because of severe corrosion and holes. The design life of ASTs is usually 30 years. The average corrosion rate of the roof is usually 0.5 mm/year. Corrosion products that formed on the inner side of roofs consist mainly of iron sulphides. Hydrogen sulphide, as in the gas oil AST, causes severe corrosion of roofs. The second region exhibiting corrosion in the heavy fuel oil tanks is the bottom (Figure 3.16). Sludge is formed during storage of the fuel oil. Large quantities of heterotrophic microorganisms and SRB are found in the sludge that made contact with the bottom. These microorganisms cause localized attacks on the bottoms.

3.6.4 *Crude Oil Storage Tanks*

The main corrosion problem in these tanks generally occurs on the inner surface of the roofs and on the bottoms. Usually the roofs are changed every 15 years because of severe internal corrosion. The second region exhibiting corrosion in the crude oil AST is the bottom. A large quantity of sludge is typically formed during storage of the crude oil. For example, 2 m of sludge was found after 18 years of crude oil AST

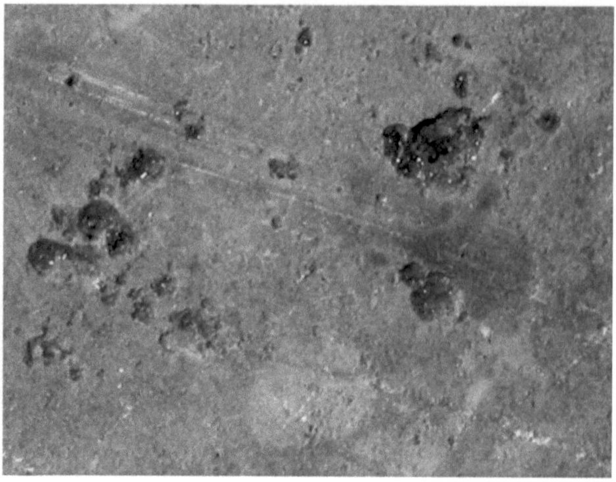

Fig. 3.16 Corroded bottoms of the fuel oil AST after 10 years of service. (For a full color version of this figure, see the Color Section)

service. Large quantities of SRB were found in the sludge. These microorganisms were responsible for localized attacks in the bottoms (see Figures 2.14b and 3.17). Corrosion products consisted of 80 wt% of inorganic (iron sulphide and rust) and 20 wt% of organic substances. The corrosion rate of the bottom varies from 0.3 to 0.5 mm/year.

3.6.5 Anti-Corrosion Techniques for ASTs

It has been shown that various parts of the inner surfaces of ASTs containing crude oil, gasoline, gas oil, and fuel oil need protection from corrosion. The following anti-corrosion techniques exist: technological measures, coatings (organic and metallic), and corrosion inhibitors. We shall discuss each of these techniques.

3.6.5.1 Technological Measures

Technological measures include draining the tank and cleaning the bottom from sludge, filling an AST with fuels to the maximum, dehumidifying incoming material through air "breathing" vents, inert gas purging (usually dry nitrogen), storage of fuels under pressure of inert gas free of water vapors and oxygen, and the treatment of fuels by removing aggressive corrosive components such as hydrogen sulphide, water, and oxygen through the use of scavengers and coalescers.

Fig. 3.17 Corroded bottom of a crude oil AST. The morphology of the surface and microbiological examination showed MIC. (For a full color version of this figure, see the Color Section)

3.6.5.2 Organic Coatings

Organic coatings remain the most prevalent technique for anti-corrosion protection of an AST. The API RP 652 standard recommends two types of coating systems: thin (<500 μm) and thick (>500 μm). The recommended thin coating systems are coal tar epoxy, epoxy phenolic, epoxy amine, epoxy polyamide, and epoxy polyamidoamine. The coal tar coatings are harmful to people and the environment. Thick-film coatings consist of a glass-reinforced lining based on polyesters (isophthalic, bis-phenol-A, and vinyl ester) or epoxy resin. Glass reinforcement includes flake, chopped strand, mat, and roving. For new tanks or for older tanks where only internal corrosion is occurring, 900 to 1,400 μm thick coatings may be used. For older storage tanks where bottoms have corroded both internally and externally, 2,000 to 3,000 μm thick glass-reinforced coatings are often used. The following coatings were examined and recommended for anti-corrosion protection of an AST: inorganic zinc silicate (150 μm), epoxy (200 to 1,500 μm), silicone-epoxy (250 μm), epoxy-phenolic (300 μm), epoxy novolac (300 to 400 μm), polysiloxane (300 μm), polyurethane (500 μm), epoxy reinforced with glass and mineral flakes (500 μm), glass-filled epoxy with rust converter, inhibitor, and passivator (600 μm), vinyl ester with acrylic copolymer (1,250 μm), epoxy vinyl ester (1,500 μm) and vinyl ester (1,500 μm).

3.6.5.3 Metallic Coatings

Zinc and aluminum arc-sprayed (metalizing) coatings are used for the protection of the inner surface of the AST from corrosion. Hot dip aluminized steel and zinc-nickel galvanic coatings are recommended for tanks containing fuels with oxygenates. The recommended thickness for zinc coatings is 100 μm, and for aluminum it is 300 μm. *Porosity is the ratio of free volume (cavities) in the coating to the total geometric volume of the coating on a metal surface*. The porosity of zinc coatings is 1 to 3% and is less than the porosity of aluminum coatings (5 to 15%). The adhesion of zinc coatings to steel surfaces is higher than that of aluminum coatings. Zinc coatings are not resistant to H_2S if the latter is present in fuels, and the electrode polarity of zinc and iron may change during the cleaning of inner surfaces of an AST with hot water at 90°C. This phenomenon can cause the dissolution of iron instead of zinc if cracks are present in the zinc coating. Aluminum coatings are resistant to H_2S and hot water, but are susceptible to sparks, if a steel object falls inside fuel tanks containing flammable hydrocarbon gases. Therefore, the same requirements must be carried out during repair and maintenance work inside steel tanks with and without aluminum coatings. Zinc does not cause sparks in such cases. However, it is not resistant to hydrogen sulphide, and zinc dust formed during the metalizing process is more dangerous for people's health, than aluminum dust. Aluminum is resistant to hydrogen sulphide. Hot water treatment of aluminum coatings is favorable for decreasing their porosity because of the formation of aluminum hydroxides in the pores of the coatings. The porous surface of arc-sprayed coatings is a good base for the penetration of liquid paints and the formation of combined metalizing-paint coatings on steel surfaces. Therefore, combined coatings were studied and recommended for use. The approximate lifetime of aluminum-epoxy coatings in fuels is 30 years.

3.6.5.4 Corrosion Inhibitors

Liquid phase: Many corrosion inhibitors, such as amines, amides, acetates, and sulphonates, dissolved in the hydrocarbon phase, are known but they have not found wide use in fuels. Small quantities of water in fuels (200 to 1,000 ppm) can cause severe corrosion of carbon steel. Inorganic corrosion inhibitors (nitrites and phosphates) injected in concentrations of 100 to 200 ppm to mixtures of gasoline and water effectively protect carbon steel from corrosion. These inhibitors are more effective under agitating conditions than during stagnation. Fuels can be deteriorated by microorganisms in as little as 6 months. About 30 types of bacteria, 80 types of filamentous fungi and 12 types of yeasts were found in fuels. Only 20 to 40% of these microorganisms are capable of using hydrocarbons for their proliferation. Bacteria can cause filter plugging, deteriorate fuel quality, and induce corrosion of AST bottoms. To inhibit or prevent the bacterial deterioration of fuels and MIC in the bottoms of ASTs, bacteriocides are sometimes injected. Isothiazolone (10 ppm), isothiazolin (200 ppm), quaternary ammonium compounds, organoborinanes, pyridinethione, hexahydrotriazines, imidazolcarbamate, and others are used. If microbial sludge has already formed at the bottom, bacteriocides would not prevent the development of MIC. Bottoms must be cleaned from any sludge. The best method of controlling microbial sludge formation is through periodical drainage (once a week) of water and microbiological control of the water and fuel phases. The important parameter of the total bacteria count (TBC) is its growth tendency rather than its absolute value. If TBC = 10^3 bacteria/ml in the aqueous phase, and is constant with time (it is recommended checking once a month), and TBC = 0 in the fuel, the latter is not contaminated by microorganisms. The pH of the aqueous phase in this case may be 5 to 7. Any changes of conditions can result in a proliferation of microorganisms and deterioration of fuels.

Vapor phase: Vapor (or volatile) phase inhibitors (VPIs) (see Section 3.2) are used for the protection of the inner surface of the upper parts of an AST which have contact with the gaseous phase containing hydrocarbon vapors, air, water vapors, and H_2S emitted from the liquid fuels. Various organic substances are recommended for VPIs such as dicyclohexylamine nitrite $[(C_6H_{11})_2NH_2NO_2]$, cyclohexylamine carbonate $[(C_6H_{11}NH_2)CO_2]$, monoethanolamine $[NH_2(CH_2)_2OH]$, and others. VPIs may be used as solids (granules, tablets, powder) and liquids. The efficiency of VPIs depends on their vapor pressure, the airtightness (hermeticity) of the AST, temperature, and water vapor content. It is impossible to close an AST tightly from the atmosphere because of the "breathing" process a tank undergoes. Because of the "breathing", vapors are emitted from the tanks during filling, and air with water vapors enters during emptying of the tanks. A VPI must be introduced in the vapor zone of the tanks throughout their use, as organic molecules of VPI also prevent additional corrosion of rusted surfaces of the AST. This method is especially cost effective in small tanks.

3.6.5.5 Internal Cathodic Protection

It is impossible to use cathodic protection inside AST containing fuels because the latter are not electrolytes. Zinc sacrificial anodes can be used on the tank bottom if aqueous phase (with salt content >0.3 wt%) is also present on the bottom. All anti-corrosive techniques described in this section are compatible for the protection of the inner surfaces of underground storage tanks.

Recommended Literature

1. API RP 652, Lining of Aboveground Petroleum Storage Tank Bottoms, 1997, 21 pp.[1]
2. API Standard 653, Tank Inspection, Repair, Alteration, and Reconstruction, 2001, 116 pp.
3. Groysman, A., Corrosion of Aboveground Fuel Storage Tanks, *Materials Performance*, **44**(9), 2005, pp. 44–48.
4. Groysman, A., Anticorrosion Technique for Aboveground Storage Tanks, *Materials Performance*, **44**(11), 2005, pp. 40–43.
5. Groysman, A. and Belaschenko, V., Study of Anticorrosion Properties of Metal Arc – Sprayed Coatings on Carbon Steel for Use in Petroleum Products. In *Proceedings of 12th International Corrosion Congress*, Vol. 1, 19–24 September 1993, NACE International, Houston, Texas, USA, 1993, pp. 63–76.

[1] API RP = American Petroleum Institute Recommended Practice.

6. Bregman, J.I., *Corrosion Inhibitors*, McMillan Co., New York, 1963.
7. Videla, H.A., *Manual of Biocorrosion*, CRC Lewis Publishers, Boca Raton, 1996, 273 pp.
8. Rosenfeld, I.L. and Persiyantseva, V.P., *Inhibitors of Atmospheric Corrosion*, Nauka, Moscow, 1985, 278 pp. [in Russian].

3.7 Corrosion and Corrosion Control in the Presence of Naphthenic Acids

Crude oils consist of many organic substances which are not usually corrosive to metals. Some crude oils contain varying amounts of *naphthenic acids* (NA) possessing high corrosiveness under particular conditions (at temperatures from 180 to 360°C and a high flow velocity). NA are probably the products of natural long-time crude oil oxidation. They are mainly cycloaliphatic mono-carboxylic acids derived from "*naphthenes*". The latter term relates to cycloalkanes – saturated monocyclic hydrocarbons (with or without side chains), particularly cyclopentane, cyclohexane and their alkyl derivatives. The carboxylic group (—COOH) is usually attached to a side chain rather than directly to the cycloaliphatic ring (Figure 3.18).

The terms *naphthenic* and *naphthenes* seem to be obsolete, except in the petroleum industry. In fact, NA are the mixture of different polycyclic compounds possessing a carboxylic group (—COOH) and may have unsaturated bonds, aromatic rings, and hydroxyl and phenol groups. In other words, NA can be regarded as cyclopentane or cyclohexane – substituted acetic acids with higher molecular weight acids being bi-, tri-, or polycyclic. Sometimes all organic acids are called by the generic name NA. We have to differentiate NA from acyclic aliphatic (paraffinic or fatty, tall oil, hydroxyl, dibasic, and aromatic) acids. According to IUPAC (International Union of Pure and Applied Chemistry), the term "*organic acids*" applies to a broad range of organic compounds which contain the organic acid group —COOH:

- aliphatic (fatty) acids RCOOH, where R is a straight $CH_3(CH_2)_n$ or branched chain;
- aromatic acids: ArCOOH, where Ar is a benzene ring or substituted benzene rings;
- naphthenic acids: XRCOOH, where X is a cycloparaffinic ring; the chemical formula of NA is $X(CH_2)_nCOOH$ ($n \geq 0$).

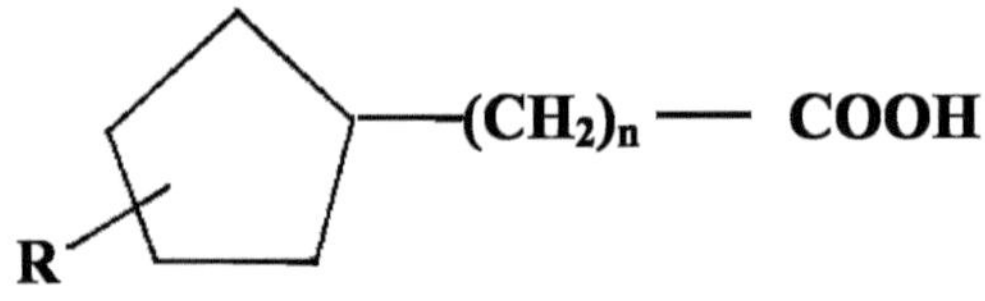

Fig. 3.18 Naphthenic acids. **R** represents multiple alkyl groups or fused cycloaliphatic rings.

All of these three groups can be present in crude oils. It is very difficult to isolate NA separately from the acyclic aliphatic acids RCOOH and aromatic acids ArCOOH. The Russian chemist Vladimir Vasilyevich Markovnikov (1838–1904) was the first to isolate naphthenic acids from the Caucases crude oil and to study their physical and chemical properties. He determined a new class of organic substances – *naphthenes*, therefore the acids received the name "*naphthenic*". Aromatic olefinic, hydroxy, and dibasic acids are present as minor components of NA. While the major use of NA has been in the production of metal soaps, they also can react to form esters, amine salts, amides, imidazolines and many other derivatives.

The following factors influence naphthenic acid corrosion: the thermodynamic possibility of reactions between metals and NA, but we are not able to calculate the thermodynamic function (Gibbs energy), because, in most cases, we do not know the chemistry of NA and corrosion products; the chemical and physical properties of NA and their concentrations, temperature, sulphur content in crude oil, and their distillation fractions, flow velocity, metallurgy, and time of contact between NA and metals. Corrosion aggressiveness of NA depends on their type: its structure, molecular weight, boiling point, and temperature of use. Sometimes di- and tri-cyclic acids can undergo thermal decomposition to produce lower molecular weight naphthenic and aliphatic acids. The physical and chemical properties of NA depend on the nature and source of the crude oil. NA are liquids with low freezing points and a wide boiling point range (216 to 350°C). They are completely soluble in organic solvents and oils but insoluble in water. The molecular weight of NA may change from 114 till 350. Chemically, NA behave like typical carboxylic acids with an acid strength similar to those of the higher aliphatic acids, but are slightly weaker than low molecular weight carboxylic acids such as acetic acid. Some properties of seven representatives of NA and four aliphatic acids are shown in Appendix H, Table H.5. An increase in the molecular weight of NA results in an increasing boiling point temperature. We can compare these properties with those of some low molecular fatty acids: formic, acetic, propionic (propanoic), and butanoic (butyric) acids. These acids are released from crude oils with a high content of NA. Similar to NA, an increase of molecular weight of aliphatic acids results in higher boiling points as in all classes of molecular compounds.

NA are distributed differently in various petroleum distillated fractions, and as a result, their corrosiveness is usually also different. *Total Acid Number* (TAN) values (also called *neutralization numbers*) are used to evaluate the corrosiveness of crude oils and their distillate fractions. TAN is a measure of the numbers of milligrams of potassium hydroxide (KOH) needed to neutralize 1 g of crude or its distillate fraction. TAN = 220 to 320 mg KOH/g for most NA. Some authors emphasize that there is no correlation between TAN of crude oils/ distillates and their corrosiveness. In spite of this, some rules of thumb exist. It was suggested using a threshold of 0.5 mg KOH/g of crude oils and of 1.5 mg KOH/g of distillate. Certainly, these values should not be used as absolutes.

The chemical structure of NA significantly influences their corrosiveness. Not all NA are corrosive, even some of them (high molecular weight NA), can inhibit corrosion of carbon steel under certain conditions. Therefore, it is difficult to find a correlation between the NA content in crude oils (or in distillate fractions) and their corrosiveness. Usually NA are concentrated in heavy gas oils (boiling temperatures 220 to 320°C), but sometimes the kerosene fraction of narrow boiling temperatures 190 to 210°C) can also be highly corrosive. Distillation towers, pipelines, and other equipment containing crude oil, kerosene, light diesel, heavy diesel, atmospheric residue, light and heavy vacuum gas oil and vacuum residue operating at temperatures above 190°C are possible areas for NA attack. In spite of naphthenic acid corrosion being known since 1930s, the corrosion mechanism as well as the chemical reactions describing it is not known because of the complex influences on acid type corrosion, various constituents in the crude oil (especially sulphur-containing compounds), temperature, and fluid flow. Some authors assume the following corrosion reaction:

$$2XRCOOH_{(l)} + Fe_{(s)} \rightarrow Fe(XRCOO)_{2(l)} + H_{2(g)}. \tag{3.20}$$

This mechanism suggests the formation of iron naphthenates and hydrogen gas, which allows controlling NA corrosion via the evolving hydrogen. Iron naphthenates dissolve in oil, therefore corrosion products do not remain on the metal surface.

Corrosion damage at high temperatures in the oil refining industry may be attributed not only to NA but also to other sources such as high temperature sulphidic corrosion and non-naphthenic acids. NA result in corrosiveness of crude oils and their fractions before, during, or after the distillation processes.

NA boil and condense together with hydrocarbons of crude oil on suitable trays in the distillation tower, therefore they are most corrosive at their boiling points, namely during evaporation and condensation on a metallic surface, which cause impingement of liquid droplets and severe corrosion in the form of pits (see Section 2.3). Pure NA are more corrosive than light aliphatic organic acids, but less corrosive than aromatic organic acids at their boiling points. If the corrosion rates of carbon steel in contact with boiling low molecular weight organic acids are 25 mm/year, the corrosion rates of carbon steel in pure boiling NA are 100 to 400 mm/year.

Sulphur-containing compounds can inhibit corrosion by NA as a result of the formation of an iron sulphide protective scale. In contrast to high temperature sulphidic corrosion, no protective scale is formed, therefore it is not simple to define that NA were responsible for corrosion. NA corrosion was not detected at >360°C, probably because of the formation of a protective coke layer or the inhibitive properties of high molecular NA. The influence of metallurgy is well known. The higher the molybdenum (Mo) content in steel, the lower the corrosion of stainless steel in the presence of NA. Flow velocity is a very important factor and usually the higher TAN, the greater the sensitivity to velocity. There is a wide spectrum of mitigation measures.

3.7.1 Mitigation Measures

(a) The *blending* (dilution) of crude oils/distillates having high TAN (above 0.5 mg KOH/g) with crude oils/distillates having low TAN, respectively, to a target value of less than 0.5 mg KOH/g.
(b) *Neutralizing*. This is one of the earliest methods which includes the addition of caustic soda (NaOH) in the crude oil. Injection of NaOH may cause caustic embrittlement of equipment and fouling formation on downstream processing.
(c) *Materials selection*. The greater the content of molybdenum (Mo) in stainless steels, the better their resistance to NA corrosion. Stainless steels SS 316 and SS 317 are often used if an NA attack is expected. The molybdenum content in steels must be above 2.5 wt%. Aluminum and titanium are also resistant to NA.
(d) The use of *corrosion inhibitors*. Various organic corrosion inhibitors based on phosphate esters and thiophosphorus compounds were developed for use in hydrocarbons at high temperatures. At 50 to 100 ppm of phosphate esters, corrosion inhibitor high protection (94 to 100% efficiency) occurs. Effectiveness is enhanced with pre-treatment. The latter is the injection of a large quantity of corrosion inhibitor (for example, 100 ppm) in the non-acidic stream (crude oil or distillate) during 1 to 3 days. Corrosion inhibitors form a dense protective layer on a metal surface during pre-treatment. Then injected concentrations may be diminished to 15 ppm in the acidic streams. In the case where phosphate esters corrosion inhibitors are used, deterioration of catalyst and kerosene may occur. Sulphur-based inhibitors are less effective than the phosphorus ones, but do not deteriorate the catalyst activity and fuel quality. Some organic acids (tannic and gallic), alcohols, heavy vacuum gas oil fraction, and high molecular weight NA (with boiling point >300°C) can also inhibit corrosion. Amine- and amide-based corrosion inhibitors are ineffective in the high-temperature environment of NA corrosion.
(e) *Use of coatings*. Aluminized steel, niobium and niobium oxide thermal spray coating showed good resistance to high temperature NA corrosion.

Note. NA are treacherous in the oil refining industry because of their high and unpredictable corrosiveness. However, NA derivatives are widely used in the treating of wood (preservatives), corrosion inhibitors, emulsifiers, defoamers, paint and ink driers, tire cord adhesives, fuel and lubricant additives, heat/UV stabilizers, cutting oils, and vinyl stabilizers.

Recommended Literature

1. Korb, L.J. and Sprowls, D.O., *Metals Handbook, Vol. 13: Corrosion*, ASM International, USA, 1987, pp. 1271–1273.
2. Brient, J.A., Wessner, P.J. and Doyle, M., Naphthenic Acids. In *Kirk-Othmer Encyclopedia of Chemical Technology*, John Wiley & Sons, Inc., 1995.

3. Slavcheva, E., Shone, B. and Turnbull, A., Review of Naphthenic Acid Corrosion in Oil Refining, *British Corrosion Journal* **34**(2), 1999, pp. 125–131.
4. Groysman, A. et al., Corrosiveness of Acidic Crude Oil and Its Fractions, *Materials Performance* **44**(4), 2005, pp. 34–39.
5. Zeitlmeisl, M.J., Harrel, J.B. and Campbell, J., Naphthenic Acid Corrosion Control, *Hydrocarbon Engineering*, October, 2000, pp. 83–85.
6. Wu, X.Q., Jing, H.M., Zheng, Y.G., Yao, Z.M. and Ke, W., Study on High-Temperature Naphthenic Acid Corrosion and Erosion-Corrosion of Aluminized Carbon Steel, *Journal of Materials Science* **39**, 2004, pp. 975–985.
7. Groysman, A., Penner, J., Brodsky, N. and Skorodinsky, S., Naphthenic Acids Combat Naphthenic Acid Corrosion. In: *Proceedings 17th International Corrosion Congress*, Las Vegas, USA, October 6–10, 2008, Paper No. 2257, 2008, 16 pp.
8. Groysman, A., Penner, J., Brodsky, N. and Skorodinsky, S., New Inhibitors for Preventing Naphthenic Acid Corrosion. In: *Proceedinds EUROCORR 2008, The European Corrosion Congress*, Edinburgh, Scotland, 7–11 September 2008, 17 pp.
9. Groysman, A. et al., Naphthenic Acid Corrosion Study. In: *Proceedings EUROCORR 2007, The European Corrosion Congress*, Freiburg im Breisgau, Germany, 9–13 September 2007, 23 pp.

Chapter 4
Corrosion Control Measures

The main cause of corrosion of metals and alloys is their thermodynamic instability in the environment. Only some of them, gold, silver, and platinum are resistant to the environment of the Earth's crust and exist as pure metals. Many "attractions" are present around the metals: water and aqueous solutions of electrolytes, gases (oxygen, ozone, sulphur oxides, nitrogen oxides, and hydrogen sulphide), salts, acids, alkalis, organic substances, and microorganisms. The conditions around the metals, such as high temperatures and temperature changes, high velocities of liquids or their stagnation, also contribute towards corrosion. Metals are not able to be apathetic to the environmental "attractions" and conditions. Our aim: to keep metallic structures in a good state, to prevent their oxidation, deterioration, loss of functional properties, damage, and failure. In order to select the correct measures of corrosion control, we have to study the corrosion mechanism, how metals react with the environment, the factors of metallic corrosion, how metals behave in different media (in water, in the atmosphere, in the presence of various salts and gases) and, of course, corrosion phenomena (general, pitting, galvanic, erosion, cavitation, MIC, and others).

If a person feels ill, he goes to the doctor. What does a doctor do? He carries out all the necessary physical and biochemical analyses. Only on the basis of these results, can a doctor come to a conclusion about the illness and may provide the solution for the remedy. A similar situation occurs with the "illness" of metals – their corrosion. If we know the causes and factors of the corrosion of metals, we can select the correct methods of prevention and control. It is very important to predict the occurrence of possible corrosion problems and to plan correct anti-corrosion measures accordingly. For example, if we design new equipment with different alloys, it is important to select those with close electrode potentials, or electrically isolate different alloys, or to increase an anodic area towards a cathodic one, and to carry out all this – in order to exclude or to diminish the probability of galvanic corrosion. If the main factor of corrosion is the flow velocity of a liquid, we have to take measures for its change (increase in the case of low velocity and stagnation, or decrease in the case of turbulence), or to select a suitable alloy or coating resistant to erosion or cavitation. If it is impossible to use a corrosion-resistant alloy

A. Groysman, *Corrosion for Everybody*, DOI 10.1007/978-90-481-3477-9_4,
© Springer Science+Business Media B.V. 2010

or suitable coating (metallic, ceramic or polymeric) we can select the proper corrosion inhibitor, neutralizer, oxygen scavenger, use cathodic protection, or keep the temperature above the dew point. We differentiate all the corrosion control methods into six groups:

(a) *The use of coatings* (organic, inorganic, and metallic).
(b) *The use of electrochemical methods*: cathodic and anodic protection.
(c) *Change the environment*: the use of corrosion inhibitors; removal of the aggressive components, such as oxygen (deaeration), hydrogen sulphide, chlorides, ammonia; neutralization (injection of alkalis into acidic solutions, or acids into alkali solutions); drying the atmosphere (removal of water vapors); use of biocides, etc.
(d) *Correct selection of materials*: corrosion-resistant metals and alloys, polymeric materials, ceramics, glasses, and composites. We have to remember that there is no universal metal or other material resistant to all media and under all conditions. Polymeric materials are not resistant to high temperatures and high mechanical stresses. Ceramics and glasses are brittle. Composite materials are not resistant to high temperatures, some chemicals, and are relatively expensive.
(e) *Correct design*. Metallic structures and equipment must be designed in such a manner that they would be convenient for drainage, cleaning, surface preparation, and painting; not to use different alloys in a general electrolyte solution, in order to prevent galvanic corrosion; not to use sharp elbows; etc.
(f) *Technological measures*, namely, changes of process conditions. For example, keeping the temperature 20 to 30°C above the dewpoint in order to prevent condensation of corrosive substances (for instance, H_2O with dissolved HCl or H_2SO_4); decrease flow velocity of liquid in the case of erosion, or increase its flow velocity in the case of stagnation, in order to prevent formation of deposits (fouling and, as a result, a possible under deposit corrosion).

Combining methods of corrosion control exist too. For example, we can use corrosion inhibitors in paints or in concrete; cathodic protection together with coatings (and corrosion inhibitors); keeping the temperature above the dewpoint and injection corrosion inhibitors and neutralizers in the processing streams. As we see, there is wide spectrum of corrosion control methods. We shall become familiar with some of them in the following sections in order to make the correct selection and to prevent corrosion of metallic structures.

4.1 Use of Coatings

The use of any coating is based on the fact that metal must be isolated from an aggressive environment. Probably, use of coatings is one of the most ancient anticorrosion and waterproof techniques. We can mention the story of Noah, the ark and the flood from Genesis. Noah's ark was waterproofed with pitch (bitumen, asphalt, or tar). The reed basket that carried the infant Moses into the Nile River in Egypt was waterproofed with pitch. The Babylonians (7th century BC) and the Nabateans ("Oilmen of the Dead Sea", 4th–3rd centuries BC) used bitumen for protection of various constructions.

All coatings can be differentiated into three groups according to the nature of the basic material: organic, inorganic, and metallic. Organic coatings may be paints, polymer materials, greases, and other paraffinic mixtures. Inorganic coatings may be enamels, ceramics (among them cement), and glasses. Metallic coatings are any metal or alloy. We shall begin with organic coatings.

4.1.1 Organic Coatings

Corrosion occurs only in the presence of an oxidizer in the environment of metal. The purpose of coatings is to isolate the metal surface from any oxidizer. It is possible to understand the protection mechanism of organic coatings if we accept the electrochemical mechanism of corrosion. The corrosion rate is equivalent to the electric corrosion current (I_{corr}) formed between an anode and a cathode on the metal surface:

$$I_{\text{corr}} = (E_c - E_a)/R, \tag{4.1}$$

where E_c and E_a represent the electrode potentials of cathode and anode on a metal surface; R is the electrical resistivity of the region close to the metallic surface (in electrolytes) between cathode and anode in the outer electric circle. Organic coatings have high electrical resistivity R_{coat} relating to the resistivity of electrolyte $R_{\text{electrolyte}}$. Therefore, electric corrosion current I_{corr} decreases according to (4.1) or is even excluded fully in the presence of organic coatings. This mechanism relates to the protection by any non-conductive coating. Organic coatings are produced after the drying of liquid paints. From which components do organic coatings, or paints, consist of?

Paint is a multi-complex system which consists of a mixture of resinous binder, pigment, additive, and solvent. Every paint component has its own purpose.

Resinous binder can be called resin, or binder; it is the main component and essence of the paint which establishes most of its chemical and physical properties. It may range from the natural egg-white to the synthetic organic polymers such as epoxies, urethanes, or vinyls. A binder determines the drying characteristics of the paint and controls the chemical and physical properties of the hardened coating.

The *pigments* are the substances of high dispersity (powders or fillers), non-soluble in resins and water, added to the paints for the formation of a thick coherent film, for imparting color, for protection of the binder by absorbing ultraviolet light and for corrosion inhibition in primer paints. The pigments are differentiated into protective (anti-corrosive, or inhibitive pigments), decorative (coloring pigments), and special (bactericidal, anti-foulant, luminous, etc.). The pigments can be natural and synthetic. Typical pigments are oxides, sulphides, and salts of transient metals (Fe, Co, Cr, Ni, and others), powders of non-ferrous metals (Al, Zn, Cu, Ni) and alloys (bronze, brass), and soot (carbon black – a dark powdery deposit of unburned fuel residue). The following pigments are used in paints. White pigments: lithopone (a mixture of barium sulphate $BaSO_4$ and zinc sulphide ZnS), rutile and anastase (TiO_2), zinc oxide (ZnO). Red pigments: haematite (Fe_2O_3), minium (red lead Pb_3O_4). Yellow pigment: ochre (Fe_2O_3 + alumosilicates). Black pigment: soot. Green pigments: Cr_2O_3, emerald green ($Cr_2O_3{\cdot}H_2O + B_2O_3$). The color of some pigments can significantly differ, for example, of ultramarine from green to violet. Some pigments are toxic, for example, white lead $2PbCO_3{\cdot}Pb(OH)_2$, red lead (Pb_3O_4), calcium plumbate ($CaPbO_3$) and zinc chromates ($ZnCrO_4$). Many steel structures have been painted with primers based on these toxic pigments for the last 100 years. The use of toxic pigments based on lead salts and chromates is now forbidden. It is interesting to emphasize that France, Belgium, and Austria banned white-lead interior paints in 1909.

Varnishes are paints that do not contain pigments, therefore they are transparent.

The *additives* (sometimes called *extenders*) are substances of high dispersity (powders, or fillers) added for improving the mechanical and maintenance properties of coatings, to replace part of the pigment content, and for reducing the cost of organic coatings. Typical additives are talc (magnesium silicates), clay, or kaolin (aluminum silicates), soot (carbon black), quartz (SiO_2), graphite (C, carbon), limestone, or chalk ($CaCO_3$), dolomite ($CaCO_3{\cdot}MgCO_3$), baryte ($BaSO_4$), woolastonite ($CaSiO_3$), and mica (silicate minerals).

The *solvents* are the liquid components used for attaining the appropriate viscosity to paints for facilitating their application to metal surfaces. Traditional solvents contain volatile organic compounds (VOC), namely, hydrocarbons (white spirit, xylene, toluene), alcohols (methanol, ethanol), ketones (acetone), and esters. The selection of solvents includes cost, volatility, low toxicity, and acceptable fire hazard. Some other components can be added to paint for modifying the rheology of media, for assisting the dispersion of pigments, and for the acceleration of drying.

To sum up, *paint is a solution or dispersion of resinous binder with solid discrete phase of pigment and additives in a volatile solvent. Resin* dissolved or dispersed in a volatile *solvent* is called the *vehicle*; this is glue holding the paint cohesively together and adhesively to the metal surface. During and after application, the solvent evaporates and the dried film consisting of the non-continuous pigmentary phase in a continuous phase of solidified binder, is formed on the metal surface. The solidification or curing process depends on the chemical nature of the vehicle. We have to differentiate *coating* from *paint*. After painting and drying (solidification or curing), a hard coating (film) is formed on the metal surface. When paint is drying, the hard film is called a *coating*.

Protection efficiency (as a result, a duration) of organic coatings is defined by their physico-chemical and mechanical properties, metal nature to be protected, surface preparation, environmental media type (atmosphere, water, acids, alkalis), and environmental conditions (temperature and their changes, stresses, wear). Thus, physico-chemical and mechanical properties of organic coatings provide their chemical and mechanical resistance. Chemical resistance means resistivity to penetration of aggressive species, such as water, oxygen, hydrogen sulphide, chlorides, acids, alkalies, organic solvents; resistivity to temperature and its changes, to solar energy, and keeping the color. Mechanical resistance means resistivity to wear, erosion, impingement, to be elastic and flexible, to have enough hardness, and of course, to maintain high *adhesion* to the metal surface. The last property is so important that we shall describe it. *Adhesion is the pull-off strength between a coating film and metal surface needed for film removing. Adhesion is defined as the greatest perpendicular force that a surface area can bear before a plug of material is detached.* Therefore, adhesion is measured in values of pressure. It usually decreases with an increase of the exposure time of coatings in the environment. We have to differentiate *adhesion* from *cohesion*. The latter is the bond strength between the particles of paint ingredients (inside the volume of the paint). Adhesion is one of the main coating properties defining the duration of the coating, and depending on the quality of the surface preparation, type of paint, coating thickness, and nature of a metal. Penetration of aggressive species through coating films from the environment to the metal surface depends on adhesion, and the latter, in its turn, depends on the penetrating properties (chemical resistance) of the coatings. The efficiency of coatings to protect the metal constructions from corrosion (resistance and duration) depends on their *thickness, adhesion, uniformity*, and *porosity*. All these parameters are interrelated. For example, the thicker the coating, the fewer the quantity of aggressive components that penetrate through the coating film and, correspondingly, the better its chemical resistance (Figure 4.1). Inner tensile stresses appear and are enhanced in the volume of organic coatings when their thickness increases. These tensile stresses cause weakening of the bonding strength between coating and metal surface and, as a result, adhesion decreases. Thus, the greater the thickness of a coating, the lesser its mechanical resistance (properties). As a result, dependence between protection efficiency and coating thickness can be described by two curves, as shown in Figure 4.1. There is an optimal thickness for every organic coating if we take into consideration the chemical and mechanical properties influencing the resistance (duration) of organic coatings. Thus, the assumption that greater thickness produces greater protective properties and duration is not valid.

The general protective mechanism of organic coatings includes physical barrier to aggressive substances, inhibition by pigments (oxides, phosphates, and other salts), and galvanic action of zinc or aluminum powder if they are added to paints. Usually several layers are used in order to decrease the *porosity* of coatings. The first layer of coatings is sometimes called a *primer*. If the primer contains phosphoric acid, it reacts with the metal surface and chemical bonds are formed between metal surface atoms and the molecules of the primer. Many organic coatings are based on chemical organic substances containing polar groups such as hydroxyl (–OH), carboxyl (–COOH), and imine (–NH). If wetting the metal surface with paints is good, physical contact is formed between coating and metal surface. Any contaminants (salts, oil, grease, dust, dirt, water, old paint) remaining on the metal surface will decrease the adhesion of coatings. Thus, adhesion depends significantly on the surface preparation.

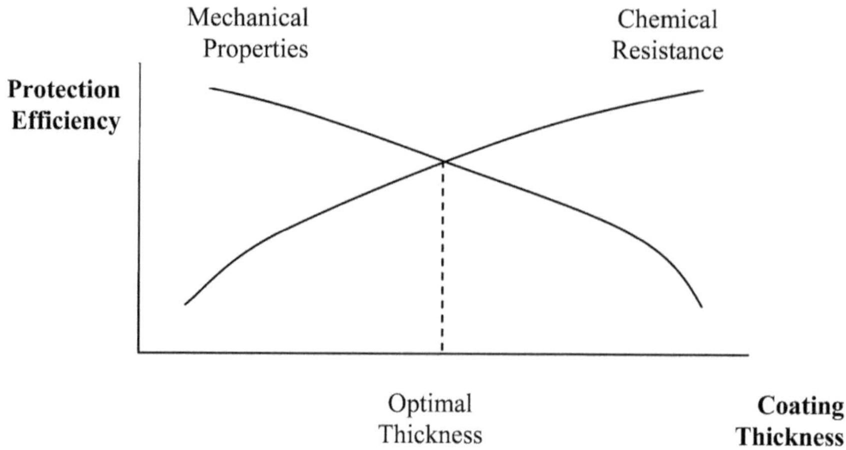

Fig. 4.1 Schematic dependence between protection efficiency (duration) and coating thickness.

4.1.2 Surface Preparation

We can categorize surface preparation methods into three groups: *mechanical* (*physical*), *chemical*, use of rust converters, and *thermal*. Mechanical energy is used in physical methods for removing contaminants from the metal surface. In chemical methods, active chemicals are used for dissolving corrosion products and removing contaminants. Thermal methods use high temperatures (thermal energy) for transformation of corrosion products into dense inert layers, burning old paints and contaminants on the surface of the metal.

4.1.2.1 Mechanical Methods

Mechanical (*physical*) *methods* include

(a) Metallic blast-cleaning abrasives: cast steel and cast iron shot and grit, crushed slags of copper and nickel.

(b) Non-metallic abrasives: ceramic materials (SiO_2, Al_2O_3, Fe_3O_4, Si_3N_4, SiC), glass bead, salts [$NaHCO_3$, garnet – (Ca, $Mn)_3(Fe,Al)_2(SiO_4)_3$, zircon – $ZrSiO_4$, novaculite – siliceous quartz rock, aluminum silicate, and iron silicate], crushed slags (electric power generating ash), agricultural shell products (walnut shells, cherry pits, and corncobs).
These abrasives may be used with pressurized air (abrasive air blast cleaning with a pressure of about 7 atm) or with pressurized water (water blast cleaning, sometimes called *hydroblasting*). The latter may be carried out independently with pressures up to 140 atm (low pressure), up to 1,700 (high pressure), and above 1,700 atm. (ultra high-pressure waterjetting).

(c) Hand and power tool cleaning includes wire brushes, non-woven abrasive pads, scrapers, chisels, knives, and chipping hammers.

The productivity of mechanical cleaning methods is slow: 10 to 20 m^2/hour. Therefore, it is impossible to clean at once all large surfaces needed for painting. Large surfaces are partly mechanically cleaned and must then be immediately painted. Dust remains on the metallic surface after cleaning. Industrial vacuum cleaners are used for removing such dust before painting.

4.1.2.2 Chemical Cleaning

Chemical cleaning includes the use of:

(a) *Organic solvents* – cleaning surface from oil, grease, wax, and other organic contaminants. They include petroleum distillates (kerosene, naphtha, mineral, or white spirit) or chlorinated solvents (trichloroethylene or perchloroethylene). The latter can be hydrolyzed in the presence of water and dangerous hydrochloric acid may form on the metal surface. Organic solvents are volatile, harmful, and dangerous to the environment and people. They are restrictedly used, especially in closed spaces (tanks, vessels, tankers, and reactors). Chlorinated solvents are banned for use in most countries because of their very high toxicity. Therefore, water-based cleaners, alkaline or acidic, are preferentially used nowdays.
(b) *Alkaline solutions* – cleaning surface from grease and acidic salts. Alkaline cleaners contain sodium or potassium hydroxide, alkaline salts (silicates and carbonates), surfactants, inhibitors, and soaps.
(c) *Acidic solutions* – cleaning surface from rust, metal oxides, mill scale. Acidic cleaners contain acid (usually sulphuric acid), surfactants, and inhibitors. They are primarily used to remove corrosion products from the metal surface. The cleaning process using acidic solutions is called *pickling*. Corrosion organic inhibitors are added to acids during cleaning (see Section 4.3). The disadvantage of pickling is hydrogen formation as a result of the cathodic process, and then possible blistering of coatings and hydrogen damages. Therefore, the metal surface should be carefully washed with water from the acidic solution after cleaning.
(d) *Detergents* – cleaning surface from polymeric contaminants. Detergent cleaners contain mixtures of surfactants, dispersants, inhibitors, and soaps.

Chemical treatment can be applied manually, by immersion, spraying, or suspension in the vapors of the medium. Sometimes use of alkaline cleaners is called *degreasing* because they remove grease from metal surfaces. High-pressure steam can be used for this purpose too.

Now we know how to prepare a metal surface before painting. Sometimes it is impossible to use mechanical or chemical cleaning to remove rust from complicated structures or inside vessels and tanks, pipes, cars, window grating, or iron-barred railings. What to do in these cases? Can we convert and modify rust without removing it from the metal surface? Next we shall discuss rust converters and surface tolerant coatings.

4.1.2.3 Rust Converters and Surface Tolerant Coatings

The use of rust converters and surface tolerant coatings relates not to chemical cleaning, but to chemical surface modification. Rust on a carbon steel surface usually consists of three layers. The outer layer (far from the metal surface) is thick (from several hundreds μm to several mm thickness), loose, and may be easily removed by means of any hand tool instrument. The second (intermediate) layer is not so thick (10 to 100 μm), and may be removed by means of various mechanical tools. The inner layer, which is close to the metal surface, is chemically bonded with it, very thin (about 5 μm thickness) and can be removed only by means of bending of the metallic element.

Rust converters work on the inner and sometimes on the second (intermediate) rust layers. It is necessary to remove loose rust (outer thick layer) mechanically, and then to treat the thin rust (thickness <50 μm) bonding to the metal surface with rust converter. Rust is a complicated mixture of corrosion products of iron and various contaminants which were present in the environment. Salts containing anions (Cl^-, SO_4^{2-}, CO_3^{2-}) and cations (Ca^{2+}, Mg^{2+}) may be present in rust (if the rust is formed in water solutions); water, dust, dirt, soot, fats, oil, and sand may be present (if the rust is formed in the atmosphere). Thus, rust may be different in composition, thickness, and adhesion to the steel surface. We must convert rust consisting of ferric and ferrous hydroxides and oxides into inert non-soluble salts, and to neutralize aggressive species (Cl^- and SO_4^{2-}). Rust converters containing acids can react with rust and modify it. The first rust converters appeared in 1950s and consisted of phosphoric acid (H_3PO_4) and zinc (Zn) powder. When the rusted (thickness less than 50 μm) surface is wetted with an acidic mixture, ferric and ferrous hydroxides react with phosphoric acid and zinc, and inert non-soluble phosphate salts are formed on the steel's surface: $FePO_4$, $Zn_3(PO_4)_2$, $Fe_3(PO_4)_2$. Any rust converter chemically modifies rust which tightly adheres to the steel surface. Rust converters cannot convert mill scale (iron oxides FeO, Fe_3O_4 and Fe_2O_3). Usually the rust converters contain acids (phosphoric or tannic – resulting in the formation of non-soluble salts), complex salts [$K_4Fe(CN)_6$, or $K_3Fe(CN)_6$, forming complexes with Fe^{2+} and Fe^{3+} cations of rust], corrosion inhibitors, surface agents (surfactants), and alcohols for better penetration of rust converter compounds through the rust layer, and barium salts for precipitation SO_4^{2-} anions. Rust converters on the basis of yeasts and polyvinylacetate dispersion were developed too, although they did not find wide application.

Rust converters chemically react with rust, act as primers and cannot stop and prevent rust from returning. They are sometimes called *rust killers*. They result in the formation of an inert layer on the steel surface, and should be topcoated. If this layer is not topcoated, leaching could occur if exposed to atmospheric humidity or water. Rust converters based on acids (phosphoric or tannic) are liquids and must be carefully washed with water to remove the residue of non-reacted acidic components after their application. After applying, most rust converters are not compatible with topcoats at temperatures above 40°C. They may be used for restoration of antique implements. Converters not only for rust, but also for corrosion products formed on copper, zinc, and aluminum alloys were developed too. We have to differentiate between *rust converters* and *surface-tolerant coatings*. Surface-tolerant coatings are coatings that are used on hand-cleaned surfaces, old coatings, and surfaces cleaned by high-pressure (~150 atm) waterjetting. Surface-tolerant coatings can be applied over non-perfectly cleaned rusted surfaces. They do not contain acids, and were created on the basis of epoxy, polyurethane, and other resins, which can be used to treat dense rust (with thickness less than 100 μm) remaining on steel surface, and can be served as independent coatings. The mechanism of usage of surface-tolerant coatings is based on their penetration through rust and inertization of all rust components. Brush and airless application are the best methods for assuring that the right substances penetrate the surface-rust layer. After drying, surface-tolerant coatings form a protective thick film on rusted constructions, and are used under particular conditions in the atmosphere, water, and fuels.

We may ask what kind of cleaning and surface preparation is better, the mechanical or chemical method? There is no ideal cleaning. When chemical cleaning is used, chemical compounds (for example, acids) remain on the surface of the metal. When mechanical cleaning is used, it is difficult to remove salts which are always present in rust. High-pressure water blast has proved itself to be a good surface preparation method and is widely used nowdays in industry. Mechanical methods such as shot and sand blasting remove scale and roughen the surface providing a useful surface for good adhesion of paints.

4.1.2.4 Thermal (Flame) Cleaning

The third group of surface preparation methods is *thermal*, or *flame, cleaning*. Various lamps and flame cleaning are used for burning old coatings, oil, and grease, and detaching scale by differential expansion on the metal surface. Flame cleaning is usually followed by wire-brushing. A laser can be used for transforming ferric hydroxide into ferric oxide in rust. A dry surface is formed in this case which can be painted. Thermal (flame) cleaning has not found wide use in industry.

Ultrasonic cleaning uses swept resonance frequency technology, and is used for the surface preparation of small metallic elements which are usually immersed in a bath with liquid for cleaning.

4.1.3 Selection of Coating System

The main question is how to *select the coating system?*

First of all, we have to define the service conditions and requirements for the coating system to be used: temperature (and its changes), chemicals (type and their concentration) which will be in contact with the metal surface, immersion or splash, sunrays, wear, etc. Then it is important to carry out accelerated tests of the recommended coating systems under laboratory conditions. For example, if we need to select the right system for the protection of the bottom inside of gas oil storage tanks, we have to immerse the panels coated with the recommended systems in three phases: in gas oil (organic phase), in an aqueous electrolyte solution (imitating water separation on the bottom), and in the gaseous phase (included hydrogen sulphide, because gas oil can contain this gas). The examination accelerated test period must not be less than three months. We may examine at ambient (20°C) or at higher (40°C or 90°C) temperatures for enhancing corrosive conditions. We may add some salts (3% NaCl and 0.2% $NaBO_3$) to water. The latter compound is an oxidizer and promotes the aging of coatings. It is important to check the references where particular coating systems have yet been used. Then, to check the safety conditions for use of the coating systems recommended. Most paints contain organic solvents which are harmful to the environment and people. In order to overcome these problems, *water-based* paints, *powder* paints, and *solventless* (*solvent free*) paints were developed. Water-based paints have water as the solvent, they dry purely by evaporation, while oil-based paints have chemical drying agents added. The greatest advantages of water-based paints are that they are "friendly" for the environment and brushes and rollers can be washed out in water. Powder paints are applied in powder form at a high temperature when they are melting or chemically cross-linked on the metal surface. Solventless paints do not contain organic solvents, and consist of 100% solids, therefore they have many benefits. A binder in solventless paints is a liquid. One layer of solventless coatings may be very thick: from 400 to 1000 μm. In order to reach such a thickness with conventional paints containing solvents, we have to produce 10 to 20 layers. As VOC (volatile organic compounds) is nearly zero in solventless paints, hazardous organic compounds do not evaporate into the atmosphere, and there is no danger for health, spark and explosions. When the solvent evaporates, pores remain in the coating film after drying. The porosity of solventless coatings is very low, significantly less than that of paints containing solvents. The curing process is faster for solventless coatings. Such additives (fillers) as aluminum oxides, fiber glasses, and chips in coatings increase their resistance to erosion and wear.

To sum up, the protective properties of organic coating systems and, consequently, their durability are defined by the correct selection of the coating system based on laboratory examination and checking the references, and the quality of the surface preparation and application. Suitable and qualifiable inspection must be carried out at all these stages. Only then will we enjoy and get satisfaction from using the most ancient method of corrosion control.

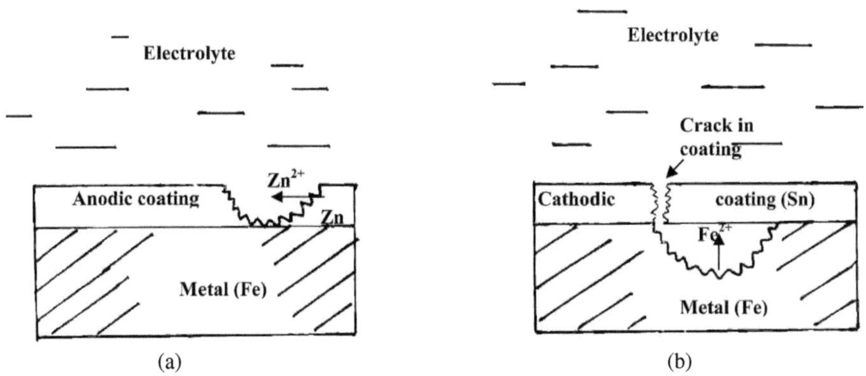

Fig. 4.2 (a) Anodic and (b) cathodic coatings on metal surface. Metal does not corrode under an anodic coating, but corrodes under a cathodic one, if there are pinholes in the latter.

4.1.4 Metallic Coatings

Throughout the whole of the book we have talked about corrosion of metals, and now we intend to talk about the paradox that metals may be protected with metals. Yes, we know that carbon steel structures and equipment are widely protected from corrosion and aesthetically coated with other metals and alloys. What is the protection mechanism in this case? There are less and more noble metals regarding to iron (see Tables 1.3 and 2.3). In relation to this, there are two main types of metallic coatings according to their protective mechanism. When a metallic coating is less noble (more negative electrode potential) than the metal to be protected, the coating is called anodic (it will corrode as an anode) and provides cathodic protection (see Section 4.2) in the case of scratches, edges, pores and any defects, that is, it protects non-covered metal in the presence of electrolytes (Figure 4.2a).

For the protection of carbon steel, the anodic metallic coatings are zinc, aluminum, magnesium, and their alloys. Cadmium is also anodic to carbon steel and was widely used as metallic coating in the past, but it is an extremely toxic metal and is not recommended for use. The more electrical conductivity of an electrolyte, the better the efficiency of the protective properties of anodic coatings when its uniformity is destroyed and pores appear in the coatings. Typical galvanic corrosion of a less noble metal occurs in this case. Zinc and aluminum are widely used as anodic coatings for carbon steel structures and equipment.

If a coating metal is more noble (more positive electrode potential) than the metal needing protection, a coating is called cathodic. Cathodic coatings are tin, chromium, nickel, copper and its alloys, stainless steels, and silver on carbon steel. If scratches, edges, pores, or any other defects appear in cathodic coatings on carbon steel, the latter will be anodic towards metallic cathodic coatings, and will corrode under these coatings (Figure 4.2b). This situation is very dangerous because steel will corrode under the coating and we will not be able to immediately

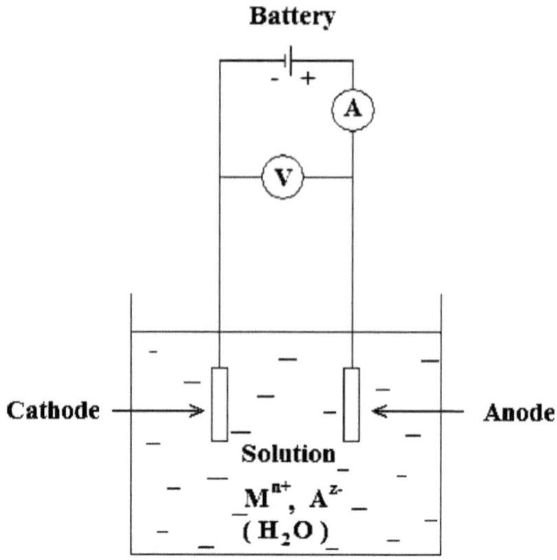

Fig. 4.3 *Electroplating* – producing electrolytic coatings, A – amperometer; V – voltmeter; M^{n+} – cations; A^{z-} – anions. Cathode – metal to be protected.

detect the rust. Rust formed under cathodic coatings will cause a decrease of the adhesion and delamination of the coating from the steel surface.

Metallic coatings may be divided into several groups according to the way they are prepared: *electrolytic*, *hot-dip coatings*, *metal spraying*, *chemical plating*, *vapor-deposited coatings*, *diffusion treatment*, and *coatings prepared by mechanical methods*.

Electrolytic coatings are produced by the electrochemical process that has several names: *electrogalvanizing*, *electrodeposition*, and *electroplating*. We can use any of these names. Thus, electroplating is the producing of a metallic coating on a metal surface by the action of an electric current (Figure 4.3). The object of a metal or alloy (for example, carbon steel) to be protected with a metallic coating (electrolytically plated) is connected to the cathode (negative pole of battery) of an electrolytic cell. This metallic object (cathode) is immersed in a solution which contains a salt of the metal to be deposited. For example, if we need a nickel coating, we may take $NiCl_2$ salt in water. This dissociates in water and Ni^{2+} cations are formed. Organic and even fused salt electrolytes may be used for electrodeposition too. A metallic electrode made of nickel (anode) is immersed in the same solution and connected to the positive pole of the electrolytic cell. When an electric current flows between the anode and cathode from the direct current power supply (battery), positively charged Ni^{2+} ions are attracted to the negatively charged object (the cathode). It provides the electrons to reduce them (Ni^{2+}) to pure nickel on a carbon steel object according to the reaction:

$$Ni^{2+}_{(aq)} + 2e^- \rightarrow Ni_{(s)}. \qquad (4.2)$$

Table 4.1 The melting temperatures of metals used for producing hot-dip coatings.

Metal	Melting temperature, °C
Zn	419
Al	660
Sn	212
Pb	327

This reaction is opposite to corrosion, and a nickel coating is formed on the steel object. The positively charged nickel electrode (anode) is oxidized to the Ni^{2+} cations, which are applied to the solution. In this case, corrosion of nickel anode is a beneficial phenomenon. The thickness of the electrolytic coating formed on a metal is determined by the duration of the plating. Usually the thicknesses vary from 0.1 to 40 μm. *Electrolytic* coatings became possible after the invention the "*voltaic pile*" (electric battery) by Alessandro Volta in 1797. His friend the Italian chemist Luigi Brugnatelli was probably the first to plate silver medals with gold using the electric Volta pile in 1805. The Emperor of France, Napoleon Bonaparte (1769–1832) was personally interested in electrolytic coatings. Russian ambassadors in Paris and London at the beginning of the 19th century, had to report about this issue every month to the Russian government. Probably, "scientific spying" during that period played a significant role, and at the end of the 1830s scientists in both Russia and England had devised electrolytic coatings of copper for printing plates. By the middle of the 19th century, electroplating of nickel, tin, brass, and zinc were applied for commercial purposes. One of the disadvantages of electroplating is that uneven current distribution results in a non-uniform metallic coating on the object. An even current distribution is reached by auxiliary anodes and current screens. Use of various organic additives in electrolytic solutions also helps to form a uniform coating on the objects.

Hot-dip coatings are produced when a metal object (for example, made of carbon steel) is dipped into a bath of molten metal (for instance, zinc). Metals with low melting temperature are used for the hot-dip coatings: zinc, aluminum, their alloys, and tin (Table 4.1). Lead was also used in the past, but because of its high toxicity is now banned from use. Zinc is mostly used as a hot-dip coating and in this case such a process is called hot-dip galvanizing. Similarly to the electrolytic coatings, the thickness of hot-dip coatings is defined by the duration that the object is left in the bath with molten metal. Usually the thicknesses of hot-dip coatings vary from 10 to 400 μm.

Metal spraying is the process of producing metallic coatings on metal surfaces by means of spraying with compressed air metals or alloys after their melting. History tells that the Swiss engineer Max Ulrich Schoop from Zurich liked to go hunting. This was probably around 1910 when he payed attention to the fact that after shooting, small bullets of lead penetrated into a tree and might remain in its bark. He asked "why could I not do the same with a metal object instead of a tree?" Thus,

Schoop invented the first flame spraying "gun" for metalization in 1913. Such a process has been called the "*Schoop process*" for many years, and it has received many synonyms: *flame spraying, thermal spraying, metalizing*, or *metal spraying*. It can be carried out by three methods: *flame spraying*, *electric arc spraying* and *plasma spraying*. *Flame spraying* is the process of melting metal with an oxygen–acetylene flame and dispersing with compressed air onto the metal object to be coated. The metal used as a coating material may be wire or powder form. Molten particles of metal or alloys move with compressed air onto the metal surface to be protected, impact and flatten. Molten particles of metals are oxidized by the air during their moving from the "gun" to the metal surface. The distance between the "gun" and the metal surface to be protected is usually about 1 to 2 meters. Therefore, a finished coating consists of a mixture of melted metal and its oxides which are solidified. The requirements for preparation of the metal surface are similar to those before painting. *Electric arc spraying* is the process of metal melting with an electric arc between two wires (connected to plus and minus electric poles) and dispersing with compressed air. *Plasma spraying* is the process of powder melting by a plasma beam and dispersing with compressed air. The temperature in the last method is very high: about 15,000 to 20,000°C; therefore, high melting metals, alloys, and ceramics (oxides, carbides, and nitrides) are used for plasma spraying. All three variations of metalizing require a spraying "gun". Robotic "gun" systems have been developed for metalizing large surfaces. Metalizing equipment is mobile, appropriate for many complex shapes, and not limited by size. The thicknesses of metal spray coatings usually vary from 40 to 500 μm. Metal spraying is used for coatings with aluminum, zinc, their alloys (85%Zn–15%Al), aluminum alloy with 5% magnesium (Al–Mg5), and stainless steels. A severely corroded carbon steel surface may be metalized with carbon steel for repair, namely, for increasing the thickness of the corroded metal. The main drawback and "advantage" of metal spraying coatings is their high porosity which depends on the type of metal spraying (its density) and process type. For example, aluminum coatings have a higher porosity (5 to 15%) than zinc coatings (1 to 3%) (see Section 3.6). The lowest porosity may be received by plasma spraying: 0.5 to 2%. Corrosives in the environment can penetrate through pores to the metal surface under a metalized coating. Because of the different porosity of various metal spray coatings, the minimum thickness needed for metal protection is also different. The minimum thickness of the coating is the thickness needed for closing of all the pores in the coating, that is, the thickness where electrolytes would be unable to penetrate through the metallic coating to the base metal surface to be protected. Thus, the minimum thickness for a zinc coating is 100 μm, for aluminum and stainless steel coatings is 300 μm (see Section 3.6). The lifetime of metallic coatings depends on their thickness, namely, on their mass. Adhesion of metalized coatings is higher than that of paints. Metalizing may be used alone or in combination with organic coatings. It also increases drastically the adhesion (because of relatively high porosity of metal spray coatings) of topcoats when they are applied over metalized coatings. Such "sandwich" coatings (metal spraying following by organic coatings) may serve for approximately 30 years (see Section 3.6). The process of filling the pores of metal sprayed coatings with paint is called *sealing*. Treatment of aluminum metalized coatings with water at 90 to 100°C results in good sealing because of the formation of aluminum hydroxides in the pores of the coatings. Good adhesion of zinc and aluminum metalized coatings to steel allows the shaping of constructions (for example, sheets for tanks) in different forms without delamination. Aluminum and zinc coatings are used for protection inside and outside surfaces of tanks used for the storage of fuels and water (see Section 3.6). The advantage of metal sprayed coatings is that sheets with such coatings can be welded and then coated (sealed) with paints.

Chemical plating is the process of producing metal coatings on a metal surface by means of chemical reaction. The miners in the Middle Ages knew that if an iron object came into contact with the "vitriol water" ("blue vitriol") of the copper mines (an aqueous solution of copper sulphate formed when copper sulphide ores have been oxidized), the surface of the iron will be covered by a red copper layer. The following oxidation – reduction reaction occurred:

$$Fe_{(s)} + CuSO_{4(aq)} \rightarrow \text{"Red Iron"}, \tag{4.3}$$

$$Fe_{(s)} + Cu^{2+}_{(aq)} \rightarrow Fe^{2+}_{(aq)} + Cu_{(s)}. \tag{4.4}$$

This is the chemical plating or metal deposition of copper on iron (steel) surface. Miners observed this metal plating many times. The copper coating is very thin (about 1 μm), porous, and is not well attached to the steel surface.

Chemical plating of nickel was developed in the 1940s. A carbon steel object is immersed in a bath containing an acidic water solution of nickel cations Ni^{2+} and hypophosphite anions $H_2PO_2^-$. The following reaction occurs:

$$Ni^{2+}_{(aq)} + H_2PO^-_{2(aq)} + H_2O_{(l)} \rightarrow Ni_{(s)} + H_2PO^-_{3(aq)} + 2H^+_{(aq)}. \tag{4.5}$$

Nickel coatings are produced by the deposition of nickel cations as a result of chemical reduction, and this process is similar to reaction (4.2), but without an outer electric current. Therefore, this process is called *electroless plating* and it has many advantages over *electrolytic plating*. One major advantage of electroless plating is that a uniform coating is formed even on objects having crevices and complicated forms, such as corners and others.

Thicknesses of electroless coatings vary from 5 to 125 μm. When reaction (4.5) takes place on a carbon steel surface in the bath, nickel coating contains phosphorus (2 to 13%) and this element determines the mechanical properties of the coating.

Diffusion treatment is the process of surface changing of metallic objects by diffusion of protective metal. When treating carbon steel with carbon powder in a furnace to 900°C, carbon diffuses into the steel surface. This process is called *carburizing*. If nitrogen diffuses into the surface of a metal, the process is called *nitriding*. These diffusion processes have been used for many years to improve the mechanical properties of steels. Similar diffusion may be carried out with any metal. Probably, the Englishman Sherard Cowper-Coles was the first to heat zinc dust together with a carbon steel object to 350–400°C in 1904. The melting temperature of zinc is 419°C. Zinc atoms diffuse at a high temperature into the upper layer (surface) of steel and partly react with iron. An iron–zinc alloy and the outer layer of pure zinc are fused together as a result of diffusion of zinc atoms into the iron surface. The coating thickness depends on the reaction time and varies between 10 to 50 μm. Such a process of the formation of a protective zinc coating on steel objects is called *sherardizing* and was introduced in the 1920s. If aluminum is used for the diffusion treatment, the process is called *calorizing* and if chromium is used, *chromizing*. In *calorizing*, aluminum is usually used for diffusion treatment together with alloy ferro aluminum (Fe–Al) or other alloys, alumina (Al_2O_3), and salt ammonium chloride (NH_4Cl). In this case, high-temperature wear resistance of the formed coating is increased significantly. Diffusion treatment provides coatings of a uniform thickness.

Vapor-deposited coatings are the coatings producing by processes in vacuum by *physical vapor deposition* (PVD) or *chemical vapor deposition* (CVD). The species are deposited in the form of individual atoms or molecules in these processes. They are carried out in special vacuum boxes, therefore are not damaging to the environment as are some other metal coating processes, for example, electroplating or metal spraying.

The PVD process is differentiated into *evaporation*, *sputtering*, and *ion plating*. Historically, *evaporation* was the first PVD process. Practically all metals can be evaporated, but aluminum, chromium, and stainless steels are widely used. The thicknesses of metal coatings obtained by evaporation are several microns. The disadvantages are non-uniformity and relatively low adhesion. Such coatings are usually used for decorative and optical applications, and are not used as anti-corrosion coatings.

Sputtering takes place in an inert gas (argon) at low pressure. The argon ions formed by electric discharge, beat the solid coating material connected to the minus electric pole of the power supply. Sputtered atoms of the coating material move to the metal object to be coated. Thin metal coatings of 0.1 μm thickness may be formed with good adhesion. High-chromium alloy coatings on turbine blades are produced by sputtering. This method was developed in the 1970s for semiconductors. The drawbacks of sputtering are thickness limitation and difficulties in forming uniform coatings on complicated objects.

Ion plating is based on evaporation (or sputtering) and a glow discharge. The application of ion plating is the formation of nitrides or carbides on the steel surface. Therefore, this method is not exactly PVD. In ion plating, nitrogen or carbon ions are obtained from the ionization of nitrogen or hydrocarbon gases. Then these ions react with iron on the steel surface forming nitrides or carbides. Therefore, these coatings have the best adhesion, structure, and fewer imperfections comparing with coatings obtained by evaporation and sputtering.

CVD is based on chemical reactions. For example, if chromium is needed as coating, it should be converted into a gas compound. For this purpose, ammonium chloride as the source of halide is used in the hydrogen atmosphere at about 1000°C. A source of chromium is usually chromium powder. As a result of chemical reactions chromium converts into a volatile $CrCl_2$ compound which in its turn reacts according to the following reactions:

$$CrCl_{2(g)} + Fe_{(s)} \rightarrow Cr_{(s)} + FeCl_{2(g)}, \tag{4.6}$$

$$CrCl_{2(g)} + H_{2(g)} \rightarrow Cr_{(s)} + 2HCl_{(g)}, \tag{4.7}$$

$$CrCl_{2(g)} \rightarrow Cr_{(s)} + Cl_{2(g)}, \tag{4.8}$$

and chromium atoms deposit on the carbon steel surface. Chromium atoms diffuse into the carbon steel surface under high temperature ($\approx$ 1000°C), and the content of the chromium reaches 13 to 30% in the carbon steel surface, which increases the corrosion resistance and mechanical properties of steel. Sometimes CVD with chromium is called *chromizing*. The greatest disadvantage of this technique is that halide gases at high temperatures are highly corrosive and corrosion-resistant materials must be used. CVD coatings contain chromium, aluminum, titanium, manganese, boron, and silicon. These are used as anti-corrosion coatings (including high temperature oxidation resistance), and improve wear, abrasion, and friction properties of the protected surfaces. CVD coatings are thick (up to 6.5 mm), dense with good anti-corrosion properties, and their cost is usually lower than that of PVD coatings.

Titanium nitride is very hard, stable, has good corrosion resistance in the atmosphere, and has a pleasing gold appearance which allows the use of titanium nitride as a coating on steel instead gold, for example on the domed roofs of churches (see Figure 6.8).

Mechanical methods of coating are the processes that use physical bonding of one metal to another for the creation of metal coatings. Soldiers during World War I (perhaps even before) observed that shrapnel from disintegrated metal shell casings traveling at high velocities, sometimes bonded with steel stanchions and other metallic surfaces that they stuck. Then the American engineer L.R. Carl reported in 1944 about the welding of brass discs under high velocity impact, and concluded that the weld was not a fusion weld but that it had been formed by a solid-state mechanism. In the 1960s explosion welding was used for producing the clad metals of Cu–Ni/Cu/Cu–Ni needed by the US Mint for new coinage. As the vice president of one company producing clad metals noted, “people made lots of money making money”. Mechanical methods of coating are divided into the following methods.

Cladding is the method of cold or hot rolling of coating high resistant metals or alloys onto the metal to be protected. Clad products include stainless steels, titanium, copper alloys, aluminum, vanadium, tantalum, and zirconium to steel. This method is used for producing the protective metal coatings inside pressure vessels, autoclaves, reactors, distillation columns, condensers, tubesheets, and shells of heat exchangers.

Explosive bonding (*explosion welding*) is the method of welding by a controlled explosion of the coating metal (high corrosion resistant) to the metal to be protected.

Overlay welding is the welding of a metal highly resistant to corrosion to the surface of a metal to be protected.

Extrusion is the process when the coating metal and the metal to be protected are extruded together. Generally speaking, extrusion is the process by which long, straight metal parts may be produced. Extrusion is done by squeezing the metal and coating together in a closed cavity through a tool known as a die using either a mechanical or hydraulic press.

Mechanical plating is a process for impact deposition of some ductile metals from a powder on a steel object. Ductile (soft) metals are zinc, aluminum, copper, tin, and their alloys. This method relies on the principal that if there are two pure metals hitting each other, the softer one will cold weld to the harder one. Other names for mechanical plating are *peen plating, impact plating, mechanical* or *cold galvanizing*. This method was developed in the early 1950s and takes place in a rotating barrel containing small carbon steel objects, powder of the coating metal (for example, zinc) suspended in an aqueous solution, and glass beads. Rotated glass beads "hammers or peens" the metal powder onto the objects. An adhered uniform layer of coating is formed with thicknesses of 2 to 100 μm.

Recommended Literature

1. Hare, C.H., *Protective Coatings*, Technology Publishing Company, Pittsburgh, Pennsylvania, USA, 1994, 514 pp.
2. Mattsson, E., *Basic Corrosion Technology for Scientists and Engineers*, Second Edition, The Institute of Metals, UK, 1996, pp. 107–125.
3. Keane, J.D. (Ed.), *SSPC, Vol. 1, Good Painting Practice*, Second Edition, SSPC, Pittsburgh, USA, 1989, 580 pp. [SSPC – Steel Structure Paint Council].
4. Revie, R.W. (Ed.), *Uhlig's Corrosion Handbook*, Second Edition, Wiley-Interscience, Inc., 2006, pp. 1023–1059.
5. Talbot, D. and Talbot, J., *Corrosion Science and Technology*, CRC Press, USA, 1998, pp. 185–191, 219–226.
6. Mühlberg, K., Surface-Tolerant Coatings – Some Experiences, *Protective Coatings Europe (PCE)*, December, 2001, pp. 13–21.
7. Korb, L.J. and Sprowls, D.O., *Metals Handbook, Vol. 13: Corrosion*, Ninth Edition, ASM International, USA, 1987, pp. 456–458.
8. von Baeckman, W., Schwenk, W. and Prinz, W. (Eds.), *Handbook of Cathodic Corrosion Protection*, Third Edition, Gulf Publishing Company, Houston, Texas, USA, 1997, 567 pp.

4.2 Electrochemical Methods of Corrosion Control

The electrochemical mechanism of corrosion in electrolytes allows the use of electric current and electric potential in order to protect metals from corrosion. Therefore, electrochemical methods work only in solutions of electrolytes. *Electrochemical methods* are a general term for cathodic and anodic protection. They are different and they should not be mixed. In *cathodic protection*, metallic equipment is connected to the *negative* pole of the power supply and turns completely into a cathode, which does not corrode. *Anodic protection* is based on the phenomenon on passivity (see Appendix D). In anodic protection, metallic equipment is connected to the positive pole of the power supply, and the electric potential changes in the positive direction, in order to reach the passive state. Thus, anodic protection is only applied for metals and alloys capable of being in a passive state in a particular media.

4.2.1 Cathodic Protection

There are three ways to show the principles of cathodic protection.

Chemical explanation. How can we prevent the anodic dissolution, or corrosion process, of iron immersed in an electrolyte solution?

$$Fe_{(s)} \leftrightarrow Fe^{2+}_{(aq)} + 2e^{-}. \tag{4.9}$$

Chemists know that one of the ways to change the direction of this reaction is to increase the concentration of ferrous cations Fe^{2+} in the electrolyte or, alternatively, to increase the number of electrons in iron. It is simple to realize the latter. Thus, if we connect the iron to the negative pole of a direct current power supply, electrons will "flow" to the iron, and reaction (4.9) would slow down to a negligible value or even to stop it.

Thermodynamic explanation of cathodic protection is based on the Pourbaix diagram (Figure 4.4). It shows three realms of the possible existence of iron in water: *corrosion* (Fe^{2+}, Fe^{3+} and FeO_2H^-), *passivity* (Fe_3O_4 and Fe_2O_3), and *immunity* (Fe). The electrode potential of iron and carbon steels in neutral aqueous solutions is about −0.44 V regarding SHE. One can detect from the Pourbaix diagram that if we diminish this potential under −0.52 V (the arrow AB), we enter the realm of immunity, where a reaction opposite to (4.9) takes place, that is, the anodic reaction of iron dissolution will be suppressed.

Polarization curves can also explain the principle of cathodic protection. Let us represent the Evans diagram for corrosion of iron in water (Figure 4.5). For the reversible iron electrode we can write

$$Fe^{2+}_{(aq)} + 2e^{-} \underset{\overleftarrow{\mathbf{i}}_1}{\overset{\overrightarrow{\mathbf{i}}_1}{\longleftrightarrow}} Fe_{(s)}. \tag{4.10}$$

Current densities $\overrightarrow{\mathbf{i}}_1$ and $\overleftarrow{\mathbf{i}}_1$ are the rates of the cathodic and anodic processes, respectively, for the reversible reaction (4.10). Cathodic ($\overrightarrow{\mathbf{i}}_1$) and anodic ($\overleftarrow{\mathbf{i}}_1$) curves for this reversible iron electrode are shown by curves 1 and 2, respectively, in Figure 4.5. For the oxygen reversible electrode we can write

$$O_{2(g)} + 2H_2O_{(l)} + 4e^{-} \underset{\overleftarrow{\mathbf{i}}_2}{\overset{\overrightarrow{\mathbf{i}}_2}{\longleftrightarrow}} 4OH^{-}_{(aq)}. \tag{4.11}$$

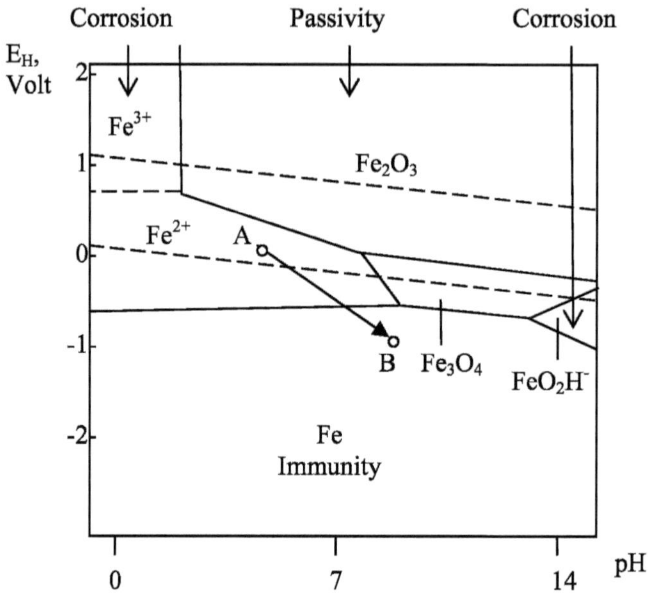

Fig. 4.4 Potential–pH (Pourbaix) diagram for Fe–H_2O at 25°C.

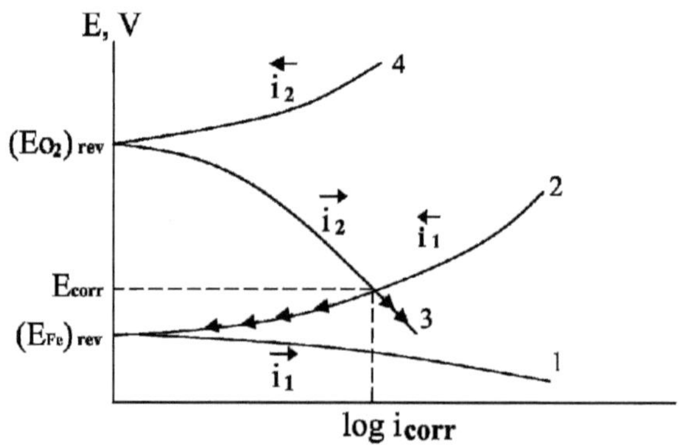

Fig. 4.5 The Evans diagram for corrosion of iron in water. $(E_{Fe})_{rev}$ and $(E_{O_2})_{rev}$ – reversible potentials for reactions (4.10) and (4.11), respectively.

Current densities $\overrightarrow{\mathbf{i}}_2$ and $\overleftarrow{\mathbf{i}}_2$ are the rates of the cathodic and anodic processes for (4.11), respectively. Cathodic ($\overrightarrow{\mathbf{i}}_2$) and anodic ($\overleftarrow{\mathbf{i}}_2$) curves for the oxygen reversible electrode are shown by curves 3 and 4, respectively, in Figure 4.5.

The corrosion current i_{corr} on iron immersed in water occurs at corrosion potential E_{corr} (cross-section of anodic curve 2 for iron dissolution and the cathodic curve 3 for oxygen reduction on the iron surface). We connect iron to the negative pole and diminish its electrode potential, that is we move, according to the polarization anodic curve 2, in the negative direction. In other words, cathodic polarization of the iron electrode (change of electrode potential in the negative direction) results in retarding of the anodic dissolution of iron ($\overleftarrow{i_1}$ value diminishes), acceleration of the cathodic reduction of dissolved oxygen ($\overrightarrow{i_2}$ value increases), and formation of hydroxide ions OH^- on the iron surface. According to Figure 4.5, in order to completely prevent the anodic dissolution of iron (curve 2), we have to diminish its potential to the value less than the reversible potential $(E_{Fe})_{rev}$ of the iron electrode in the electrolyte solution. Usually polarization of iron equipment in the cathodic direction occurs up to 100 mV more negative than the corrosion potential E_{corr}. Therefore, in order to carry out correctly the cathodic protection by means of cathodic polarization of the potential in the negative direction, we have to measure E_{corr} of the iron equipment or structure under real conditions, and then to polarize 100 mV more negative than E_{corr}.

Two ways exist for cathodic polarization in practice: connecting the main metal to be protected (for example, iron) to a less noble metal (aluminum, zinc, magnesium, or their alloys), or connecting to the negative pole of the outer power supply (rectifier or battery). The first method is based on the use of *sacrificial anodes* because they are sacrificed by being dissolved as an anode and turn the metallic construction to a cathode which does not corrode. Principally, any metal may be protected by using suitable sacrificial anodes in a solution of electrolytes, and the electromotive force series (Table 1.3) can help select them. Sometimes this method is called a *passive* one (there is no relationship to *passivity*!), because we connect equipment to be protected to a *sacrificial anode*, and "forget" about corrosion for some time. The second method of cathodic protection is based on the connection to the negative pole of the rectifier and use of an impressed electric current. This method is sometimes called an *active* method of cathodic protection.

The more electrical conductance of media, the better cathodic protection works. Therefore, sacrificial anodes work better in cooling water of high hardness (more calcium and magnesium salts, and corresponding conductance), than in low hardness waters. Cathodic protection is one of many methods of corrosion control of underground and undersea metallic structures and equipment. It is not simple to define the border of the electrical conductance when sacrificial anodes work well or not applicable. Of course, the metallic structures in the atmosphere, in fuels and demineralized water (or other media of high electrical resistance) will not be protected by cathodic protection. Any system for cathodic protection includes anode, cathode, general electrolyte and electric current. We shall discuss separately two types of cathodic protection: *sacrificial anodes* and *impressed current* use.

4.2.1.1 Sacrificial Anodes

The use of sacrificial anodes is one of the ancient methods of corrosion control (see Section 6.1) and is based on the principles of galvanic corrosion which has positive application in this case for the saving of metallic structures. In order to

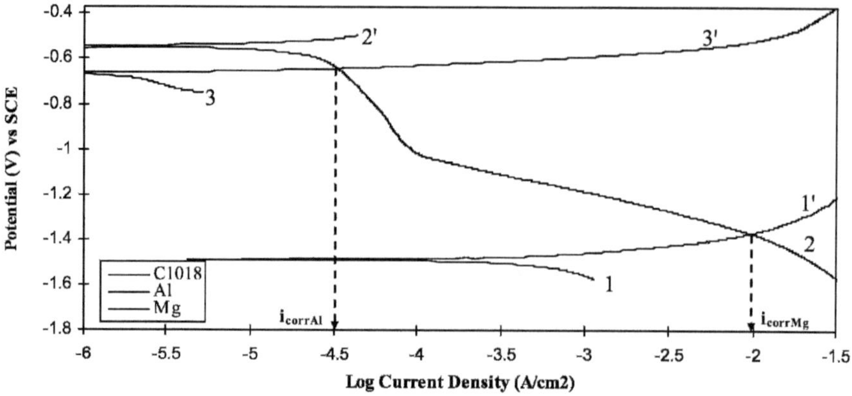

Fig. 4.6 Polarization curves of mild steel (C1018), magnesium (AZ61) and aluminum (6063) alloys in 0.5% NaCl aqueous solution. 1, 2, 3 – cathodic curves; 1′, 2′, 3′ – anodic curves. 1 – Mg; 2 – Mild steel; 3 – Al. The chemical composition of alloys is given in Appendix F.

select the type of anode we have to define the electrochemical characteristics of the anode and the cathode in specific media. Therefore, we have to measure not only the electrode potentials of the anode and the cathode, but also their polarization curves. Electrode potentials of sacrificial anodes and cathodes may be changed with time, therefore thermodynamic electrochemical characteristics may not always give the correct forecast for use of sacrificial anodes.

How to select the material of sacrificial anodes for protection of carbon steel structures? The selection depends on the properties of the environment (its electrical conductance) and electrochemical characteristics of anodes. Magnesium as a sacrificial anode is not used for the protection of carbon steel in media with high electrical conductance: sea water, cooling water and other electrolyte solutions (with conductance $>$3,000 μS/cm). This is explained by the large difference in electrode potentials of iron and magnesium (about 1.0 V – not a bad battery!), and appropriate polarization characteristics (Figure 4.6).

Polarization curves of magnesium and mild steel show the very high corrosion rate of a magnesium anode in cooling water and 0.5% NaCl aqueous solution: 200 mm/year, or 350 $kg{\cdot}m^{-2}{\cdot}year^{-1}$! Magnesium anodes will sacrifice (dissolve) quickly and we often have to change them. Magnesium anodes are used in hot water (about 80°C) for the protection of inner surfaces of boilers (we may check their presence in the boilers on the roof!), and in soils with low electrical conductance.

Zinc anodes are used in sea water, non-brackish waters, and in soils. Aluminum has one essential disadvantage: it has a very good passive film of Al_2O_3 which is not conductive to working well with aluminum as a sacrificial anode. Such elements as tin, mercury, indium, and antimony are added to aluminum, in order to de-passivate it. Thus, aluminum alloys Al–Zn–Hg, Al–Zn–In and Al–Zn–Sn are used as sacrificial anodes. Aluminum anodes are sometimes used for the protection inside bottoms of tanks and tankers containing crude oil and fuels when sea water is used as ballast.

In order for sacrificial anodes to work well, the content of salts in aqueous solutions should be above 0.3 wt%. Aluminum anodes in such tankers may fall and cause sparking in the presence of explosive gases (hydrocarbons, etc.). Zinc anodes do not result in sparking when falling on the carbon steel surface. Therefore, zinc is preferred for use in tankers in the presence of flammable and explosive gases.

Sacrificial anodes are used for the protection of the outer surfaces of pipelines for short distances, outer and inner surfaces of tanks in contact with soil (outside) and aqueous solutions of electrolytes (inside), inner surfaces of heat exchangers with cooling water, supports, and other carbon steel structures in different aqueous solutions and soils.

The efficiency of sacrificial anodes depends on the impurities contained in them. Any cathodic inclusions, for example, iron in zinc anode, result in diminution of efficiency because of the existence of inner galvanic cells and electric currents between iron and zinc. In this case, zinc will be destroyed quicker. Therefore, any anode material has restrictions regarding quantities of cathodic impurities. For example, a magnesium anode must not have iron above 0.003 wt%. If a magnesium anode has 0.01% of iron, efficiency diminishes from 100% to 74%.

The voltage between anode and protected metallic structure depends on the type of material and environment, and is usually about 1 V. Sacrificial anodes must be checked and periodically changed. Usually they are installed for 4 to 5 years.

4.2.1.2 Impressed Current System of Cathodic Protection

Although use of sacrificial anodes is a relatively simple method of corrosion control, there are disadvantages. The first one is that they only work for a limited period, corrode and should be replaced. Usually sacrificial anodes do not work for more than 4 to 5 years. The second drawback is that they only work in media of significant electrical conductance. The third one is that corrosion products formed as a result of the active work of sacrificial anodes in the chemical and food industry can deteriorate the products (chemicals) through non-desired contaminants. If sacrificial anodes are used for the protection of long underground pipelines, it will be very complicated to replace them every few years.

An *impressed current system* can decide all these problems. Such a system consists of a transformer of alternating electric current in a direct one. The metallic structure requiring cathodic protection is connected to the negative pole of the electric power source (Figure 4.7). In order to close up the electric current auxiliary anodes, called ground bed anodes, are connected to the positive pole of the rectifier.

4.2.2 Criteria for Cathodic Protection

Intensive use of an impressed current for the protection of underground gas pipelines began in 1920–1930s. It was important to define the criterion for cathodic protec-

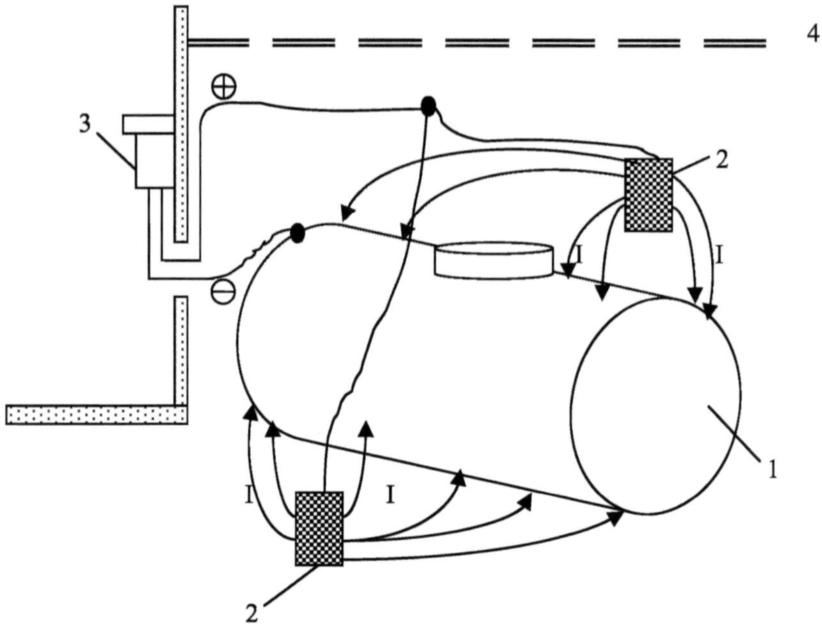

Fig. 4.7 Cathodic protection of an underground steel tank with impressed current: 1 – the tank to be protected; 2 – auxiliary anodes; 3 – rectifier; 4 – ground level). *I* – protective electric current.

tion. Two criteria are used. The first one was suggested by the American electrical engineer Robert J. Kuhn in 1928–1933. The Belgian scientist Marcel Pourbaix invented his diagram 10 years later which allowed the definition of what electric potential must be applied to iron for its complete protection. Kuhn knew nothing about the criteria of thermodynamic – the Pourbaix diagram, which showed that iron became immune to corrosion at an electric potential of less than −0.53 V with respect to SHE. He practically proved that the electric potential of a protected buried pipeline in soil should be less than −0.85 V with respect to a saturated copper/copper sulphate reference electrode. As $E_{Cu/CuSO_4} = +0.317$ V with respect to SHE, and if we take the criterion value −0.85 V, suggested by Kuhn,

$$E = -0.85 + 0.317 = -0.533 \text{ V with regard to SHE,}$$

and shows good agreement between thermodynamic and practical values! Electrical engineers who dealt with cathodic protection thought that the impressed current method should completely protect underground structures. Unfortunately, electrochemical principles show that it is impossible to provide complete protection of buried structures by means of cathodic protection. According to polarization curves (see Figure 4.5), small anodic currents remain on the steel surface. A bare steel surface requires large electric current values for cathodic protection. In order to decrease them, we have to use organic coatings together with cathodic protection.

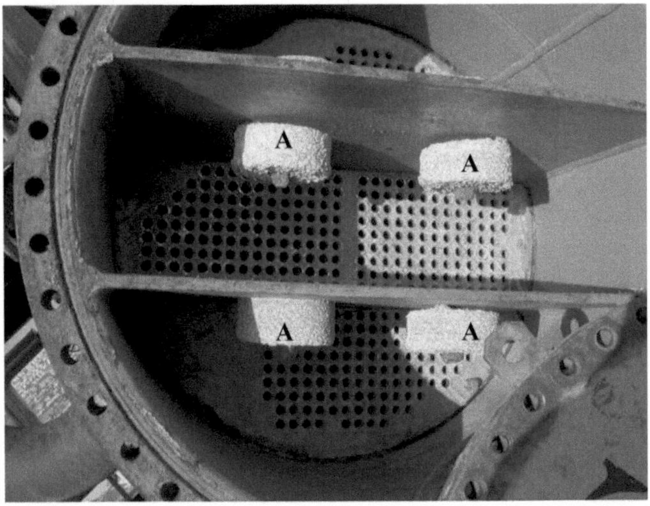

Fig. 4.8 Aluminum sacrificial anodes (A) installed on the inside of heat exchangers coated with epoxy paint (green). The cooling water flows inside the tubes. (For a full color version of this figure, see the Color Section)

4.2.3 Use of Organic Coatings Together with Cathodic Protection

Organic coatings possess high electrical resistance (R) and the electric current (I) required for cathodic protection of coated structures drastically decreases according to Ohm's Law: $I = E/R$. If the electric current diminishes, the number of required sacrificial anodes also decreases and their duration increases correspondingly. As an example, aluminum anodes are used in cooling water for the protection of the insides of heat exchangers coated with epoxy paint (Figure 4.8). Cathodic protection alone cannot provide complete corrosion control of underwater and underground structures. From another side, there are no organic coatings without defects in the form of pores, cracks, and discontinuities. Every coating system has its duration. Let us first apply a coating on the metallic structure and then use the cathodic protection in order to protect the metallic surface against defects in the coatings. If cathodic protection is used for an uncoated structure, the electric current should reach all bare surfaces of the structure. If cathodic protection is used for a coated structure, the electric current must not reach all the surface of this structure but only places with defects: pores, cracks, and other discontinuities.

4.2.4 Limitations and Disadvantages of Cathodic Protection

People are generally well informed about the side-effects of many medicines. Use of some chemicals (corrosion inhibitors, passivators, biocides) and paints (coal-tar, minium, etc.) can cause undesired effects on the environment and the health of people. Use of cathodic protection is not excluded. Cathodic protection has a negative side, and there is a philosophical approach to its use (see Section 6.2). Impressed current systems may cause stray current corrosion (if metallic constructions are in the region of the activity of cathodic protection and are not connected to an impressed current), coating debonding and hydrogen embrittlement (in the case of overprotection and as a result of water decomposition and hydrogen evolving), and corrosion of aluminum (because hydroxyl ions forming during cathodic protection are harmful for aluminum).

Recommended Literature

1. Korb, L.J. and Sprowls, D.O., *Metals Handbook, Vol. 13: Corrosion*, ASM International, USA, 1987, pp. 466–477.
2. von Baeckman, W., Schwenk, W. and Prinz, W. (Eds.), *Handbook of Cathodic Corrosion Protection*, Third Edition, Gulf Publishing Company, Houston, TX, USA, 1997, 567 pp.
3. Uhlig, H.H. and Revie, W.R., *Corrosion and Corrosion Control: An Introduction to Corrosion Science and Engineering*, Third Edition, John Wiley & Sons, USA, 1985, pp. 217–232.
4. Mattsson, E., *Basic Corrosion Technology for Scientists and Engineers*, Second Edition, The Institute of Materials, UK, 1996, pp. 95–103.
5. Shreir, L.L., Jarman, R.A. and Burstein, G.T. (Eds.), *Corrosion, Vols. 1 and 2*, Third Edition, Butterworth Heinemann, UK, 1994, pp. 10:1–10:154.
6. Revie, R.W. (Ed.), *Uhlig's Corrosion Handbook*, Second Edition, Wiley-Interscience, 2006, pp. 1061–1087.
7. Peabody, A.W., *Control of Pipeline Corrosion*, Second Edition, Edited by Ronald L. Bianchetti, NACE International, USA, 2001, 347 pp.

4.3 Change of Chemistry of the Environment

The environment for solid metal may be solid, liquid, gaseous, and two- or three-phase media. For instance, concrete is an example of a solid medium, cooling water is a liquid one, the atmosphere is a gaseous medium, fuel with drainage water is a two-phase medium, and soil is an example of a three-phase mixture of liquid water or aqueous solution, solid salts and oxides, and gases in pores of the solid phase. Water vapors (steam) may flow together with various gases. Liquid water may be mixed with air or suspended particles when it flows inside tubes. We can change the chemistry of any environment around the metal but, as a rule, water (or aqueous solution), organic liquids (usually, oils, and fuels), atmosphere, and concrete are treated or changed. Change of corrosive soil of high electrical conductance for non-corrosive sand is one of the methods for changing the chemistry of the environment around a metal, causing diminishing corrosiveness of the environment. Deaeration, or removal of dissolved oxygen from water, is another example. Regulation and control pH are based on neutralizing the acidic or alkaline species in aqueous solutions, and we can decrease their corrosiveness in such a manner. Purification of water from organic substances, inorganic ions, and dissolved oxygen is the main procedure for producing non-corrosive boiler feed water. Otherwise, boilers would undergo severe corrosion and explosions. Softening is the procedure of diminishing the concentration of calcium (Ca^{2+}) and magnesium (Mg^{2+}) ions in cooling water. Hard water (high concentrations of Ca^{2+} and Mg^{2+}) can give rise to carbonate deposits on heat exchanger tubes. Soft water (low concentrations of Ca^{2+} and Mg^{2+}) can cause corrosion. Any addition of chemicals into the environment results in its chemical change and, as a result, influences its aggressiveness, diminishing or increasing the corrosion of metals. One of the most widespread methods of changing the chemistry of the environment is the use of *corrosion inhibitors*.

4.3.1 Corrosion Inhibitors

> In my childhood, I heard a story about a small mouse (a mouse probably is always small) which penetrated into an ear of a big elephant, and the latter died. Everybody has suffered from the sting of a small gnat (mosquito). Small non-visible bacteria can kill any person. Homoeopathists use very small concentrations of medicines for treatment of many illnesses. Thus, we are familiar with the powerful influence of very small objects on the existence of big ones.

Inhibitors are the general name for substances that, being present in *suitable* concentrations, decrease the rate of chemical reactions, diminish growth in biology, and impede the proceeding of physiological processes or even stop them. Inhibitors got their names from the Latin word *inhibere* which means *to suppress*, *to hold*, or *to retain*. *Corrosion inhibitors* are chemicals that, when present in very low concentrations in a corrosive environment, retard the corrosion of metals. Concentrations of corrosion inhibitors can change from 1 to 15,000 ppm (0.0001 to 1.5 wt%). Inhibitors are substances that work oppositely to catalysts, namely, slow down the reactions. Catalysts are not spent in chemical reactions but inhibitors are spent in electrochemical corrosion reactions. Corrosion inhibitors can be solids, liquids, and gases, and can be used in solid, liquid, and gaseous media. Solid media can be concrete, coal slurries or organic coatings (paints). Liquids may be water, aqueous solutions, or organic solvents. A gaseous medium is an atmosphere or water vapor. There is no clear classification of inhibitors. Most authors divide them into anodic (passivators), cathodic, and mixed inhibitors according to the mechanism of the influence on electrochemical reactions occurring on a metallic surface resulting in corrosion. Corrosion inhibitors can be divided into those which depend on using a medium, or type of protected metal. The first division concerns the inhibitors which are used in water and aqueous media (cooling waters), in acids, in organic media, in the atmosphere, in paints, and in concrete. Another classification can relate to inhibitors protecting iron and steel, copper and its alloys, aluminum and its alloys, zinc and its alloys, etc. Corrosion inhibitors can be classified according to their nature: organic and inorganic. We shall describe the inhibitors according to their use in specific media: in water and aqueous solutions of electrolytes (for instance, cooling waters), in acids (for example, aqueous solutions of sulphuric acid, hydrochloric acid, or citric acid), in two-phase media (hydrocarbons–water), in atmosphere, in paints, and in concrete.

4.3.2 Corrosion Inhibitors in Water and Aqueous Solutions of Electrolytes

When we talk about water, we have to point out a particular kind of water: cooling water, potable water, boiler feed water, water in extinguishing lines. Of course, corrosion inhibitors may be different for various kinds of water. For example, for most cooling water systems, corrosion inhibitors must work together with biocides and anti-scaling agents; for potable water, corrosion inhibitors must be non-toxic for people; for boiler feed water, inhibitors (oxygen scavengers) must react with small concentrations of dissolved oxygen at high temperatures; for water in extinguishing lines, inhibitors must be effective under stagnant conditions; for engine coolant waters, inhibitors must be applicable for many different alloys (carbon steel, cast iron, copper alloys, aluminum alloys) which are present in the system. Even if we talk about cooling waters, they may differ significantly in chemical composition, for example, be of medium, hard, or soft hardness. This fact influences drastically the efficiency of corrosion inhibitors. Those for iron and carbon steels, copper and its alloys, aluminum and its alloys in water and in aqueous solutions are summarized in Table 4.2.

4.3.2.1 Iron and Carbon Steels

Corrosion inhibitors for iron and carbon steels can be of the anodic, cathodic, and adsorbed (organic) type.

Anodic inhibitors are compounds that suppress the anodic electrochemical reaction of dissolution of metals. They are chromates, phosphates, silicates, nitrites, carbonates, molybdates, borates, hydroxides, benzoates, dissolved oxygen, and peroxocompounds. Potassium chromates (K_2CrO_4) and dichromates ($K_2Cr_2O_7$) have been used for many years, but are now forbidden for use as they can cause cancer.

Phosphates. Inhibitive properties of phosphates are based on their hydrolysis and increase pH above 8.0, and on formation of ferric and ferrous phosphate layers (called *passive films*) on an iron surface.

$$Na_3PO_{4(s)} + H_2O_{(l)} \rightarrow 3Na^+_{(aq)} + HPO^{2-}_{4(aq)} + \mathbf{OH}^-_{(aq)} \text{ (pH > 7; 11 to 12)}, \tag{4.12}$$

$$Na_2HPO_{4(s)} + H_2O_{(l)} \rightarrow 2Na^+_{(aq)} + H_2PO^-_{4(aq)} + \mathbf{OH}^-_{(aq)} \text{ (pH > 7; 8 to 9)}. \tag{4.13}$$

Phosphates protect carbon steel at temperatures up to 80°C, but their efficiency decreases compared with the efficiency at low temperatures (about 20°C). In waters of

Table 4.2 Corrosion inhibitors of some metals and alloys in water.

Corrosion inhibitor	*Chemical formula*	*Metal/alloy protected*	*Recommended concentration of corrosion inhibitors*		*Type of protection*
			ppm	$M \cdot 10^3$	
Di-sodium-phosphate	Na_2HPO_4	Fe, mild steel, cast iron, Cu	80	0.563	Anodic
Tri-sodium (or potassium)-phosphate	Na_3PO_4 K_3PO_4	Fe, mild steel	650 940	3.96 4.43	Anodic
Sodium pyro-phosphate	$Na_4P_2O_7$	Fe, mild steel	120	0.451	Anodic
Peroxodiphosphate potassium	$K_4P_2O_8$	Fe, mild steel	25 to 100	0.072 to 0.288	Anodic
Sodium peroxocarbonate	$Na_2C_2O_6$	Fe, mild steel	> 12,500	> 75.3	Anodic
Sodium and potassium peroxoborates	$NaBO_3$, KBO_3	Fe, Mild steel	> 11,000 > 14,000	> 134 > 143	Anodic
Phosphoric acid	H_3PO_4	Fe, mild steel	200	2.04	Anodic
Hexa-meta-phosphate	$(NaPO_3)_6$	Fe, mild steel	15 to 20 ppm PO_4^{3-}	0.158 to 0.211 PO_4^{3-}	Anodic
Sodium Nitrite	$NaNO_2$	Fe, mild steel, cast iron, Sn alloys	400 to 1,000	5.79 to 14.5	Anodic
Sodium Molybdate	Na_2MoO_4	Mild steel	150 to 1,000	0.728 to 4.85	Anodic
Sodium Borate	$NaBO_2$	Fe, mild steel, Zn, Cu	480 to 960	7.29 to 14.6	Anodic
Sodium tetra Borate	$Na_2B_4O_7$	Fe, mild steel, Zn, Cu	660 to 2,640	3.28 to 13	Anodic
Silicates	Na_2SiO_3 $Na_2Si_2O_5$ $Na_2Si_3O_7$	Fe, mild steel, cast iron, Zn, Cu, Al	50 to 150 ppm SiO_2	0.83 to 2.5 SiO_2	Anodic
Sodium carbonate or bicarbonate	Na_2CO_3 $NaHCO_3$	Fe, mild steel, cast iron	> 5 g/l > 20 g/l	> 47 > 238	Anodic
Sodium hydroxide	NaOH	Fe, mild steel, cast iron	> 800	> 20	Anodic
Potassium hydroxide	KOH	Fe, mild steel, cast iron	> 1,500	> 26	Anodic

greater hardness, the efficiency of phosphates is higher because of the formation of mixed phosphates of calcium, magnesium, and iron on the protected metal surface. The concentrations of inhibitors needed to protect metals depend on the chemical composition of the aqueous solution, temperature, and flowing regime. We should be cautious regarding the use of phosphates! Mono-phosphates (NaH_2PO_4) do not protect at all because they reduce the pH of water as a result of their hydrolysis and they can even stimulate corrosion of carbon steel.

Table 4.2 Continued.

Corrosion inhibitor	*Chemical formula*	*Metal/alloy protected*	*Recommended concentration of corrosion inhibitors*		*Type of protection*
			ppm	$M \cdot 10^3$	
Ammonium hydroxide	NH_4OH (25% wt. aqueous)	Fe, mild steel, cast iron,	> 1,500	> 11	Anodic
Sodium Benzoate	C_6H_5COONa	Fe, mild steel	> 20 g/l	> 139	Anodic
Organic compounds: tannins, benzoates, carboxylates and other salts of carboxylic acids, amines, fatty polyamines, lignins, ligno-sulfonates		Fe, mild steel, cast iron, Zn, Cu, Al	50 to 300		Film forming
Calcium bicarbonate	$Ca(HCO_3)_2$	Fe, mild steel			Cathodic
Zinc salts	$ZnSO_4$, $ZnCl_2$	Fe, mild steel	40 to 80 ppm $Zn^{2+}_{(aq)}$	0.61 to 1.22 $Zn^{2+}_{(aq)}$	Cathodic
Phosphorus-containing organic compounds: phosphonates, phosphonic acids, phospho-alkyl-carboxylic acids		Fe, mild steel	100 to 200	0.34 to 0.97 phosphonic acids	Anodic-cathodic

Notes: ppm – parts per million; M – mol/Liter.

$$NaH_2PO_{4(s)} + H_2O_{(l)} \rightarrow Na^+_{(aq)} + HPO^{2-}_{4(aq)} + \mathbf{H_3O^+_{(aq)}} \quad pH < 7;\ 5 \text{ to } 6. \quad (4.14)$$

When calcium and magnesium ions are present in water, di- and tri-phosphates at high pH (above 8) and temperatures above 50°C may result in scale deposits [$Ca_3(PO_4)_2$ and $Mg_3(PO_4)_2$] on heat exchanger tubes and diminish the ability of heat transfer. Another problem is that phosphates are favorable to bacteria growth in water. The last problem with the use of phosphates is that there are restrictions on phosphates concentrations in sewage waters. In most countries, the allowable concentration is 1.5 ppm of phosphorous P (4.5 ppm PO_4^{3-}) in sewage waters. If phosphates are used in cooling water systems, generally the concentration is less than 0.2 ppm P (0.6 ppm PO_4^{3-}) in sewage water (after biotreatment).

Silicates: Different types of silicates can be used as corrosion inhibitors. They differ by the ratio [Na_2O]/[SiO_2]: sodium meta-silicate Na_2SiO_3 (or $Na_2O{\cdot}SiO_2$), sodium di-silicate $Na_2Si_2O_5$ (or $Na_2O{\cdot}2SiO_2$), and sodium tri-silicate $Na_2Si_3O_7$ (or $Na_2O{\cdot}3SiO_2$). Silicates are effective inhibitors in waters at higher temperatures (60 to 80°C). The inhibitive properties of sodium silicates are explained by the increase of pH and formation of protective films incorporating SiO_3^{2-} anions on steel surfaces, for example, $FeSiO_3$. These films even protect carbon steel after the removal of sodium silicate from the water solution. Sodium silicate can also protect a rusted steel surface. Complex compound $Fe_3O_4 \cdot FeSiO_3$ is formed on the surface in this case.

Sodium nitrite ($NaNO_2$) is a good corrosion inhibitor in closed cooling water systems and two-phase gasoline–water systems. The concentration of $NaNO_2$ needed for protection of carbon steel depends on the salt content in water, temperature, and flow velocity, and usually varies between 400 to 1,000 ppm.

Dissolved oxygen in water of high purity may be used as an anodic inhibitor (passivator). Oxygen at high concentrations passivates metallic surfaces and thus decreases anodic dissolution of carbon steels in pure water at high temperatures. This principle is used at power stations.

If *anodic* inhibitors are added in insufficient concentrations they may result in localized corrosion because of an increase of the ratio of cathodic to anodic surface. Thus, anodic reaction is concentrated on small anodic sites and results in pitting corrosion.

The presence of chloride ions in water influences the protective properties of all inhibitors and accelerate corrosion of iron. Therefore, more concentrations of anodic inhibitors are needed in the presence of chlorides.

Cathodic inhibitors are the chemical compounds that suppress cathodic electrochemical reaction on metallic surfaces. They do not cause local attack if their concentration is less than the value needed for effective protection. Calcium bicarbonate $Ca(HCO_3)_2$, zinc salts ($ZnSO_4$, $ZnCl_2$), polyphosphates, and phosphonates are cathodic inhibitors. Some cathodic inhibitors form non-soluble substances with products of cathodic reaction hydroxyl ions (OH^-) after oxygen depolarization on cathodic sites of metallic surfaces and thus prevent access of oxygen to cathodic regions.

$$Ca(HCO_3)_{2(aq)} + OH^-_{(aq)} \rightarrow \mathbf{CaCO_{3(s)}} + HCO^-_{3(aq)} + H_2O_{(l)}, \tag{4.15}$$

$$Ca^{2+}_{(aq)} + 2HCO^-_{3(aq)} \rightarrow \mathbf{CaCO_{3(s)}} + CO_{2(aq)} + H_2O_{(l)}, \tag{4.16}$$

$$ZnSO_{4(aq)} + 2OH^-_{(aq)} \rightarrow \mathbf{Zn(OH)_{2(s)}} + SO^{2-}_{4(aq)}. \tag{4.17}$$

Polyphosphates are the generic name of inorganic polymeric substances with the general formula $(NaPO_3)_n$. Main representatives of polyphosphates are sodium tripolyphosphate ($Na_5P_3O_{10}$) and sodium hexa-meta-phosphate $(NaPO_3)_6$. The latter is hygroscopic and after being kept in the atmosphere, is transformed into pyrophosphate ($Na_4P_2O_7$) and orto-phosphate (Na_3PO_4). Hexa-meta-phosphates are not only used as corrosion inhibitors but also for softening of water and to remove scale from steam boilers and washing machines.

Phosphonates are the organic compounds having direct bonds between phosphorus and carbon atoms which are resistant to hydrolysis. Typical representatives of phosphonates are amino-tris-methylene phosphonic acid (AMP) and 1-Hydroxyethane-1,1-diphosphonic acid (HEDP). Phosphonates influence both anodic and cathodic electrochemical reactions on metal surfaces. Cathodic inhibitors are usually less effective than anodic ones.

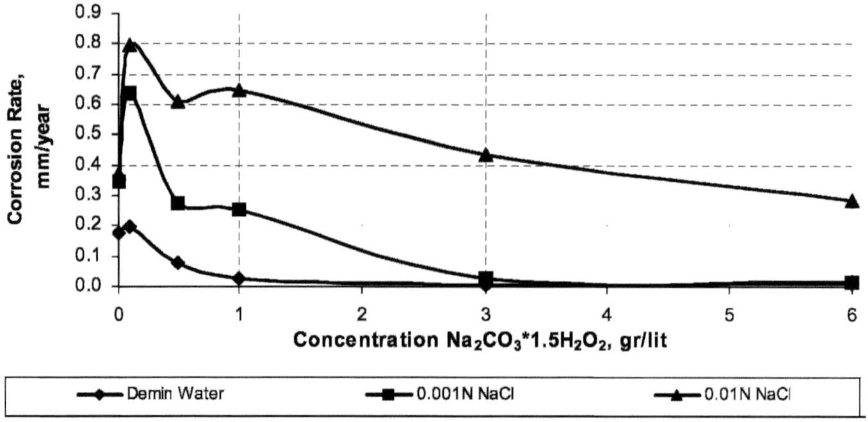

Fig. 4.9 Corrosion rate of mild steel in different solutions of NaCl versus concentrations of sodium peroxocarbonate ($Na_2CO_3 \cdot 1.5H_2O_2$).

Peroxo-compounds. Probably, the reader is familiar with the bleaching properties of hydrogen peroxide (H_2O_2) for textiles (wool and silk). If women want to change their hair color to white, they have to use hydrogen peroxide. It also has antiseptic properties. The presence of the bond (—O—O—) in the molecule H_2O_2 is responsible for its oxidizing properties. Peroxo-compounds, such as peroxo-borates ($NaBO_3$, KBO_3), peroxo-carbonates ($Na_2C_2O_6$ or $Na_2CO_3 \cdot 1.5H_2O_2$), peroxo-diphosphates ($K_4P_2O_8$) are very strong oxidizing agents as they contain similar peroxide (—O—O—) groups. They may decompose in aqueous solutions with the formation of active oxygen atoms, which are able to passivate metal surfaces. *Passivity* relates to the formation of passive oxide, hydroxide, or salt layers or films on metallic surfaces (see Appendix D). Peroxo-compounds can work as corrosion inhibitors because of their oxidizing capability and formation of a non-soluble film and thus retarding anodic reaction on metal surfaces. Similar to other anodic inhibitors, peroxo-compounds increase the corrosion rate of metals at lower concentrations (0.1 to 0.5 g/l) and decrease the corrosion rate at higher concentrations (6 to 10 g/l) (Figure 4.9). Thus, critical minimum concentrations of peroxo-compounds are needed which must be kept in a solution for the effective protection of metals from corrosion. Peroxo-diphosphates ($K_4P_2O_8$) do not possess such properties of anodic inhibitors and protect carbon steel surfaces without an increased corrosion rate at low concentrations. This fact may be explained by the direct oxidation of iron and immediate formation of phosphates on carbon steel surfaces.

Organic corrosion inhibitors. Benzoates, carboxylates and other salts of carboxylic acids, amines (high molecular weight, liquids), amides, tannins, and lignins are used as corrosion inhibitors in aqueous media. They are adsorbed on metallic surfaces and protect them from aggressive components. Sometimes amines are called organic filmers, and form films on metal surfaces that generally are substantially thicker than films established with the proper application of inorganic inhibitors.

Synergistic mixtures of corrosion inhibitors diminish both anodic and cathodic reactions on metal surfaces. Sometimes such mixtures are called *synergistic*, as together they decrease the corrosion rate of metals more effectively than inhibitors being present separately. *Synergism* is the principle of some substances improving the performances of others. Many organic and inorganic compounds relate to this type. Zinc salts are usually used synergistically with phosphates, polyphosphates, phosphonates, silicates and molybdates. If the concentration of zinc salt alone must be about 60 ppm Zn^{2+} as the corrosion inhibitor of carbon steel in aqueous solutions, the concentration must be 0.5 to 2 ppm in a mixture with phosphate and phosphonate. If the concentration of phosphate as a lone corrosion inhibitor of carbon steel in aqueous solutions should be about 60 to 120 ppm PO_4^{3-}, the concentration of phosphate must be 5.5 to 7.5 ppm PO_4^{3-} in a mixture with zinc salt. When phosphonates are used alone, their concentration must be about 100 ppm. If they are used in synergistic mixtures, the recommended concentrations may be 15 to 20 ppm. Silicates can be synergized with polyphosphates and zinc salts. Three and more corrosion inhibitors may also be present in synergistic mixtures.

4.3.2.2 Copper

Small concentrations of copper (at least 0.1 ppm) in water can deposit on steel surfaces and cause galvanic localized corrosion (see Section 2.4). Sodium mercaptobenzothiazole (MBT), benzotriazole (BTA), and tolyltriazole (TTA) with the general name *aromatic azoles* at concentrations ~5 ppm, are usually used as corrosion inhibitors of copper in aqueous solutions. Their inhibitive properties are based on the formation of a protective film on the copper surface.

4.3.2.3 Aluminum

Hexametaphosphate, calcium bicarbonate $Ca(HCO_3)_2$, mixture of borax (sodium borate, tetraborate, or disodium tetraborate) with mercaptobenzothiazole (MBT), sodium silicate, and nitrates are used as corrosion inhibitors of aluminum in aqueous solutions.

4.3.3 Corrosion Inhibitors in Acidic Media (Pickling Inhibitors)

Aqueous solutions of inorganic and organic acids are used for chemically cleaning carbon steels, stainless steels, titanium, copper, and other alloys. Usually, organic substances having N, S, or O atoms with free electron pairs, or some quaternary ammonium compounds with no donor electrons, are used as corrosion inhibitors of metals in acidic media. Thiourea [$CS(NH_2)_2$], amines, polyamines, amides, imidazolines, orto- and para-tolylthiourea, quinoline, derivatives of thioglycolic acid and 3-mercaptopropionic acid, gelatine, casein, sorbitol, agar – agar, dekstrin, and katapin at concentrations of 0.01 to 0.2 wt% (100 to 2,000 ppm) are used in acidic media. Usually mixtures of various amines of high molecular weight are used in industry as pickling inhibitors. Some of them work as cathodic inhibitors and slow down cathodic reduction of hydrogen cations (see reaction (1.47)), others work as filmers and prevent penetration of hydrogen cations into metallic surfaces.

4.3.4 Mechanism of Corrosion Control with Inhibitors

In order to select effective corrosion inhibitors and regulate their protective properties, we have to understand their mechanism. How they protect the metals from corrosion? When we talk about coatings, the main protective mechanism is to isolate the metallic surface from corrosive media. If people do not want to get drunk they eat bread and butter before drinking vodka. Butter coats the wall of the stomach, vodka cannot reach the main organs, and a person will not become drunk. A similar principle exists when corrosion inhibitors contact a metal surface. Many corrosion inhibitors can form protective films on a metal surface and diminish possible contact with aggressive components. In order to protect metals from corrosion, inhibitors must reach the surface of metals and react with the products of electrochemical reactions, or be adsorbed. The protective mechanisms of anodic, cathodic, and adsorbing inhibitors are different. The protective mechanism of anodic inhibitors (phosphates, carbonates, molybdates, and nitrites) is based on a reaction with a metal surface and the formation of passive layers of oxides, hydroxides, or salts. These inhibitors significantly influence the corrosion potential of protected metals. The protective mechanism of cathodic inhibitors is generally based on a reaction with the products of a cathodic electrochemical reaction (OH^-). For example, Zn^{2+} ions react with OH^- ions with the formation of non-soluble $Zn(OH)_2$ at cathodic sites of metallic surfaces (see reaction (4.17)). Inhibitors change the kinetics of electrochemical reactions that cause corrosion in such a way that their rate drastically decreases. Organic inhibitors are adsorbed on metal surfaces. Presence of polar bonds C—N, C—S, C—O and C—P in organic molecules of inhibitors with free electrons on N, S, O and P atoms promote their adsorption on metallic surfaces. The inhibition efficiency of the homologous series of organic substances differing only in the heteroatom is usually in the following sequence: $P > S > N > O$. The mechanism of adsorption may be physical or chemical. When weak Coulomb forces are formed between atoms of inhibitors (for example, N) and metallic atoms, the adsorption is physical. If chemical strong bonds are formed between atom of inhibitors (for example, P or S) and metallic atoms, the chemisorption occurs. Organic inhibitors are sometimes called *non-passivating* types. They nearly have no influence on the corrosion potential of metals.

4.3.5 Factors Influencing Efficiency of Corrosion Inhibitors

Following factors influence the efficiency of corrosion inhibitors: the chemical composition of aqueous solution, pH, flow rate, temperature, and metal surface conditions (roughness, presence of corrosion products, and other compounds). Inhibitors are most effective in a definite pH range. Nitrites lose their effectiveness below a pH of 5.5 to 6.0; polyphosphates should be used between pH = 6.5 to 7.5; Zn – phosphate inhibitors are usually used at a pH = 7.8 to 8.2. The greater the concentration of chloride and sulphate ions, the greater the concentration of corrosion inhibitors needed for protection. Generally more concentrations of corrosion inhibitors are needed for protection under stagnant conditions than under the circulation of media.

The nature of a metal surface is very important. The presence of hydrocarbons (oil or grease) or corrosion products on the surface of the metal negatively affects the efficiency of the corrosion inhibitors, because the latter cannot penetrate the surface of the metal. The metallic surface must be cleaned and passivated before use of corrosion inhibitors. Most companies recommend carrying out passivation by special chemicals containing corrosion inhibitors before use in cooling water systems.

Suspended solids are detrimental in the use of inhibitors because the first can be adsorbed by the metallic surface thus preventing the corrosion inhibitors from reaching the surface. Alternatively, corrosion inhibitors may be adsorbed by suspended solids in water and more concentrations are needed.

Usually inhibitive properties are diminished with an increase of temperature. But some corrosion inhibitors, for example, silicates, are more effective at 60 to 80°C than at 25°C.

4.3.6 Inhibitor Efficiency

In order to compare various corrosion inhibitors and select the most effective one for corrosion control, we have to know how to calculate their efficiency. There are two values, E and γ, for determinating inhibitor efficiency:

$$E, \% = \frac{CR_\circ - CR_i}{CR_\circ} \cdot 100\%, \tag{4.18}$$

$$\gamma = \frac{CR_\circ}{CR_i}, \tag{4.19}$$

where $CR_\circ$ – the corrosion rate of metal in media without inhibitor; CR_i – the corrosion rate of metal in media with inhibitor. The corrosion rate can be measured by any available method: weight loss or electrochemical method (see Chapter 5). The inhibitor efficiency values E and γ are interrelated by the equation

$$E, \% = \left(1 - \frac{1}{\gamma}\right) \cdot 100\%. \tag{4.20}$$

Most scientists and engineers use E values, because they show efficiency in percents, 100% being full protection. In order to select allowable (optimal, minimum, or rational) concentrations of corrosion inhibitor, we have to study the corrosion rate of a particular metal as a function of the concentration of corrosion inhibitor in a particular solution. It is very important to compare the efficiency of various inhibitors correctly, preferably using concentration units in mol/L. Secondly, if we select the corrosion inhibitor, the question is what is the allowable E value: 60%, 70%, 80%, 90%, 99% or 100%? Let us define the minimum allowable E value. First of all, the corrosion rate of a metal in a particular solution must be determined. For example, the experimental corrosion rate of carbon steel in cooling water is 0.6 mm/year. Now we have to determine the maximum allowable corrosion rate for this metal in the solution (see Chapter 5). The maximum allowable corrosion rate of carbon steel heat exchanger tubes in cooling water is 0.1 mm/year (in the oil and petroleum industries). The minimum allowable inhibitor efficiency in this particular case is

$$E_{\text{min}} = \left(\frac{0.6 - 0.1}{0.6}\right) \cdot 100\% = 83\%. \tag{4.21}$$

This means that our target for the selection and use corrosion inhibitors is $E = 83\%$. Only inhibitors which give this value and more can be selected for use.

4.3.7 Application of Corrosion Inhibitors and Some Recommendations

Inhibitors are rarely used in the form of a single compound nowdays. For example, sodium nitrite ($NaNO_2$) alone at 400 ppm is an effective inhibitor of carbon steel in closed cooling water systems. The minimum concentrations of different inhibitors for efficient protection may significantly vary and depend on temperature, chemical composition of media, and flow rate. Synergestic mixtures of anodic and cathodic inhibitors are often used in cooling water systems. For example, zinc and magnesium salts are used with phosphates and phosphonates. If copper alloys are present in cooling water systems, organic azole compounds should be added. Mixtures of corrosion inhibitors containing borate, silicate, molybdate, nitrite, nitrate, and benzotriazole are used in closed water cooling systems of motor vehicles. Such mixtures can protect carbon steel, cast iron, aluminum alloys, and copper alloys. Mixtures of calcium gluconate and sodium benzoate, gluconate with tannin, calcium, and zinc gluconates are synergestic corrosion inhibitors of carbon steel in water. Inhibitors can prevent specific corrosion failures. For example, nitrates are used as inhibitors of stress corrosion cracking of SS 304. Specific corrosion inhibitors (usually organic amines) are used in the oil and gas industry, crude oil refineries, in lubricating oils and hydraulic liquids, and for protecting metals in the atmosphere, in paints, and in concrete.

Corrosion inhibitors in paints are usually *pigments*. Chromates ($SrCrO_4$) and lead-based compounds (Pb_3O_4, $PbCrO_4$) commonly used in paints in olden times are toxic and are now banned from use. Zinc powder is also restricted. "Green" inorganic and organic inhibitors have been developed for use in paints. Organic inhibitors include amines and phosphate esters; inorganic inhibitors include phosphates and molybdates of zinc, aluminum and iron, Na-benzoate, Zn-gluconate, phosphites, stearates, acetates, acrylates, and silicates.

Nitrites, borates, benzoates, formiates, surface active agents, salts of organic acids, their mixtures and VCI (VPI) are used as corrosion inhibitors of carbon steel in *concrete*.

Regarding the application of VCI for protection of metallic devices and constructions in the atmosphere, the reader is referred to Sections 3.2 and 3.6.

4.3.8 Inhibitors and Ecology

There is no clear and accepted definition of "environmentally friendly", or "green" corrosion inhibitors. They must be assessed from the health, safety, and environmental point of view. Corrosion inhibitors must be low toxic (better non-toxic), biodegradable, with low bioaccumulation, and should not contain harmful elements and components. One example is the prohibition of chromates which have been used for many years as effective corrosion inhibitors, or restricted use of phosphates in sewage waters. For instance, Zn^{2+} cations, as all in nature, must be present in water in optimal, or no-risk, concentrations. These values are between 0.5 to 50 μg/l. When the concentration of Zn^{2+} is less than 0.5 μg/l, zinc is deficient; when the concentration of Zn^{2+} is above 50 μg/l, zinc is toxic for many kinds of fish. The criterion for selecting corrosion inhibitors in the 20th century changed significantly. Before the 1960s it was efficiency, then until the 1980s it was economy, now the main criterion is ecology. Intensive research has been carried out into the substitution of toxic inhibitors for less toxic or non-toxic ones. For example, plant extracts aloe leaves and peels from oranges, mangoes, and pomegranates, tobacco, black pepper, castor oil seeds, acacia gum, and lignin are recommended for protection of steel from corrosion by acids. The first oxygen scavengers for treating boiler feed water were natural products, extracts of vegetation, that included catechol, pyrogallol, and tannins. Then they were replaced by efficient, but toxic hydrazine compounds. Tannin and other natural products are now back in favor. If we are talking about inhibitors in paints, lead-containing (Pb_3O_4) and chromates ($SrCrO_4$) are toxic, and Zn-containing compounds [$Zn_3(PO_4)_2$ and ZnO] possess some level of ecotoxicity. New "green" inhibitor magnesium oxy-amino-phosphate is recommended for use in paints.

4.3.9 Conclusions

1. There is no theoretical way of forecasting inhibitive efficiency of known or newly synthesized inhibitors in a particular medium. Only experiments can produce results and make conclusions about the possible use of inhibitors in a medium.
2. Use of corrosion inhibitors is influenced by regulations that have been developed because of toxicity and environmental effects resulting from industrial effluents. Therefore, it is necessary to analyze the requirements for water, fuels or atmosphere when we begin to use corrosion inhibitors, even that which has been used for many years in other places, because new information about their toxicity may be available.

Recommended Literature

1. Shreir, L.L., Jarman, R.A. and Burstein, G.T. (Eds.), *Corrosion, Vol. 2*, Third Edition, Butterworth Heinemann, UK, 1994, pp. 17:10–17:65.
2. Revie, R.W. (Ed.), *Uhlig's Corrosion Handbook*, Second Edition, John Wiley & Sons, Inc., USA, 2006, pp. 1089–1105.
3. Bregman, J.J., *Corrosion Inhibitors*, MacMillan, New York/London, 1963.
4. Rosenfeld, I.L., *Corrosion Inhibitors*, McGraw-Hill, New York, 1977, 327 pp.
5. Mattsson, E., *Basic Corrosion Technology for Scientists and Engineers*, Second Edition, The Institute of Materials, UK, 1996, pp. 103–107.
6. Korb, L.J. and Sprowls, D.O., *Metals Handbook, Vol. 13: Corrosion*, ASM International, USA, 1987, pp. 478–486, 494–497.
7. Rosenfeld, I.L. and Persiyantzeva, V.P., *Inhibitors of Atmospheric Corrosion*, Nauka, Moscow, 1985, 278 pp. [in Russian].
8. Uhlig, H.H. and Revie, R.W., *Corrosion and Corrosion Control: An Introduction to Corrosion Science and Engineering*, Third Edition, John Willey & Sons, USA, 1985, pp. 263–277.
9. *Corrosion Inhibitors*, The Institute of Materials, London, 1994.

Chapter 5
Corrosion Monitoring

Now we know that it is impossible to prevent corrosion of many metals used by us because they are thermodynamically unstable in the environment. If corrosion of metals is inevitable, we want, of course, to know how long equipment or structure will serve safely and efficiently under particular conditions. In other words, if we know the corrosion rate and corrosion type, how may we predict the life longevity of the equipment? When should we carry out shutdown, repair, or replace it? To answer these questions, *corrosion monitoring* is carried out. Most corrosion damage occurs during relatively short periods of time, usually during startup, changes, upsets, or shutdowns. Therefore, *corrosion monitoring* methods must supply the real-time (immediate) information. To use the metallic constructions without corrosion monitoring is similar to living and working with closed eyes and ears. Why else do we have to monitor corrosion of metal equipment and structures in industry? Various electronic devices are used. Their reliable functioning depends on the quality of electric contacts made of different metals and alloys. If corrosive species are present in the environment, electric contacts may corrode, reliability of the data received by these devices diminishes, and corrosion may result in malfunctioning and unpredicted disasters. This is especially important in aircraft, space vehicles, military technique, etc. Therefore, we should monitor corrosive aggressiveness of the environment where electronic or other devices are in contact with it. The third reason for corrosion monitoring relates to preventive anti-corrosion measures. Various methods are used for corrosion control. Corrosion inhibitors, neutralizers, anti-scaling inhibitors, biocides, demulsifiers, and oxygen scavengers are used for prevention of or decreasing corrosion. We must know the efficiency of these anti-corrosive chemicals. When cathodic protection is used for corrosion control of underground and underwater equipment and structures, we should know the efficiency of cathodic protection. If coatings are used for corrosion control, we have to know their efficiency and protective properties. In other words, *corrosion monitoring* helps us to define the efficiency of the corrosion control methods. To sum up, there are three important questions in industry connected with *corrosion monitoring*. The first question is a "qualitative" one. Does corrosion occur? Is there any equipment or structure suffering from corrosion? Many side issues exist. What is the form of the corrosion?

A. Groysman, *Corrosion for Everybody*, DOI 10.1007/978-90-481-3477-9_5,

© Springer Science+Business Media B.V. 2010

Is the corrosion uniform or localized? Here is a wide spectrum of corrosion phenomena: pitting, crevice, SCC, galvanic corrosion, dealloying, intergranular corrosion, erosion, cavitation, MIC, hydrogen damages, and fretting corrosion. We have to define the form of the corrosion in order to take correct preventive measures. The second question relates to the previous one. What are the reasons for and factors influencing corrosion? Certainly this question is connected with the selection of anticorrosion measures. The third question is a "quantitative" one. What is the corrosion level? What is the *corrosion rate*? The latter is the *corrosion of material occurring in the unit of time*. This is not always the corrosion product quantity. Corrosion may change with time and with distance on a metal surface. Even if we know the corrosion level, that is if we received a corrosion rate of 0.01 mm/year, or 0.1 mm/year, or 1 mm/year, the following questions remain. Is this value low or high? Is the situation dangerous or may we exist with such a corrosion rate? May we continue to use the equipment? Or it will be better to stop using it, to repair, or to take measures for corrosion control? Certainly, the attitude to all these things is always relative. If one hair remains on our head, this is little. If one hair is floating in a plate with soup, this is many! A similar situation exists with the evaluation of corrosion rate. Sometimes a corrosion rate of 0.1 mm/year is a high value. Sometimes a corrosion rate of 0.5 mm/year is allowable. For example, for heat exchanger tubes made of carbon steel with a wall thickness of 2.3 mm and a life design period of 15 years, a corrosion rate of 0.1 mm/year is a critical or allowable value. For carbon steel pipes with a wall thickness of 9 mm and a life design period of 15 years for water supply, a corrosion rate of 0.5 mm/year is an allowable value. This means that the third ("quantitative") question concerns the classification of corrosion rates for a particular industry, for specific particular equipment and structures. We may find different classifications for maximum corrosion rates in the chemical and oil industries. For example, the allowable (acceptable, maximum, critical) corrosion rates shown in Table 5.1 are recommended for particular equipment in the chemical industry.

A low allowable corrosion rate of 0.025 mm/year is used for equipment producing medicines and food, because of restrictions concerning contaminants in these products. If equipment made of expensive metals, such as platinum, tantalum, etc., is used, the maximum corrosion rate is also very low. If such equipment as valve seats, spray nozzles, orifices, etc., with critical dimensions is used, it is forbidden to change these dimensions and the allowable corrosion rate is also very low. We have to define the critical corrosion rate for every particular case, and we will show how with the following example.

The second critical corrosion rate in Table 5.1 is 0.25 mm/year (10 mpy), and is recommended for thin-walled equipment. This value is not always correct and depends on the type of industry. For example, the widespread wall thickness of heat exchanger carbon steel tubes in oil refining and the petrochemical industry is 2.336 mm. This means that if we take the maximum corrosion rate of 0.25 mm/year, after 9.3 years nothing will remain of the heat exchanger tubes.

$$\frac{2.336\ \text{mm}}{0.25\ \text{mm/year}} = 9.3\ \text{years}, \tag{5.1}$$

Table 5.1 Classification of corrosion rates for equipment in the chemical industry.[a]

Maximum corrosion rate, mm/year[b]	Characteristics and uses
0.025	Very low corrosion; for services where product contamination is a problem
0.25	Low corrosion; for thin-walled equipment
0.50	Fairly low corrosion; can be considered to be the "normal" maximum allowed in chemical equipment
1.25	High corrosion
>1.25	Excessive corrosion

[a] Landrum (1989).
[b] 1 mm/year = 40 mpy (1 mpy = 1 mils per year)

The design life of heat exchangers is 15 years in the oil refining industry. This means that the value of 0.25 mm/year is not suitable for heat exchanger tubes. In addition, it is impossible to use heat exchangers up to the formation of holes, and we should take some safety factor, for example, 0.7. This means that we may use only heat exchangers up to a corrosion wear of 30%. Thus, we have to replace the heat exchangers after

$$9.3\ \text{years} \cdot 0.7 = 6.5\ \text{years}. \tag{5.2}$$

If we use the recommended allowable corrosion rate of 0.25 mm/year, the life period of a heat exchanger (6.5 years) is less than half of the design life (15 years). Therefore, we have to define the allowable corrosion rate for any metallic equipment and structures in a particular industry. The allowable corrosion rate in the abovementioned case is

$$\frac{2.336\ \text{mm} \cdot 0.7}{15\ \text{years}} = 0.11\ \text{mm/year (5 mpy)}, \tag{5.3}$$

that is, nearly half of the recommended value of 0.25 mm/year in the chemical industry. We are talking about uniform corrosion. Different values may be received for pitting corrosion. Recommended values for the maximum rate for pitting corrosion are relative. Thus, a pitting corrosion rate of 0.11 mm/year (5 mpy) may be allowable for thick tubes and dangerous for thin ones. Therefore, it is not easy to choose the allowable corrosion rate for one system with equipment of different wall thicknesses. For example, large tubes made of carbon steel with a wall thickness of 9 mm and small ones with a thickness of 2 mm can be used in open and closed cooling water systems. In addition to carbon steel tubes, copper alloys are used in cooling water systems. Of course, different allowable corrosion rates must be accepted for various alloys in open and closed cooling water systems (Table 5.2).

Different authors recommended different values for allowable corrosion rates of carbon steel and copper alloys. The common practice is that these values are smaller for closed cooling water systems than for open ones and, for the copper alloys, smaller than for carbon steel. Unfortunately, there are no recommendations regard-

Table 5.2 Maximum (allowable) corrosion rates[a] (mm/year) in cooling water systems.

Open system		Closed system	
Carbon Steel	Copper Alloys	Carbon Steel	Copper Alloys
< 0.013	< 0.0025	0.0013 to 0.0075	< 0.0025
< 0.075	< 0.0050	0.0025	0.00005
0.013 to 0.025	0.0005		
< 0.11[b]	< 0.005[b]		

[a]Recommended in *Materials Performance* **38**(10), 1999, p. 57.
[b]Our recommendations. These values concern general corrosion.

ing allowable rates for pitting corrosion, and we should define them for specific equipment in every particular case.

In conclusion, the allowable corrosion rate must be chosen in accordance with the demands of industry, type of processing, type and metallurgy of the equipment, and the type of corrosion phenomenon. Usually corrosion rates for carbon steel less than 0.11 mm/year (5 mpy) and for copper alloys under 0.005 mm/year (0.2 mpy) are acceptable.

All who deal with designing equipment have to decide about corrosion allowance. Unfortunately, design engineers usually select a voluntary corrosion allowance value (without technical support). What is this and how to select the correct corrosion allowance?

5.1 Corrosion Allowance

Design engineers define the wall thickness D_{mech} of equipment and structures in accordance with the mechanical features of the materials (strength, tension, stress, etc.). All metallic constructions corrode at a definite corrosion rate. Therefore, design engineers add a value called a *corrosion allowance* to the equipment's wall thickness defined in accordance with the demands of the mechanical characteristics. *Corrosion allowance is the thickness in mm that the engineer has to add to the design thickness* D_{mech} *for equipment because of corrosion.* The following formula may be used for the definition of the corrosion allowance value:

$$D_{\mathrm{ca}} = \mathbf{k} \cdot t \cdot 2, \tag{5.4}$$

where D_{ca} is the corrosion allowance, mm; $\mathbf{k}$ is the corrosion rate (real or estimated), mm/year; t is the design life of the equipment, year; 2 represents a safety factor. We should know the corrosion rate value $\mathbf{k}$. We may define this value experimentally or find it in the literature for similar metal equipment under similar conditions. If it is difficult or impossible to do this, we may take the allowable corrosion rate, recommended or practically found, for the D_{ca} calculation. For example, if the design life of some equipment is 15 years, and its corrosion rate is 0.1 mm/year, in a certain medium, the corrosion allowance is:

$$D_{ca} = 0.1 \text{ mm/year} \cdot 15 \text{ years} \cdot 2 = 3 \text{ mm}. \tag{5.5}$$

Now the design engineer may add this value to the thickness calculated in accordance with the mechanical demands, and obtains the wall thickness of the equipment needed:

$$D = D_{mech} + D_{ca}. \tag{5.6}$$

5.2 Corrosion Monitoring Methods

There is no strong division of corrosion monitoring methods. They range from non-direct/non-intrusive to direct/intrusive technique. Sometimes one method (for instance, pH measurement) may be intrusive (on-line pH meter), or non-intrusive (if pH is measured for a sample obtained through an existing valve). Sometimes corrosion monitoring techniques are divided into direct (weight loss – coupons and electrical resistance probes; corrosion current – linear polarization probes; remaining wall thickness – ultrasonic or eddy-current technique) and indirect (radiography and chemical analytical methods). An intrusive technique requires entry into the process stream (coupons, electrochemical sensors, pH-meters, and electrodes). A non-intrusive technique does not require entry into the stream (non-destructive technique – eddy current, acoustic emission; hydrogen flux probes).

Three main factors influence corrosion: metal type, environment, and conditions at the border between a metal and the environment. Therefore, we can use the properties of a metal, of an environment, and of a border metal–environment for corrosion monitoring. In the light of this, corrosion monitoring methods can be divided into three groups: control of a metal condition, of the environment, and of a border metal–environment. *Corrosion monitoring methods are the control methods of corrosion situation, namely, control of metal behavior under particular environmental conditions.* Control of a metal condition is based on the physical properties of metals: mass, thickness, and electrical resistance of the metal sample. Control of the environment is based on the chemical, physico-chemical and microbiological properties, and conditions of the environment. Control of the border metal–environment is based on the detection of its physico-chemical properties.

Control of physical properties of a metal (physical methods). Historically they were the first: man's eyes or optical devices were used for the control of properties of a corroded metal surface. Different devices based on physical phenomena allow us to measure the changes in thickness of metallic equipment and structure (ultrasonic and eddy-current methods), and weight loss of the metal (weight loss "coupon" method and electric resistance measurements of a corroding metal sensor). The acoustic emission method is used for determining the local damage of metallic constructions (cracks, pits, and holes). Optical microscopes and boroscopes are used for visual analysis of a corroded metal surface. Video cameras are used for inspection of inner surfaces (including coating assessment) of tanks and voids on ships. X-ray radiographic methods are used for online wall thickness measurements of insulated pipes, detection of intergranular corrosion, and flow-accelerated corrosion.

Control of environment (chemical analytical, physico-chemical, physical, and microbiological methods). These methods are based on chemical analysis of media which are in contact with metal equipment. For instance, it is impossible to use the weight loss method in the case of corrosion of copper alloys or stainless steels in aqueous solutions because of very low weight changes of coupons made of these alloys. Various sensitive physico-chemical analytical methods can be used for the determination of the concentration of metal cations in media as a result of corrosion of metallic equipment. On the other hand, analytical methods are used for the determination of aggressive components in media: pH, Cl^-, SO_4^{2-}, O_2, CO_2, H_2S, NH_3, CN^-, Fe^{3+}, Cu^{2+}, hardness, alkalinity, suspended solids, microorganisms influencing corrosion, the presence of contaminants inducing erosion and cavitation in water and steam; and some physico-chemical properties of media (electrical conductance and redox potential). For organic liquids (fuels, solvents, acids, etc.), it is very important to measure the water content, dissolved oxygen concentration, and electrical conductance. For boiler feed water, it is important to measure the concentration of dissolved oxygen, copper and iron cations, silica, and totally dissolved solids. For steam and high temperature gases, it is very important to measure the temperature in order to eliminate water condensation and, as a result, cavitation and dewpoint corrosion. For the determination of the aggressiveness of the atmosphere, we should measure the relative humidity, the temperature and its changes, time of wetness, concentrations of sulphur oxides (SO_x), nitrogen oxides (NO_x), ammonia and its derivatives, ozone, hydrogen sulphide, and dust in atmosphere. The determination of hydrogen gas in some processing (Hydrogen Flux Monitoring) relates to the analytical methods. For the determination of corrosiveness of soil, we have to measure its electrical resistance. In some cases it is preferable to identify the microorganisms, the humidity, and pH of soil.

Microbiological analysis of water, soil, atmosphere, crude oil, and fuels is very important for the control of potentiality of the occurrence of MIC.

Control of process (technological) parameters include the measuring of flow rate, temperature, pressure, dewpoint temperature, heat transfer resistance, presence of solid particles in streams, etc.

The analysis of residue concentrations of corrosion inhibitors, neutralizers, oxygen scavengers, biocides in technological streams is very important for control of the efficiency of anti-corrosion treatment, as well as for the ecology.

Control the border metal–environment (physico-chemical methods). These methods are based on the physico-chemical properties of the border metal–environment, and may be divided into electrochemical methods, identification of corrosion products and deposits, and examination of the morphology of the metal surface.

Among electrochemical methods are corrosion-potential measurements, potentiodynamic polarization measurements (Tafel method and cyclic polarization), linear polarization resistance (LPR), electrochemical noise measurements (ENM), electrochemical impedance spectroscopy (EIS), galvanic probes (ZRA – Zero Resistance Ammeter), and Harmonic Distortion Analysis (HDA). The radioactive method "Thin Layer Activation" is used for the determination of corrosion resistance (corrosion rate) of some alloys under specific conditions. Microbiological analysis of deposits (sessile bacteria on metal surface) is also very important. Scanning Electron Microscopy (SEM), Scanning Tunneling Microscopy (STM), and Atomic Force Microscopy (AFM) are used for the examination of metal surface morphology. Energy Dispersive Spectroscopy (EDS), Energy Dispersive X-ray Fluorescence (ED-XRF), X-ray Photoelectron Spectroscopy (XPS), also known as Electron Spectroscopy for Chemical Analysis (ESCA), Auger Electron Spectroscopy (AES), Mössbauer Spectroscopy, Secondary Ion Mass Spectroscopy (SIMS), Fourier Transform Infrared Spectroscopy (FTIR), Raman spectroscopy, UV-Visible reflectance studies, and Ultraviolet Photoelectron Spectroscopy (UPS) are used for the identification of the chemical content of corrosion products, and the presence of inhibitors or any deposits on the metal surface. We may receive very useful information from the color of corrosion products formed on the metal equipment (Appendix I), but there is some uncertainty. Firstly, there is no objectivity in visual color determination. Secondly, the same color can belong to several compounds, for instance, black may be attributed to the presence of various sulphides (FeS, Ag_2S, CuS) or oxides (Fe_3O_4, CuO); white to the presence of AlOOH, $Zn(OH)_2$, or $CaCO_3$. Additional chemical examination is needed. In any case, color identification is very useful preliminary information about corrosion and its reasons. A slippery surface of brown, orange, black, green, or intermediate colors usually means the presence of biofouling.

5.2.1 Physical Methods of Corrosion Monitoring

We shall concentrate on two physical methods widely used both in laboratory and industry: the weight loss (WL) method and electrical resistance (ER) measurements, both based on a loss of corroded metal.

5.2.1.1 Weight Loss (WL) Method

When we discussed corrosion kinetics (see Section 1.4), we analyzed how we may measure the rate of corrosion according to the weight loss of a metal specimen called a "coupon". Rectangular or circular in shape, "coupons" made of a metal or an alloy representative of the material of the equipment, can be inserted into a process stream by means of retractable or retrievable holders (Figure 5.1). Corrosion rates are calculated from weight loss measurements according to formulas (1.8) and (1.9). The surface of the coupon must be carefully prepared before installation. The dimensions and initial weight of the coupons must be defined. Every coupon must be stored in a special envelope in a desiccator. Then they are installed for a

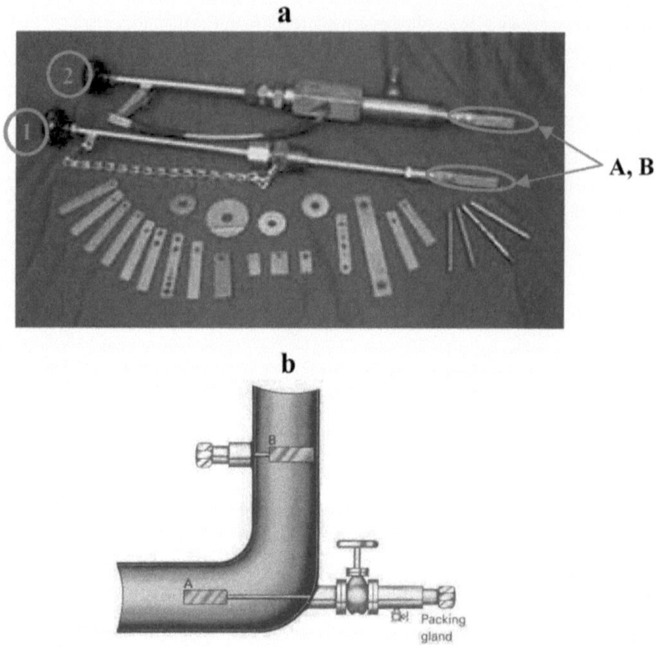

Fig. 5.1 Various coupons with retractable holders (a) for the examination of corrosion by the WL method and their installation in a process stream (b). 1 – retrievable coupon holder (up to 8 atm); 2 – high pressure or packing gland holder (up to 105 atm). A and B – coupons. (Courtesy of Metal Samples Company)

certain period depending on the corrosiveness of the process stream and coupon's material. In Section 1.4 we discussed how this period can be defined. The coupons' immersion period must be determined for every particular case. For example, this period may significantly differ in various cooling water systems. Usually 30 to 40 days are recommended. This period is correct for carbon steel in water not treated with corrosion inhibitors. Copper alloys and stainless steel are resistant to cooling water. Therefore, the immersion period for these alloys must be increased to 180 days and even more. Sometimes a period of 30 to 40 days for carbon steel coupons is unacceptable. If water is not treated with chemicals (corrosion inhibitors, anti-scaling agents, and biocides), the corrosion rate of carbon steel may be high (about 0.6 to 1 mm/year), and coupons have to be changed every 10 days. On the other hand, if chemicals are injected in cooling water, the corrosion rate of carbon steel is low (about 0.05 mm/year and less), and coupons may be changed every 90 to 120 days.

If the corrosion rate of carbon steel is measured in a two-phase system, for example, "hydrocarbons–water" in the overhead of the crude oil distillation column, the immersion period of the coupons depends on the inhibitor's efficiency and may change from 40 to 180 days, and even more. We have to keep in mind that pitting

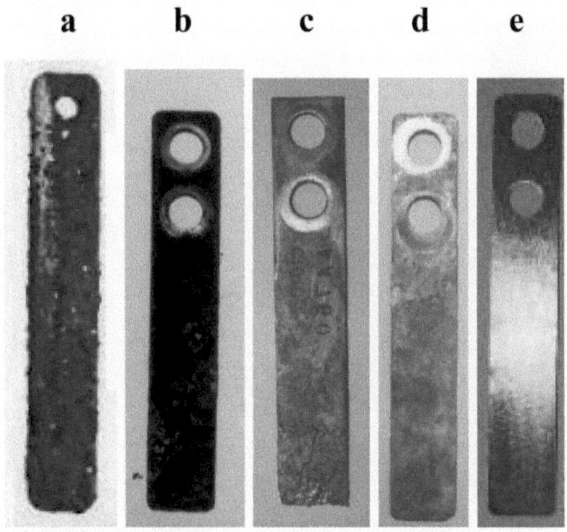

Fig. 5.2 The coupons after being exposed to cooling water (a, c) and to the process stream in the overhead of the crude distillation column (b, d): a, b – coupons with corrosion products; c, d – coupons after cleansing of corrosion products; e – original coupon. (For a full color version of this figure, see the Color Section)

corrosion is very dangerous phenomenon in any streams and an induction period (above 60 days) is needed for the determination of pitting corrosion. The coupons must be removed after a period of immersion chosen according to the goals of the study and the corrosiveness of the stream. They may be photographed with, and then without corrosion products or other deposits, and compared with the original coupon before installation. Two examples of pictures after immersion in industrial streams are shown in Figure 5.2.

Color identification of corrosion products formed on the surface of a coupon may give preliminary information about the presence of species causing corrosion (see Appendix I). If dense films of corrosion products are formed on the coupon's surface (for instance, iron oxides or iron sulphides) the thickness can be measured. Then corrosion products should be removed by means of a knife made of hard polymeric material (Teflon, polyethylene, or polypropylene) for subsequent chemical identification. Then the coupon's surface must be cleaned mechanically with an eraser or chemically with hydrochloric acid (5 to 8 wt% HCl) and injection of 0.2 wt% organic corrosion inhibitor (usually liquid amines or their blends) to decrease the loss of clean metal during chemical cleaning.

Here are some recommendations for the correct installation of coupons in equipment at the plant:

1. A coupon must be installed against flow, but in such manner so as to prevent any erosion or impingement.
2. A bypass with three and more places for a coupon's installation is used in water cooling systems (Figure 5.3), and it should be connected to the return riser to the

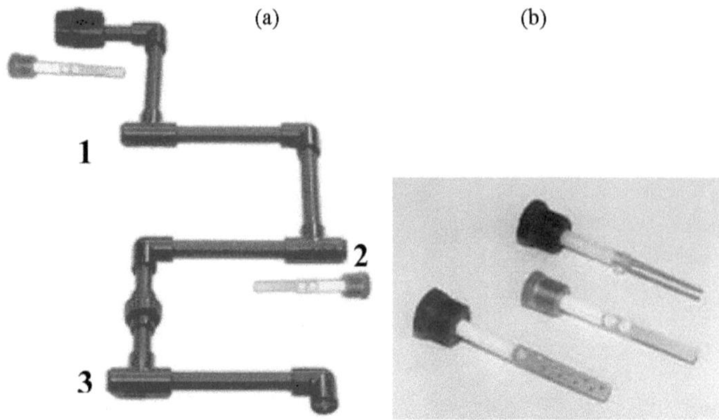

Fig. 5.3 Bypass piping assemblies (a) and coupons (b) before installation. 1, 2, 3 – places for inserting the coupons. (Courtesy of Metal Samples Company)

cooling tower or at the hottest part of the system. It is very important to install copper alloy coupons after carbon steel and aluminum ones. Do not use copper and galvanized pipes as well as brass valves for the connection of the bypass to the riser. A bypass is usually made of polymeric material (polyvinylchloride).

3. The coupons must be mounted closely to (but should not be in contact with) the surface of the equipment to be examined.
4. Flow rates 1 to 2 m/s of stream are adequate.
5. We should remember that a coupon gives information about the corrosion situation only in the place of its installation. A corrosion situation may change by moving away from the place of a coupon's installation. Therefore, the more coupons that are installed, the better the *corrosion situation map* obtained.

The benefits of the *WL method*:

1. The corrosion rate calculated by this method is precise: 3 to 5% of reproducibility. The reliability of this method is high. It is recommended to compare other methods of corrosion monitoring (for instance, electrochemical) with the WL method.
2. It is possible to measure the corrosion rate of metals in any media (without any relationship to its electrical conductance): in gases, in liquids, in solids, and under wide process conditions: up to 450°C and to 350 bar in the case of retrievable holders, and without restrictions in the case of fixed probes.
3. This is the only method from which the quantitative information can be obtained about pitting corrosion rate because the depth of pits is measured by means of an optical microscope. The electrochemical noise measurements (see Section 5.4) give qualitative information about pitting tendencies.
4. It is possible to examine a coupon's surface and corrosion products and then to identify the corrosion phenomenon and corrosion mechanism.
5. Special coupons with small holes may be used for the examination of scale formation; *disc coupons* – for the crevice corrosion examination; *stressed*

coupons – for the stress corrosion cracking examination; *rod coupons* – for erosion examination; *welded coupons* for studying the variance of corrosion behavior between welded and non-welded metals; and special *gauze* coupons for collecting biofouling (see Figure 5.13).

The limitations of the *WL method*:

1. In order to receive reliable data, in many cases we need to wait for a long time: at least 30 to 40 days, and more. Corrosion rates may be calculated only after removal of a coupon from the system. If the corrosion rate is low (about 0.001 mm/year), we need to wait longer: about 4 to 9 months. In order to receive information about pitting corrosion, we need to wait more than 60 days.
2. The coupons provide general and averaged data about corrosion over a period of time (Figure 5.4). Environmental conditions may change every moment and it is impossible to know what happened during such upsets.
3. The WL method is time-consuming. We have to treat the coupons: to clean, to weigh, and to take into account the *cleaning factor*. When we clean a coupon's surface of corrosion products by means of acid, some WL of the cleaned metal occurs. We can estimate this error in percent (called the *cleaning factor*) in the WL definition. In other words, the *cleaning factor* shows the influence of chemical cleaning on the accuracy of the WL method. In order to define the *cleaning factor*, the clean coupon may be immersed in the same acid solution with an inhibitor for a similar period as for the treated coupon and define WL. The ratio of the WL determined for the clean coupon to the WL determined for the treated coupon in percent is the cleaning factor. If the WL of a treated coupon is relatively high (10 to 100 mg), the cleaning factor is not high: about 1 to 2%. If the WL of the treated coupon is low (0.5 to 1 mg), the cleaning factor may be 50 to 100%.

Sometimes the WL method is not suitable for assessing corrosion of non-iron alloys, for example, copper alloys in water and a selection of corrosion inhibitors for them. Sensitivity and accuracy of WL of copper coupons are low in water in the presence of corrosion inhibitors. In this case Atomic Absorption and Ion Couple Plasma (ICP) methods are used for the determination of the concentration of copper ions appearing in water because of the corrosion of copper alloys (chemical analytical methods).

The main drawback of the *WL method* is that it is not always suitable for use in industry because of difficulties of installation and the coupons have to be changed every month or every few months. For example, if there are dangerous processing conditions, such as high temperature and pressure, harmful substances (hydrogen sulphide, hydrogen chloride, etc.), it is difficult or even impossible to achieve the points needed for measuring. The *WL method* does not permit on-line, real-time monitoring. Therefore we will discuss the methods that partly overcome the limitations of the WL method.

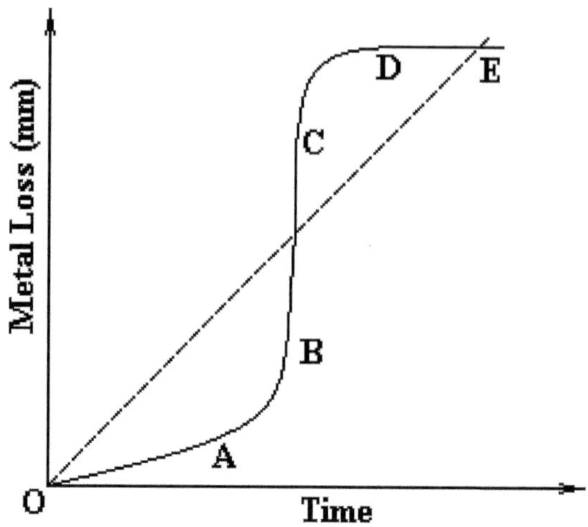

Fig. 5.4 A typical plot of metal loss versus time. OE – average corrosion rate over the total time period. OA and DE – corrosion rate for normal conditions. BC – corrosion rate at a process upset.

5.2.1.2 Electrical Resistance (ER) Method

If we take a wire, a strip, or a cylindrical metal specimen, we may use the following formula, well-known in physics, for a conducting material:

$$R = \frac{\rho \cdot L}{A}, \tag{5.7}$$

where R is the electrical resistance of conducting material, Ohm; ρ is the specific electrical resistivity of the conducting material, Ohm·m; L is the length of a specimen, m; A is the cross-sectional area of a specimen, m^2.

If a piece of a metal is immersed in corrosive media (liquid, gas, or solid), corrosion may occur and, as a result, the cross-sectional area A of a specimen will decrease. According to formula (5.7), the electrical resistance of a metal specimen will increase. Thus, a metal specimen may be used as a corrosive ER sensor. Temperature influences the electrical resistance of a metal. Therefore, an ER probe (a sensor element contacting with a corrosive medium) is mounted together with a reference element for temperature compensation, which does not physically contact a medium but is of the same temperature. Thus, a reference element is influenced by temperature in the same way as a sensor, but is not influenced by corrosive media. Practically, we measure the ratio of the electrical resistance of a sensor exposed to the environment to the electrical resistance of a reference element not exposed to the environment at the same temperature. Since temperature changes affect the electrical resistance of both the exposed and reference element equally, measuring the resistance ratio minimizes the influence of changes of the ambient temperature. Sensing elements in the ER probes are made in a variety of geometric configura-

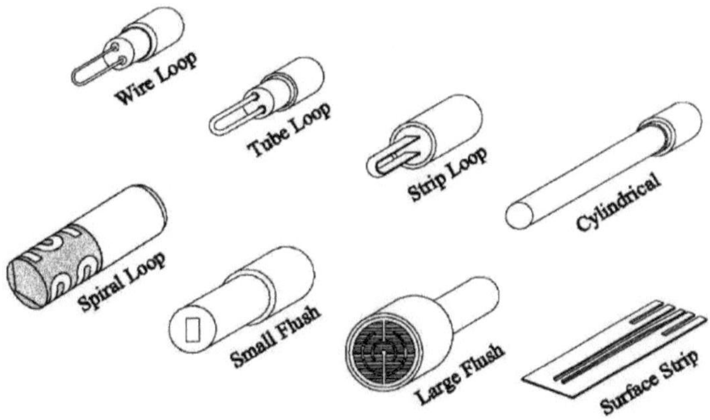

Fig. 5.5 Various types of ER sensing elements. (Courtesy of Metal Samples Company)

tions, thickness, and alloy materials (Figure 5.5). The selection of a particular element form depends on sensitivity, response time, and element (probe) "lifetime" required. Response time is the minimum period in which a measurable change takes place and governs the speed with which useful results can be obtained. A probe "lifetime", or a period required for the effective thickness of the exposed element to be consumed governs the probe replacement schedule. The lesser the element thickness, the greater its sensitivity, and the lesser its "lifetime". Since a probe's "lifetime" and response period are directly proportional to each other, the selection of an element is a compromise between sensitivity, data frequency, and probe replacement frequency. Detection levels as low as 1 nm are achieved with elements with thicknesses of several millimeters.

A portable device may be used for registration of changes of a sensor's electrical resistance. In this case the electrical resistance changes of the sensor mounted in a stream should be measured every few days. We may continually monitor by means of a transmitter connected to a distributed control system. An increase in the electrical resistance of a metal specimen (sensor) is equivalent to a decrease in thickness. Thus, the slope of a curve "dial reading of ER probe – exposure time" in Figure 5.6 shows that a metal corrodes with time and, as a result, the metal thickness and quantity diminish. We can calculate the corrosion rate according to the slope of such a curve according to

$$k = |\text{slope}| \cdot 0.001 \cdot 365\ [\text{mm/year}] = \frac{\Delta L}{\Delta t} \cdot 0.001 \cdot 365\ [\text{mm/year}]. \qquad (5.8)$$

$\Delta L/\Delta t$ is the ratio of the diminishing thickness of the sensor (change of the dial reading ER probe) in microns (μm) to the measured period in days, microns/day; 0.001 mm/μm and 365 days/year – conversion factors. The greater the slope of a line "dial reading of ER probe versus time", the greater the corrosion rate of a metal (Figure 5.6).

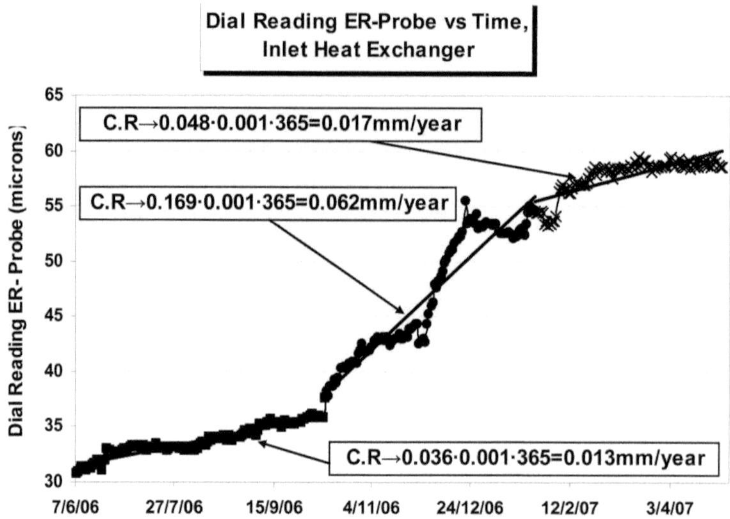

Fig. 5.6 The electrical resistance measured in the corroded thickness (microns) of a sensor element versus time. CR – corrosion rate for three various periods.

This method is used in the overhead of crude distillation units and naphthenic acid corrosion control in the oil refining industry, for inhibitor selection and optimization, for erosion control, corrosion monitoring in district heating systems, for monitoring the efficiency of cathodic protection of fuel storage tank bottoms, and internal corrosion monitoring of subsea production flowlines. The design of ER probes permits operation up to 160°C and 700 atm.

The benefits of the *ER method*:

1. We can observe from the curve "electrical resistance versus time", that the online real-time corrosion rate may be measured.
2. Because the measurements are based on the WL of metallic elements, the electrical resistance can be measured, and as a result the corrosion rate of metals in media both of high electrical conductance (solutions of electrolytes including soils of high electrical conductance) and low conductance (gases, hydrocarbons, two-phase electrolyte-non-electrolyte systems, non-conducting solvents, and soil of low electrical conductance). The ER method is similar to the WL method and is opposite to the electrochemical methods. The latter methods may be used only in media of relatively high electrical conductance.
3. The lifetime and sensitivity of ER probes depend on the geometrical dimensions of a sensor. Therefore, we may select a sensor of any lifetime and sensitivity required. Very high corrosion rates may be quickly detected (for several hours).
4. It is possible to monitor corrosion without removing the sensor and at any distance. Of course, we should replace a sensor after full corrosion. We may select a sensor of such configuration that replacement will be needed once a year or once in two years.

The limitations of *ER method*:

1. It only measures uniform corrosion. It is impossible to identify pitting corrosion and other corrosion phenomena. Pitting may result in wrong results. Therefore, we have to examine visually the surface of a sensor if sudden unexpected and sharp changes in ER data occur.
2. Some salts (FeS_2, PbS) and oxides (Fe_3O_4, MnO_2) formed on the sensor's surface are electrically conductive and give rise to wrong results.
3. Some deposits (fouling) formed on the surface of a sensor also give rise to wrong results.
4. If there is no preliminary information about corrosion rates in a particular medium, it is very difficult to select a correct sensor of a suitable sensitivity and a lifetime. Therefore, at first it is important to define the corrosion rate by means of the WL method in a particular medium, and only then to select an ER probe of a suitable sensitivity and lifetime.
5. Several days are usually needed for accurate measurements by the ER method and for receiving reliable data.

We should remember to change sensing elements when they reach 90 to 95% of their nominal "life". For instance, if an ER probe with a sensing element of thickness of 250 μm is used, we should change it when 240 μm has been corroded. However, sensing elements must be immediately changed if localized corrosion has occurred.

In order to receive reliable data, we should compare the ER data with the WL method. How are these methods used in practice? What useful information can we obtain by means of these methods?

5.2.1.3 On-Line Corrosion Monitoring at the Overhead of a Crude Distillation Unit at an Oil Refinery

Crude oil distillation is a process of separation of petroleum distillates (fuels) proceeding in a distillation tower (also called the atmospheric column, because a distillation, called a rectification, of crude oil is carried out at atmospheric pressure). The lighter petroleum fractions (gaseous methane, propane, butane, pentane, and others), water vapors, hydrogen chloride (HCl), and hydrogen sulphide (H_2S) at 120 to 130°C exit an overhead of the distillation tower (Figure 5.7).

All gases exiting the distillation tower are cooled below 100°C in an air cooler and heat exchangers, usually made of carbon steel. If the temperature goes below 100°C acidic aqueous solutions of HCl and H_2S are formed, and cause severe corrosion of carbon steel equipment. Neutralizers and corrosion inhibitors are injected into the overhead system in order to diminish or prevent dewpoint acidic corrosion. One of the objects of corrosion monitoring is to follow the efficiency of the neutralizers and corrosion inhibitors, as well as the corrosion situation in the air cooler and heat exchangers. For this, both WL coupons and ER probes are mounted in all possible places (see Figure 5.7). Usually heat exchangers have two inlets and two

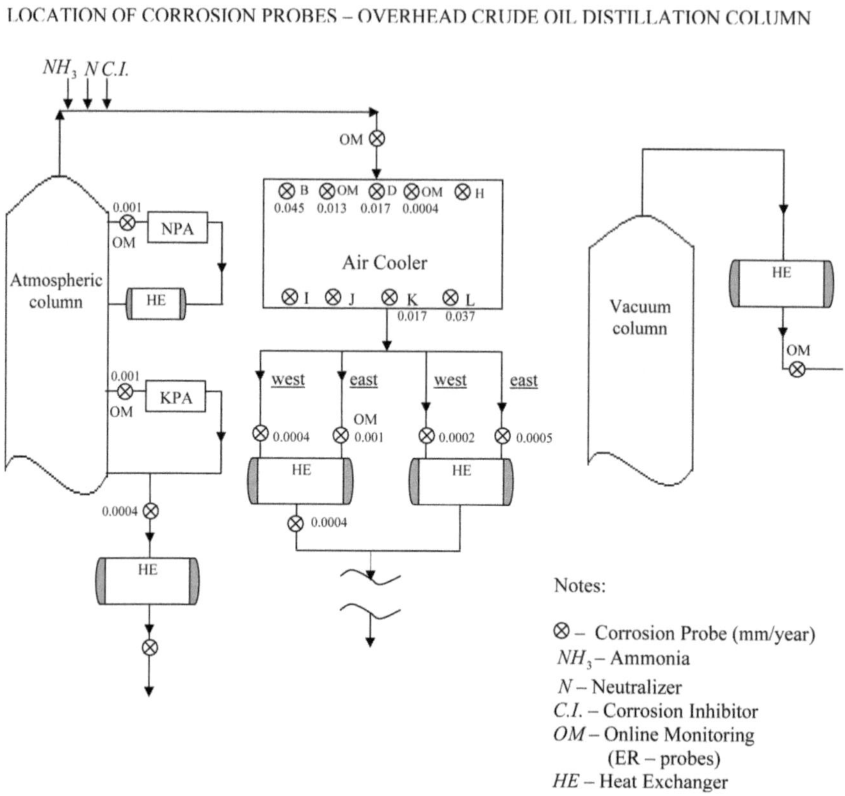

Fig. 5.7 Corrosion monitoring at the overhead of the crude oil distillation tower at an oil refinery. OM – On-line monitoring (ER probes). B, D, H, I, J, K, L – WL coupons. NPA – Naphtha Pump-Around. KPA – Kerosene Pump-Around. HE – Heat Exchanger. Corrosion rates are shown near the points of probe location.

outlets. It is very important to mount a WL coupon parallel to ER probes in such inlets and outlets.

The philosophy of on-line ER corrosion monitoring includes the data collecting system, their treatment (performing the calculation), utilization of the data, their interpretation, and comparing with the results of the WL coupons every 30 to 60 days. The analog signals (4–20 mA) sent from the ER probes are proportional to the uncorroded metal (sensor) remaining and are calibrated to give an indication of total corrosion of the carbon steel sensors in microns. The outputs of these probes are monitored via an existing distributed control system (Honeywell/TDC 3000) and are also sent to the Plant Information system (PI) where they are analyzed. Figure 5.8 shows these data as collected and stored within the PI system.

The PI system enables easy access to the history data base with any PC (personal computer) that is connected to the local net or by dialing-in from remote PCs. This means that information may be received about corrosion from almost anywhere.

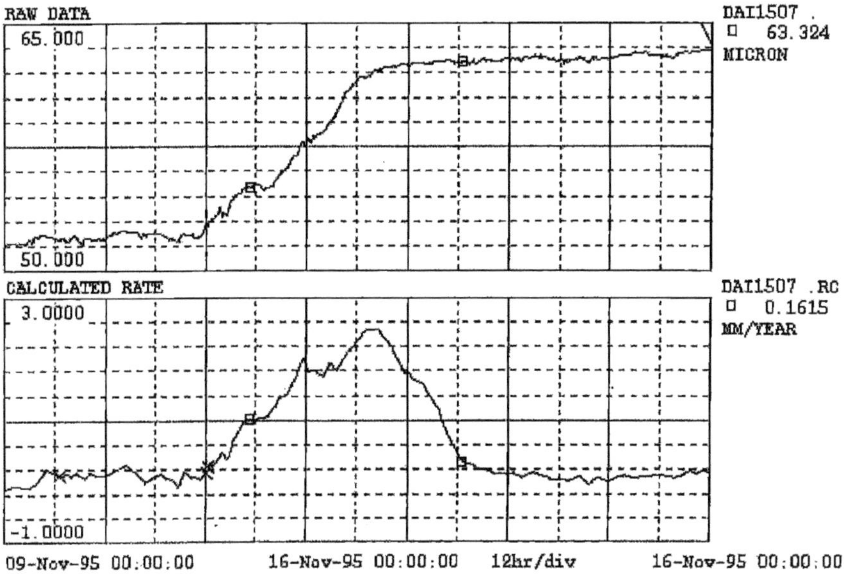

Fig. 5.8 Dial reading (micron) and calculated corrosion rate (mm/year) stored in the PI system. The upper curve shows the changes (decreasing) of a sensor's thickness (in microns) with time as a result of its corrosion; the lower curve shows the corrosion rate in mm/year (differential of the upper curve).

Performing the calculations. Based on the field instrument readings, the corrosion rates are calculated by means of differentiation of received data regarding a given period. The simplest way is to take all readings that were collected within the PI and find the regression line which best fits the data (see Figure 5.6). Another technique which is more suitable for on-line calculations is filtering by averaging over periods of time and making the corrosion calculations between separated averages. There are two parameters to adjust in the application of this algorithm: the length of the period to be averaged to give representing values and the time to separate between these two values (Figure 5.9).

The calculated corrosion rates depend on the measured period, and these data may be presented for various periods, for example, 2 hours – 2 days, 5 hours – 5 days, 7 hours – 7 days, etc. (Figure 5.10). These data show how many hours are averaged and what the distance is in days between these averages. These calculations are continuously done and the results are stored in the system.

Now we want to know about the reliability of received data. Long periods (usually longer than 30 days) give integral corrosion rate values which we should compare with the WL coupon's data. We can observe that ER data coincide with the data of the WL method (Figure 5.11), but the latter does not allow following the changes in corrosion that happen during this period.

On-line ER data allow analyzing during short periods. These data reflect process changes reflecting the corrosion and enables the experts to find the reasons for

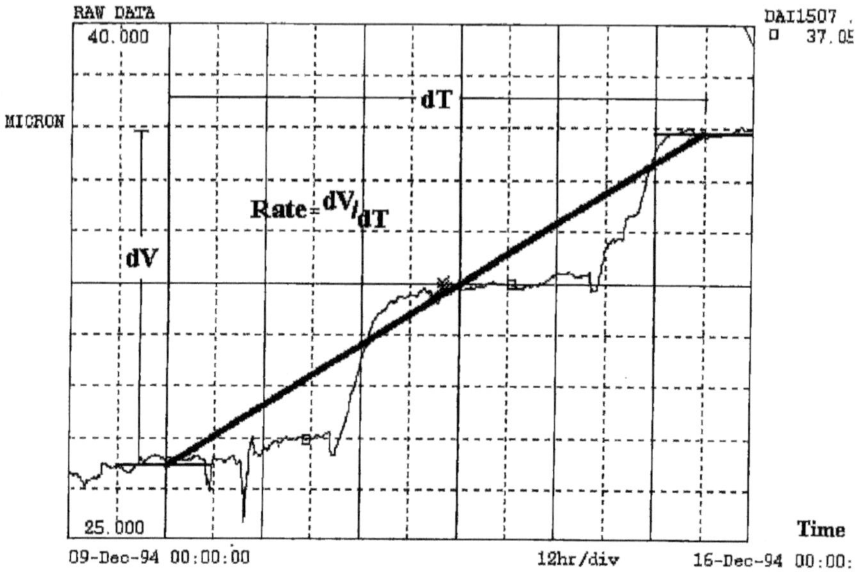

Fig. 5.9 On-line corrosion rate calculation.

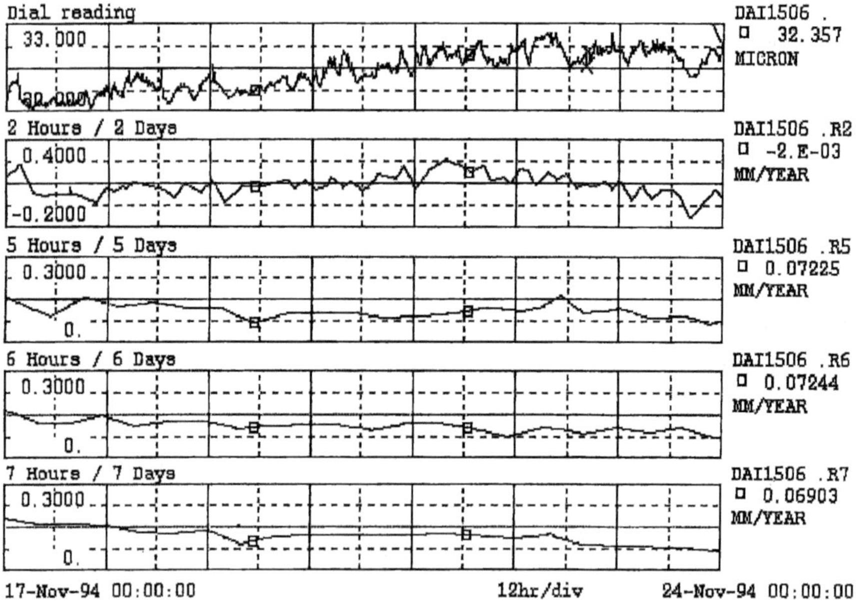

Fig. 5.10 Averaged corrosion rates versus time base.

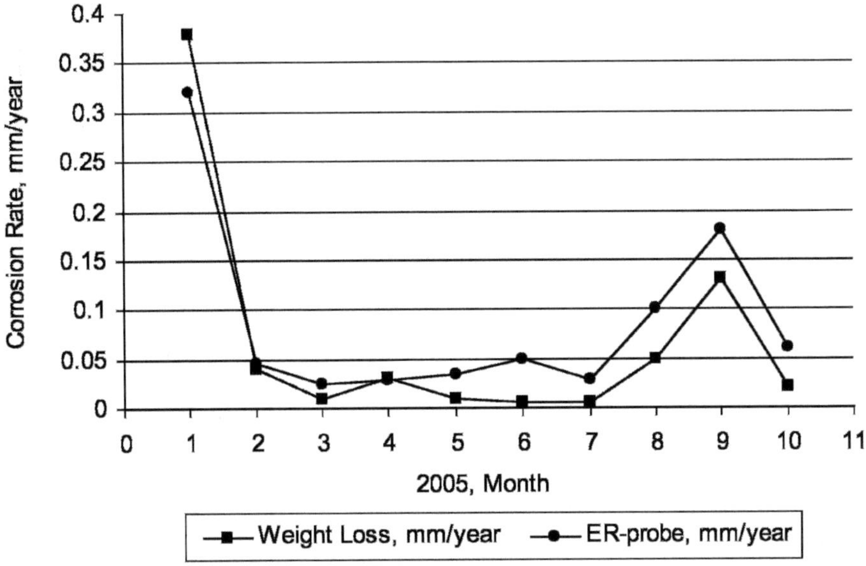

Fig. 5.11 Comparison ER and WL data.

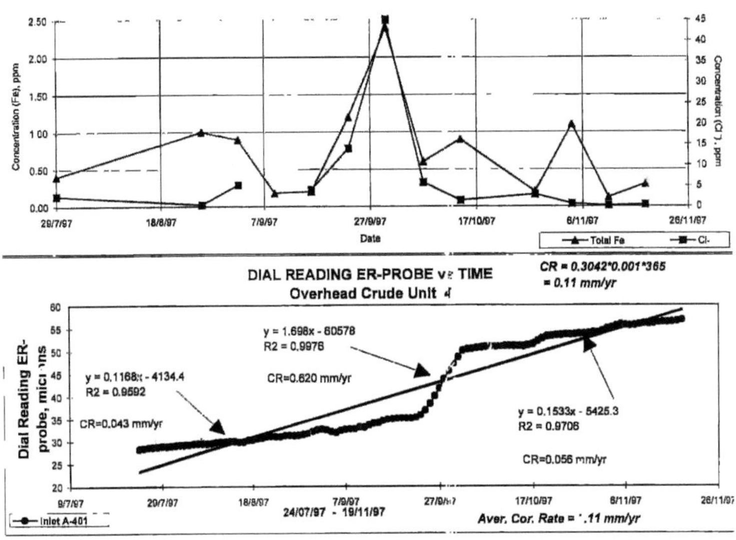

Fig. 5.12 The chemical analysis of condensed water after heat exchangers (upper graph) and the ER data (lower graph) in the overhead distillation column.

the upsets and to try to improve or eliminate the causes and minimize the corrosion damages (Figure 5.12). From this figure we can define that the high content of chlorides caused the high corrosion rate and high content of iron in the aqueous medium.

5.3 Controling the Environment (Chemical Analytical, Physico-Chemical, Physical, and Microbiological Methods)

Chemical analytical and physico-chemical methods include chemical and microbiological analysis of the environment (content of dissolved ions, gases, other chemicals, and microorganisms), and its physico-chemical parameters. These methods allow both the detection of the corrosion level (sometimes also the corrosion rate) and corrosiveness of the environment. They must be defined for specific streams at every enterprise: cooling water, condensed water, two-phase systems water-hydrocarbons, boiler feed water, chemicals, etc. All chemical treatment programs in cooling water systems require a certain water quality. This means that chemicals (corrosion inhibitors, anti-scaling agents, and biocides) work only under specific chemical and physico-chemical parameters of water. Chemical analytical methods are standardized [1, 2], and their description is beyond the scope of this book. Photocolorimetric, spectroscopic (IR – infra-red, UV – ultra-violet, visible light), atomic absorption, ion plasma couple (IPC), liquid and gas chromatography, ion-selective electrodes, fibre-optic electrodes, and pH meters are used in chemical analytical methods.

Many crude oil refineries combine WL and ER methods with chemical analytical methods (measuring of iron, copper, pH, chlorides and sulphates). The chemical analysis of iron in oil and gas production is standard [3]. Analytical measurements of iron and nickel in crude oil and gas oil are used for the evaluation of the corrosiveness of these organic liquids in the presence of naphthenic acids. Test methods for measuring the pH of soil and soil resistivity are standard [4–6]. It is recommended to measure the electrical resistivity of soil to determine its corrosiveness. The less the electrical resistivity of the soil, the more its aggressiveness. In some cases it is preferable to identify the microorganisms, the humidity, chlorides, sulphates, and pH of soil.

The question is what are the dangerous values of analytical parameters (pH, iron, copper, chlorides, microorganisms, etc.) determining the corrosion level (high or low) in a system? Usually allowable values of analytical parameters in water, fuels, other streams (one and two phases), and soil are recommended which show the possible danger of corrosion in the system.

This depends on the specific system (type of materials and media used). For instance, if equipment is made of iron and copper alloys, the total iron and copper content in cooling water systems must not exceed 1 and 0.05 ppm, respectively. For condensed water in the overhead of crude oil distillation towers at oil refineries,the recommended values are pH = 5.5 to 6.5; 0.5 ppm total iron, and 20 ppm chlorides. Otherwise, corrosion may occur. In many cases, it is important to monitor not the absolute values of chemical species concentrations, but their trend, i.e. sudden changes of these values can show the beginning of corrosion (see Figure 5.12).

Sometimes hydrogen sulphide and ammonia are measured in condensed water at the overhead, but these data do not give useful information for corrosion monitoring. Chemical analytical methods are very sensitive and allow the measurement of corrosion rates in the cases when the WL method is not applicable. For instance, copper alloys are resistant to cooling water in the presence of copper corrosion inhibitors, and the WL method does not give reliable results, but the atomic absorption method allows the measuring of a dissolved copper in solution. Any analytical method of determination of metal cations in water or any other medium after immersion of a coupon in this medium, allows the calculation of the corrosion rate of the metal (supposing that all the coupon's area dissolves and metal ions in a dissolved state are in the solution):

$$k = \frac{C \cdot V \cdot \rho_s \cdot 87.6}{A \cdot \rho_m \cdot t}, \tag{5.9}$$

where k is the corrosion rate, mm/year; C is the concentration of dissolved metal in the solution, determined by analytical method, ppm (mg of metal/kg of solution); A is the area of coupon, cm^2; V is the volume of the solution, liter; ρ_s is the density of the solution, kg/liter; ρ_m is the density of the metal, g/cm^3; t is the period of immersion of coupons in the solution, hours; 87.6 is the coefficient of inversion to mm/year. If corrosion products are formed on a coupon's surface, they must be removed and dissolved in the same solution. Formula (5.9) is applicable for laboratory examinations.

Dissolved oxygen plays an important role in the corrosion of boilers and its concentration is stringently restricted. A similar restriction concerns the content of iron, copper, total hardness, silica, total alkalinity, and electrical conductance of boiler feed water and boiler blowdown.

The benefits of the *chemical analytical methods*:

1. High sensitivity to all metals dissolved in liquids.
2. Estimation of corrosion rate of all metals in liquids under specific conditions.
3. The trend of analytical parameters in the environment can show sudden corrosion, or any danger of the occurrence of corrosion.
4. Determination of concentrations of hazardous metals (Pb, Cr, Cd, and Zn) in the environment can decide the problem of maximum contaminant levels for regulated drinking water suppliers.

The disadvantages of the *chemical analytical methods*:

1. *Quantitative* estimation of the corrosion rate is applicable for measuring only in liquids under laboratory conditions. The presence of corrosion products on a metallic surface can give rise to the wrong results.
2. The chemical content of aggressive corrosive species in the environment (water, soil and atmosphere) gives the *qualitative* estimation of its corrosiveness.
3. These methods sometimes require much time for obtaining results.

5.3.1 Monitoring of Microbiological Activity towards Metals

Microorganisms can be responsible for corrosion in many systems: cooling water, drinking water, crude oil, fuel, and water storage tanks, heat exchangers, pipes, etc.

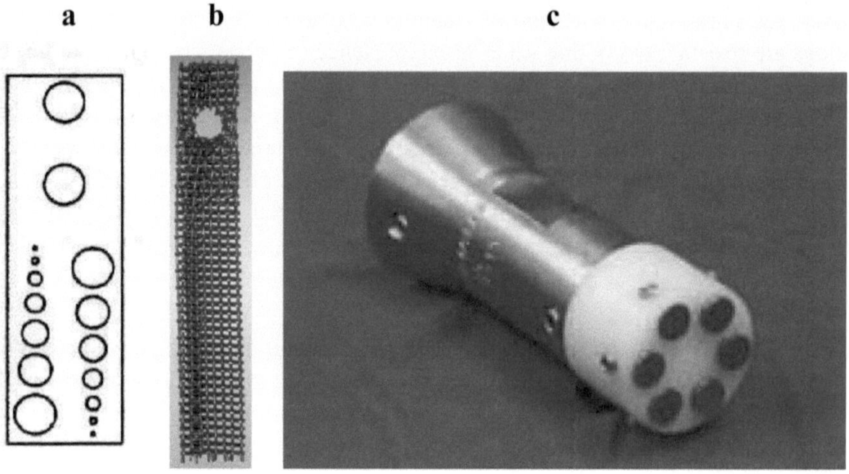

Fig. 5.13 Special metallic coupons with holes (a), Ni-Cr screen gauze (b), and bioprobe (c) for collection of biofouling and later examination. (Courtesy of Metal Samples Company)

(see Section 2.5). Different methods are used for microbiological analysis in water, soil, crude oil, fuels, and sludge [2, 7]. There is no only one method of MIC monitoring. The following techniques are used for biofilm monitoring: collection of samples on exposed studs or special coupons and afterward examination (Figure 5.13), electrochemical devices, and the measurements of the change of heat transfer over a tube array arising from the influence of biofouling, scale, and corrosion products.

Microbiological, biochemical and molecular techniques are used for identifying and quantifying microorganisms. Microbiological examination includes identification of microbial activity, that is various kinds of *planktonic* and *sessile* microorganisms, by means of "kits", biosensors, and conventional methods of microbiological analysis (bacteria growth on special artificial media). Direct visualization includes the use of fluorescent dyes that cause cells to light up under ultraviolet radiation. Use of commercial "kits" allows to quickly assay the presence of enzymes associated with microorganisms suspected of causing corrosion. The most common way to assess microbial populations in various industrial streams is to use artificial growth media for particular microorganisms associated with corrosion. In this case, we have to wait for the results over a longer time than when using a kit. Microbial activity may be determined by measuring the amount of adenosine triphosphate (ATP) in industrial streams. This metabolite drives many reactions in the cells. Detection of other substances such as organic acids, esters, hydrogen sulphide, or methane by analytical methods may indicate microbiological activity in the biofilm. Identification of proteins (for example, lipopolysaccharide) or nucleic acids may also quickly show the microbial activity. On-line, real-time determination of total microbiological activity with fluorescent bioreporters has been developed. It is recommended to determine the presence of microorganisms in deposits by means of mixing with a 2 wt% OsO_4 solution. Microorganisms intensively absorb osmium. The further determination of osmium in deposits by means of one of the physico-chemical methods (for example, EDS) may point out the presence of microorganisms. Methods for removing biofilms from a metal surface are inefficient and may cause lysis of cells. Surface analytical techniques such as SEM & EDS, ESEM (environmental scanning electron microscopy), TEM (transmission electron microscopy), CLSM (confocal laser scanning microscopy), EDIC (episcopic differential interference contrast microscopy), piezoelectric technology based on quartz-crystal microbalance, AFM, ATR-FTIR (attenuated total reflection – Fourier transform infrared spectroscopy), XPS, and AES are used for biofilm monitoring. Physico-chemical methods, such as scanning with a vibrating probe, allow the examination of electrochemical conditions on the metal surface when microorganisms are present. Spectral methods (infrared absorbance, optical measurements, and immunofluorescence) have been developed for monitoring microbiological activity.

The question is what is the allowable value of the quantity of microorganisms for using metallic equipment away from MIC danger? It is recommended that the total bacteria count (TBC) in cooling water and fuels must not exceed 10^4 bacteria/ml, and sulphate reducing bacteria (SRB) must not exceed 10 bacteria/ml. Otherwise, severe MIC may occur. It is very important to carry out the trend of the results of measurements. Recommended allowable values of microorganisms depend on a specific system (see Section 2.5).

It is very important to know how biofouling develops in real time. The most widespread method which found wide application in industry, is the method of measuring of the accumulation of a deposit (biofouling) on a metal surface.

5.3.2 Deposit Accumulation Test (Heat Transfer Resistance Method)

This method is based on the principle that microorganisms attaching inside of heat exchanger tubes (if water flows inside) result in a decrease of heat transfer. We have to measure the resistance to heat transfer from the outer surface to the inner surface. It is simple to do this if we measure the inlet and outlet water temperature and flow rate of water in the heat exchanger. Some companies manufacture such devices (for instance, corrosion and deposit monitoring system CorrDATS of Rohrback Cosasco Systems, Inc.) similar to small heat exchangers which allow carrying out the necessary measurements and to calculate the *heat transfer resistance* values (Figure 5.14). These values show the formation of all possible organic and inorganic deposits inside the heat exchanger tube surface: microorganisms, oil and other hydrocarbons, scale of carbonates, phosphates, silicates, corrosion products, silt, dirt, etc. Thus, we do not know the exact contribution of microorganisms in the resistance of heat transfer. Even if we were able to define only the biofouling formation, we would not be able to know which part and what kind of microorganisms is responsible for MIC. In spite of this substantial drawback, the heat transfer resistance method is used in industrial cooling water systems together with microbiological analysis and

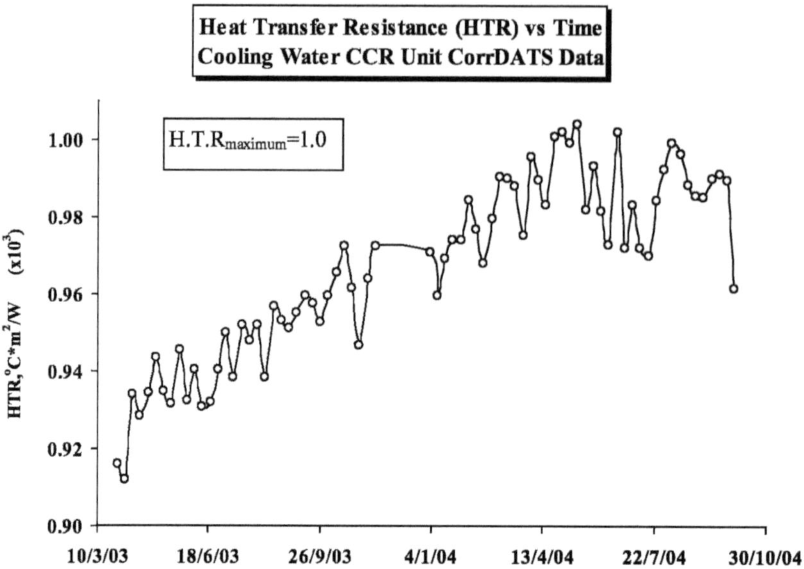

Fig. 5.14 Heat transfer resistance (HTR) versus time (cooling water system).

corrosion monitoring. Certainly, we have to wait for biofouling formation over a long time and we do not know what happens during this period.

Electrochemical methods can be used for on-line monitoring of MIC. These methods are based on the principle that MIC is an electrochemical process and, therefore, influences such electrochemical parameters of corrosion as electric current, electrode potential, and electrical resistance of metals. Electrochemical impedance spectroscopy (EIS), electrochemical noise measurements (ENM), and the scanning vibrating electrode technique (SVET) are used for this purpose. Electrochemical methods give general information about all electrochemical corrosion processes occurring on a metal surface. In other words, the changes of the measured electrochemical parameters occur as a result of all corrosion processes, even those which do not relate to MIC, and it is impossible to differentiate the MIC from other electrochemical corrosion processes.

5.3.3 Hydrogen Monitoring

Some corrosion reactions can produce hydrogen (atomic H and molecular H_2), which can penetrate into metals and alloys, and cause hydrogen damage in the form of blisters and cracks. This hydrogen can be collected either in intrusive or non-intrusive devices called *hydrogen probes*. Intrusive probes (often called "finger" probes) are inserted into a vessel or pipe. They consist of a steel sensing element that has a hollow space inside connected to a pressure device which monitors the build-up (or vacuum loss) of hydrogen pressure. Non-intrusive hydrogen probes utilize an externally applied cell, patch or gauge to monitor the rate of hydrogen egress from the outer surface of a vessel or pipe wall. An electrochemical cell is also used where hydrogen reacts electrochemically producing a current signal. This method is standardized [8]. Hydrogen monitoring is used in wet sour (H_2S) environments (oil, gas, refining, and petrochemical industries).

The benefits of *hydrogen monitoring*:

1. It provides information about potential hydrogen attacks.
2. In some cases, it provides on-line, real-time monitoring of hydrogen permeation.

The limitations of *hydrogen monitoring*:

1. It provides qualitative information.
2. Hydrogen probe data do not usually correlate with other corrosion monitoring methods, for instance, WL method.
3. The hydrogen probes can be subjected to error or delays in reading process transients, since many processes with hydrogen participation can be prone to rapid changes in the severity of hydrogen charging with plant operating conditions.

5.4 Electrochemical Methods

Electrochemical corrosion monitoring methods are based on the established fact that the corrosion of metals and alloys in solutions of electrolytes is an electrochemical process. This means that electrochemical methods are only suitable for corrosion of metals in media with high electrical conductance. Thus, if corrosion reactions have an electrochemical nature, electrode potential, electric current, and electrical resistance (as derivative of the two first parameters) formed on the metal surface are the main characteristics of corrosion. If we measure these three parameters we can characterize and monitor the corrosion of metals in electrical conductive media. Electrode potential of a metal is a thermodynamic characteristic of the corrosion process, and the formula for its definition is $\Delta G = -nFE$ (n – number of moles of electrons involved in the metal dissolution; F – Faraday constant, E – difference of electromotive forces for anodic and cathodic reactions). We have to be careful as we measure the *corrosion potential* and not a *reversible electrode potential* used in this formula (Appendix B). From the electric current one can calculate the weight loss of metal using Faraday's Law (1.37 and 5.14). The corrosion rate of a metal is proportional to the corrosion current. Based on three electrochemical parameters (*electrode potential*, *electric current*, and *electrical resistance*), electrochemical methods may be divided into qualitative and quantitative. Qualitative methods include the identification of corrosiveness of media (determination of *redox potential*) and determination of the *corrosion (electrode) potential* of a corroded metal surface.

Quantitative methods include the identification of *corrosion current* on a metal surface. Any electrochemical method giving rise to the calculation of the *corrosion current* and the resultant corrosion rate may be related to a quantitative method.

The advantages of the *electrochemical methods*:

1. They allow the quickest dynamic information about corrosion rate to be obtained in real time. Therefore, we can follow the fast changes in the corrosiveness of a system including a metal and an electrolytic environment.
2. From the engineering point of view, the electrochemical methods are convenient as they give electric signals allowing a quick reaction and allowing data to be stored in electronic devices.
3. Some electrochemical methods use polarization of metal in the environment. Such methods allow useful information to be received about active and passive regions on the metal surface, mechanism of corrosion and inhibition, as well as a corrosion rate.

The limitations of the *electrochemical methods*:

1. Electrochemical methods are based on the measurement of electric current flowing through solution. Therefore, there are limitations for their usage in a solution with low electrical conductance: in non-electrolytes (organic solvents, fuels), two-phase systems electrolyte – non-electrolyte (with large quantity of the latter), or in very pure water.
2. The corrosion rate measurements assume uniform corrosion, but this is not always correct.
3. If used in multi-phase systems, non-conductive phases may fairly quickly isolate the metal electrodes and invalidate the readings until the electrodes are cleaned again.
4. Non-corrosive electrochemical reactions occurring on a metal surface result in errors. For example, hydrogen sulphide (H_2S), hydrogen peroxide (H_2O_2), or other species in solutions take part in the oxidation-reduction (redox) non-corrosive reactions giving rise to an additive electric current non-referring to a corrosion current.
5. Not all metal surfaces corrode, but calculations relate to a current density (A/cm^2), and to a weight loss of all the metallic surface (g/cm^2). Therefore, we do not obtain a real corrosion rate.
6. So many factors influence the electrochemical processes on a metal surface and they measure so small changes in the electric current and potential that the electrochemical methods are not so reliable as the WL method, and their repeatability is bad. However, the *trend* of electric parameters may give useful information about the corrosion behavior of metals.

The results of the electrochemical methods should be compared with the WL data.

The electrochemical methods use analysis of the interface metal–electrolyte with direct current (DC), alternating current (AC), and without electric current. The measurement of redox potential and electrode (corrosion) potential relate to the technique without electric current. The DC technique is used in the linear polarization resistance (LPR), zero resistance ammeter (ZRA), potentiodynamic polarization, and electrochemical noise measurements (ENM). The AC technique is used in electrochemical impedance spectroscopy (EIS) and harmonic distortion analysis (HDA). Now we shall be familiar with some electrochemical methods which found application both in the laboratory and in industry.

5.4.1 The Measurement of Oxidation/Reduction (Redox) Potential (ORP)

Atoms of different elements may be present in solutions in various oxidized states, for instance, Fe^{2+} and Fe^{3+}, Mn^{2+} and Mn^{4+}, Co^{2+} and Co^{3+}, Cu^{+} and Cu^{2+}, S^{2-} and SO_4^{2-}, O_2 and OH^-, Cl_2 and Cl^-, etc. If such ions and molecules are present in a solution in significant quantities, and when electron exchange with the electrode is sufficiently fast, redox equilibrium is established at some inert electrode (for example, platinum):

$$\mathrm{ox} + \mathrm{ne}^- \leftrightarrow \mathrm{red}, \tag{5.10}$$

giving it a well-defined potential, or reversible *oxidation/reduction potential* (ORP):

$$E = E^{\circ}_{\mathrm{ox/red}} + \frac{RT}{nF}\ln\frac{a_{\mathrm{ox}}}{a_{\mathrm{red}}}, \tag{5.11}$$

$E^{\circ}_{\mathrm{ox/red}}$ – standard electrode potential (at $a_{\mathrm{ox}} = a_{\mathrm{red}} = 1$); a_{ox} and a_{red} – activities of oxidized (ox) and reduced (red) forms; $R = 8.314\ \mathrm{J{\cdot}mol^{-1}{\cdot}K^{-1}}$, gas constant; T – temperature in Kelvin; n – number of mol electrons participating in the redox reaction (5.10); $F = 96{,}500\ \mathrm{C{\cdot}mol^{-1}}$ (Faraday constant).

The *redox potential E* is a voltage which is proportional to the ratio of activities (concentrations) of oxidized to reduced states of specific substances in liquid solutions and is measured on a platinum electrode in relation to some reference electrode (calomel, for example). Atoms of different elements are present in various oxidized states (oxidizers and reducers) in electrolytic media, may take part in cathodic and anodic corrosive reactions, and may influence microbiological activity. *ORP* is a qualitative parameter for an identification of corrosiveness of a medium. It is useful to monitor the *trend* of *ORP* in order to define the changes in the corrosiveness of an environment. The presence of O_2, Fe^{3+}, and other oxidizing agents in a solution results in a deficiency of electrons on a metal surface and will attempt to acquire electrons. Likewise, the presence of H_2S, Fe^{2+}, Mn^{2+}, and other reducing agents in a solution results in a situation where electrons are available and will attempt to give up the electrons. In an oxidizing environment, higher ORP values exist, whereas lower ORP values exist in a more reduced environment. The polarity and value of ORP depend on the concentration of oxidizing and reducing species in the environment.

There are not so many benefits of the *ORP method*:

1. It is very simple, quick, and may be used to control the injection of oxidizing biocides or other additives.
2. It is used to control microbiological activity.
3. It can serve as a rough indication of the corrosiveness of a solution.

There are more limitations of the *ORP method*. The main drawback is that the measured ORP is referred to the buck conditions in the solution and is different from the ORP formed at the surface of equipment. Some ions which are present in various oxidized states do not take part in corrosive anodic and cathodic reactions, but influence the ORP. Deposits (for example, biofilm) may be formed on the platinum electrode, and do not allow some aggressive species (for example, O_2, Cl^-, etc.) to penetrate to the electrode's surface. Such a situation may give rise to errors in the determination of real ORP. Therefore, this method is applicable for corrosion control in boiler feed water (this is very pure water), and microbiological growth in cooling water. This is the only electrochemical method concerned with the determination of the corrosive parameter of the environment. Other electrochemical methods deal with corrosive properties of the border metal–environment.

5.4.2 The Measurement of the Corrosion Potential of Metallic Equipment

Corrosion potential is the electric (electrode) potential of a metal surface formed in contact with an electrolyte solution when an electric current does not flow through a metal. A voltmeter with high inner electrical resistance (above 10^6 Ohm) is used for this purpose. The corrosion potential measured at an open electric circuit is called an *open circuit potential* (OCP) and is used as a *qualitative* characteristic of the corrosion process on a metal surface. The corrosion potential is measured in comparison with reference electrodes such as calomel, silver/silver chloride, or copper/copper sulphate electrodes. This method is similar to the redox potential measurement, but instead an inert platinum electrode, metallic equipment or a metallic specimen similar to equipment metal is used. If the corrosion potential of a metal increases with time (changes in a positive direction), passivation occurs, and corrosion slows down. But pitting corrosion may occur! If the corrosion potential of a metal decreases with time (changes in a negative direction), the corrosion process intensifies. But it is impossible to answer how quickly corrosion may be decreased or intensified. This general rule is not always correct as many factors can shift the potential values towards more positive or negative values, but these shifts may not necessarily be related to the occurring of severe corrosion of metals. For example, a decrease of oxygen concentration at the metal surface causes more negative corrosion potential values, but they may not necessarily be associated with a high probability of steel corrosion. The presence of cathodic (or mixed anodic and cathodic) inhibitors causes the corrosion potential to shift towards a more negative value with a corresponding diminution of severity of steel corrosion. It is used to assist in the prediction

Table 5.3 Probability of corrosion according to half-cell readings.

Half-cell potential reading versus $Cu/CuSO_4$	Corrosion activity
Less negative than −0.200 V	90% probability of no corrosion
Between −0.200 and −0.350 V	An increasing probability of corrosion
More negative than −0.350 V	90% probability of corrosion

of corrosion behavior by comparison with polarization data obtained from a laboratory or site polarization scan (Appendix C). This method is used to control cathodic and anodic protection systems which hold the metal at controlled immune or passive potential respectively; for measurement of the corrosion potential of rebars in concrete (Table 5.3), for monitoring corrosivity of soil, etc. [9].

The measurements of potentials of epoxy (or other paints) coated and galvanized rebars are not suitable. There are the same advantages of *corrosion potential* measurements as for the *redox method*: it is very simple, fast, and can be used for on-line monitoring. There are more drawbacks:

1. Changes in the corrosion potential with time do not allow defining which process, cathodic or anodic, is responsible for the increase or decrease of metallic corrosion.
2. It is impossible to identify the corrosion phenomenon type, such as pitting corrosion, crevice corrosion, etc.
3. The method is qualitative and does not give information about corrosion rates.

Another electrochemical method, linear polarization resistance (LPR), is widely used for the quantitative determination of the corrosion rate of metallic equipment in aqueous solutions of electrolytes.

5.4.3 Linear Polarization Resistance (LPR) Method

This method is based on the measurement of electric current formed between two similar electrodes when the electric potential between them is about 10 to 30 mV. The propensity of the metal ions of an alloy to pass into a solution, or corrode, is inferred by the relationship between a small change in electric potential (typically 10 to 30 mV) across the corrosion interface and the increase in current density, which result from a consequent change in the flow of metallic ions into the solution. If we take two identical electrodes, put into a solution of electrolytes and begin to change the electric potential from −30 mV regarding the open circuit potential ($E_{ocp} = E_{corr}$) of a corroding electrodes with constant rate to +30 mV, a nearly straight curve "electric potential (or polarization) versus electric current density" is obtained (Figure 5.15) [10].

In other words, when the corroding electrode is polarized within a few millivolts around E_{corr}, the measured current density $\Delta \mathbf{i}$ is proportional to the corrosion cur-

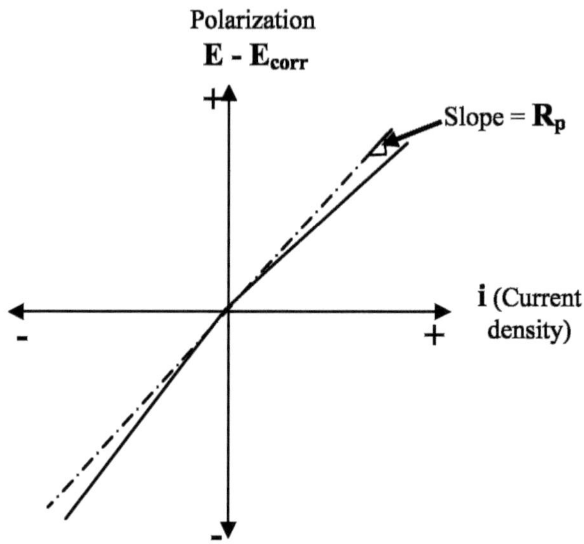

Fig. 5.15 Electric potential (polarization) versus electric current density between two similar electrodes (linear polarization resistance method).

rent density $\mathbf{i}_{corr}$. The ratio of potential change ($\Delta E = E - E_{corr}$) to current density change ($\Delta\mathbf{i}$) at the point of zero current ($i_\circ = 0$) gives the polarization resistance R_p to corrosion reaction:

$$R_p = \left(\frac{\Delta E}{\Delta \mathbf{i}}\right)_{i_\circ=0}. \tag{5.12}$$

Milton Stern, one of the inventors of the LPR method, noted that this is not a "resistance" in the usual sense. Qualitatively, we may emphasize that the greater the polarization resistance, the smaller is the corrosion rate. The theory of this method was developed by Stern and Geary who showed in 1957 that the corrosion current density $\mathbf{i}_{corr}$ (mA/cm^2) can be calculated if we know the polarization resistance R_p:

$$\mathbf{i}_{corr} = \frac{B}{R_p}, \tag{5.13}$$

where B is the Stern–Geary constant (mV) depending on the type of the metal and environment. The principle of this technique has a basis only in the evaluation of steady-state uniform corrosion. The B values are generally taken to be in the range of 26 to 30 mV for most metal-environment systems.

The use of the "polarization resistance" for measuring corrosion rates has one particularly important advantage. The potential range investigated is close to the corrosion potential and the applied currents are generally smaller than the corrosion current. Thus, the nature of the corroded surface is not changed significantly and the reactions which proceed during polarization are those which actually occur during the corrosion.

If the electrical conductivity of a solution is above 100 μS/cm ($R < 10^4$ Ohm·cm), two electrodes are used for LPR measurements. If the electrical conductivity of the solution is between 1 and 100 μS/cm ($10^4 < R < 10^6$ Ohm·cm), three electrodes are used. The third electrode is used for compensation of the IR-drop and for the measurement of real electric potential of the electrode. LPR measurements can be made both manually and automatically to produce data from 5 minutes to 24 hours.

The LPR method is used to determine internal corrosion monitoring of subsea production flowlines, corrosion rate of reinforcement in concrete, nuclear waste tanks, for evaluation of the efficacy of corrosion inhibitors in oil and gas pipelines, for measuring the corrosion rates of outer surfaces of pipes at a dig site and water cooling systems. Guidelines for on-line, in-plant corrosion monitoring using electrochemical techniques are described in [11].

Using Faraday's Law, we can convert the results of electrochemical measurements (i_{corr}, μA/cm^2) to rates of uniform corrosion (CR, mm/yr) [12]:

$$\mathrm{CR} = K \cdot i_{\mathrm{corr}} \cdot \rho^{-1} \cdot \mathrm{EW}, \tag{5.14}$$

CR – corrosion rate, mm/yr; $K = 3.27 \times 10^{-3}$, mm·g/(μA·cm·yr); ρ – density of metal, g/cm^3; i_{corr} – corrosion current, μA/cm^2; EW – equivalent weight; EW $= W/n$ (W is the atomic weight of the metal and n the valence of the metal).

The benefits of the *LPR method*:

1. It allows on-line, real-time measuring of the rate of uniform corrosion, and provides data within a few minutes.
2. The LPR method is particularly well suited to the application of aqueous electrolyte solutions so that process upsets or other corrosion conditions can be detected very quickly to enable remedial action to be taken almost immediately. The method is used for the fast estimation of performance of corrosion inhibitors in a conductive environment.
3. Some systems combine this method with electrochemical current noise measurements to determine the initiation of localized corrosion or as an indication of the stability of the inhibitor film on the metal's surface.

The limitations of the *LPR method*:

1. It is impossible to measure localized corrosion. We shall show later that a two-electrode system is used for the measuring of pitting tendency by means of electrochemical noise measurements.
2. Scale and other deposits formed on the electrode surface give rise to an increase in the electrical resistance non-referring to the change of corrosion rate of the metal.
3. The redox reactions, different from corrosion reactions, contribute to the increase of corrosion current, and can result in errors.

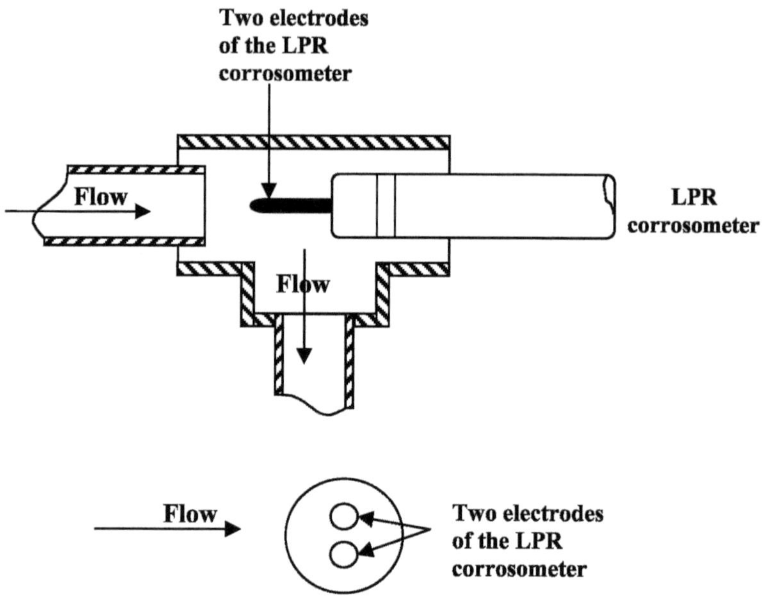

Fig. 5.16 Correct installation of the LPR electrodes.

4. The constant B in (5.13) may vary with time, and this change may result in wrong results. The method requires the prior knowledge of Tafel constants (see Appendix C) for the calculation of corrosion rates.
5. It is very important to correctly mount the electrodes in a flow stream and to prevent the "shading" of one electrode by another (Figure 5.16).
6. The main disadvantage is that the LPR method does not work in solutions of low conductivity, and it is impossible to monitor localized corrosion. Electrochemical noise measurements (ENM) have been introduced into practice in order to overcome the above-mentioned drawbacks.

5.4.4 Electrochemical Noise Measurements (ENM)

Analysis of corrosion failures shows that 70 to 90% of them result from localized corrosion. Data from the oil refining industry show that these values are between 40 to 65%. Ninety per cent of these failures occur in 10% of the time that the equipment is in service. ENM allow monitoring of localized corrosion. First of all, we have to explain what *electrochemical noise* means, a term coined in 1979. In an "ideal" world of corrosion, the electrode potential and current would be constant, metals would corrode uniformly, and all measurements would be easy. In the real world of corrosion, metals tend to corrode non-uniformly and corrosion potentials and currents vary with time. All this is connected with *electrochemical noise*. Everything in the universe is under *noise*. We know that noise in acoustics is any undesired sound. In electronics and information theory, noise refers to those random, unpredictable, and undesirable signals. Noise in radio transmission is known as static, and in television it is called "snow". Noise is everywhere, from atom to planetary orbits. All particle trajectories exhibit chaotic, fractal, and turbulent behavior. Jan Ingenhauz in 1785 watched how the grains of charcoal powder moved randomly on an alcohol film. The Scottish botanist Robert Brown reported in 1827 about the erratic "dancing" of small particles in fluids at rest. Noise is opposite to silence. As it is impossible to reach *absolute zero temperature*, it is impossible to reach *absolute silence*. Noise in many cases is undesired phenomenon. People learned to use noise in some cases. For example, have we ever thought why a child cries? A child cries if something is wrong. If a metal is "ill", a hole or crack is formed on the metal surface or inside, high energy goes out of (leaves) the active region, and we may "hear" the energy flow effects. The acoustic emission method is based on this principle. For example, tear a piece of a paper or break a tree. We will hear some noise – a crash, a crackle. This means that we hear how the bonds between atoms and molecules break. A similar situation occurs in a metallic corrosion. Metal ions leave their places in a crystal and cross into a solution in the anodic process. A metal corrodes, that is some bonds are destroyed and some bonds are formed. Some cations return to a crystal from a solution. Corrosion potential and current (important characteristics of corrosion!) are not constant. Their fluctuations occur. *Electrochemical noise* in corrosion refers to the naturally occurring fluctuations in corrosion potential and corrosion current. Adsorption/desorption processes on a metal surface, pitting, and crack initiation, etc., may be the sources of noise. If corrosion is an electrochemical phenomenon, it occurs according to a random or stochastic process. Electrochemical noise is the proof of the ideas of the Swiss physicist Auguste-Arthur de la Rive (1837) and the English corrosion scientist Uhlick Richardson Evans (1930) about the existence of anodic and cathodic sites on a metallic surface which "travel" all the time from one place to another one. Electrochemical noise may be explained by the uncertainty principle of quantum mechanics. Some uncertainties or fluctuations exist in the positions and velocities of the particles (ions, electrons, and molecules) taking part in the electrochemical process. Up to the 1970s, scientists thought that the sources of the electrochemical noise were bias and error diminishing the reliability of electrochemical results. Therefore, they did not use the electrochemical noise information. Warren P. Iverson was the first who used in 1968 the *electrochemical potential noise* measurements for the analysis of metallic corrosion. The electrochemical noise technique is defined as the analysis of the spontaneous fluctuations of the electric current and potential of a corroding electrode. Two techniques are used for measuring of the electrochemical potential and current noise. In the first technique, two "identical" electrodes are connected through a zero resistance ammeter. Changes of electric potential and current between two "identical" electrodes are measured in regards to the reference electrode with time. It is explicit that it is impossible to create two ideally "identical" electrodes. Therefore, another technique was developed with one working electrode which is connected with a counter electrode (usually made of platinum) and a reference electrode of a potentiostat. This technique is called the *potentiostatic electrochemical noise technique* where both potential and current noises are measured at the same working electrode at an open circuit potential. Current noise relates to current variations between working and counter electrodes, whereas potential noise refers to the variations in potential between a working electrode and a reference electrode. Various software programs were developed for mathematical treatment of measured values of potential and current noise.

The *electrical resistance noise* can be calculated as a derivative of a potential and a current noise according to Ohm's law, and these values are in accord with the linear polarization resistance data. The electrical resistance noise values are very important for analysis of the efficiency of protective coatings on metals.

Both electrochemical noise techniques with two "identical" and one electrode are used for monitoring of both general and localized corrosion, especially pitting and crack initiation. The main benefit of this method is that it is used for on-line measuring of all types of localized corrosion in real time. The theory and practice of ENM is described in [13, 14]. The pitting index (PI), or pitting factor (PF), is calculated according to

$$\mathrm{PI} = \frac{\sigma_I}{I_{\mathrm{corr}}}, \tag{5.15}$$

where σ_I is the current recorded from ECN (Electrochemical Current Noise) (mA); I_{corr} is the corrosion current recorded by LPR (mA) or harmonic analysis (see Section 5.4.7). Mixed quantitative-qualitative assessment is used:

a) $\mathrm{PI} < 0.01$ (general corrosion);
b) $0.01 < \mathrm{PI} < 0.1$ (intermediate zone, but still predominantly general corrosion);
c) $\mathrm{PI} > 0.1$ (localized corrosion).

The industrial devices are built on this principle for *pitting tendency* (PT) corrosion monitoring, i.e., transform raw electrochemical noise data into a process parameter that can be understood and used by operation personnel just like a temperature or pressure reading (Figure 5.17). ENM are unique among other electrochemical monitoring techniques in that they can be used to detect and distinguish between different localized corrosion phenomena such as pitting, crevice corrosion, and stress corrosion cracking. EN probes were used for measuring the pitting tendency in tanks containing high-level radioactive waste, sour oil processing facilities, and in cooling water systems [9].

5.4.5 Zero Resistance Ammetry (ZRA)

ZRA is a current-to-voltage converter, which provides a voltage that is proportional to the current flowing between the input terminals while imposing a "zero" voltage drop to the external circuit. In the ZRA technique, a macro cell current is measured between two corroded sensor elements in an electrolytic environment. It is

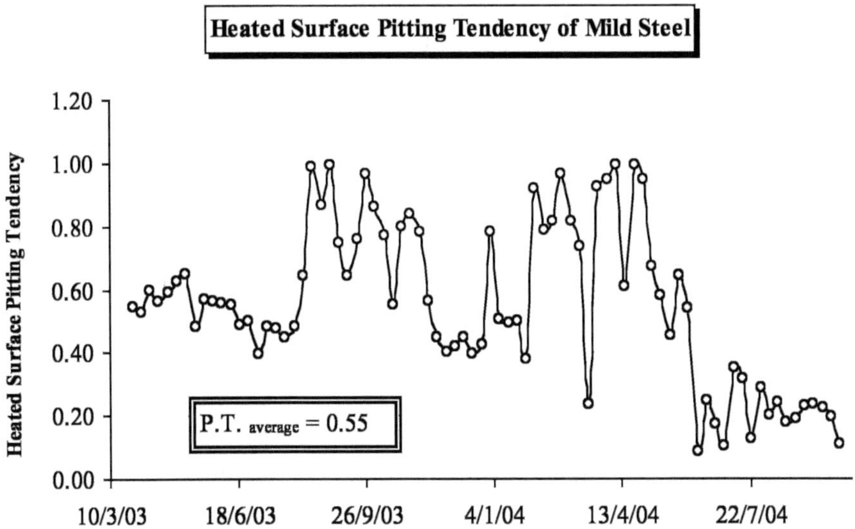

Fig. 5.17 Pitting tendency (PT) of heated mild steel surface versus time in cooling water (measured by means of the CorrDATS system of Rohrback Cosasco).

used for measuring the galvanic coupling current between two dissimilar or similar ("identical") electrodes. The resultant galvanic current is a corrosion rate value which can be obtained through Faraday's Law (1.37). This method is used also for measuring the coupling current between two "identical" electrodes to monitor electrochemical current noise. In reality, these electrodes will be slightly different and a small coupling current will exist. Such measurements are relevant in detecting the breakdown of passivity and the early stages of corrosion. The restriction is, if extensive corrosion is occurring on both electrodes, that the macro cell current measured will not accurately reflect the severity of attack.

5.4.6 Electrochemical Impedance Spectroscopy (EIS)

Everyone knows about the concept of electrical resistance – the ability of a circuit element to resist the flow of electric current, known as Ohm's law (see (1.34)). This relationship is restricted by an ideal resistor which has the following properties: it follows Ohm's law at all current and voltage values; its resistance value is independent of frequency; an alternating current (AC) and voltage signals through a resistor are in phase with each other. The real world contains circuit elements that exhibit much more complex behavior. Instead of resistance we use *impedance*, which is a more general circuit parameter. Like resistance, impedance is a measure of the ability of a circuit to resist the flow of electric current. Unlike resistance, it is not limited by the simplifying properties listed above.

Electrochemical impedance is the frequency-dependent, complex-valued proportionality factor, $\Delta E/\Delta I$, between the applied potential (or current) and the response current (or potential) in an electrochemical cell. *Electrochemical Impedance Spectroscopy* (EIS, also known as *AC impedance*, or the *electrochemical impedance technique*) is an electrochemical technique in which a low amplitude alternating potential (or current) wave is imposed on top of a potential (usually the corrosion potential and zero imposed current). In other words, EIS monitors the electric response of the border metal-environment to the applied AC signal over a frequency spectrum. Electrochemical impedance is normally measured using a small excitation signal. The relationship between the voltage and current is used to make a judgement about the rate of corrosion. The theory of the EIS based on generation of impedance spectra is described in [13, 15]. The goal of the EIS technique is to measure the impedance as a function of frequency by proper procedures and then analyze the resulting spectrum to estimate the corrosion rates and the mechanism that might give rise to the spectra. The impedance is affected by the interaction between frequency and all of the physical and chemical processes that respond to that frequency change within the electrochemical cell and across the corroding interface. In the limit of zero frequency, the impedance approaches the direct current (DC) resistance of the corroding system. Thus, the low frequency limit of EIS is the same as the LPR method.

The benefits of the *EIS technique*:

1. Rapid estimation of general corrosion rates (0.5 to 24 hours after immersion).
2. Estimation of corrosion rates in slowly changing systems (as long as stability is not compromised).
3. Estimation of low corrosion rates (less than 0.01 mpy).
4. Corrosion rate estimation in low conductive media (organic liquids, soil, and concrete), for evaluation of cathodic protection efficiency.
5. Rapid assessment of corrosion inhibitors. The EIS technique can be more effective than the LPR method.
6. Evaluation corrosion under coatings and quality assurance of coatings (pore resistance and film capacitance).

The shortcomings of the *EIS technique*:

1. Measuring an EIS spectrum takes time (usually many hours).
2. The system (metal–electrolyte) must be at a steady state. Otherwise, inaccurate results will be received. In practice, steady state can be difficult to achieve as the system can change through adsorption of solution impurities, growth of an oxide layer, build-up of reaction products in solution, temperature changes, coating degradation, etc.
3. The analysis of the EIS data is relatively complex compared to the commonly used ER or LPR techniques. The EIS method requires benchmarking with other corrosion monitoring techniques.

5.4.7 Harmonic Distortion Analysis (HDA), or Harmonic Analysis (HA)

This method is based on measuring the fundamental, second, and third harmonic currents from the current response at corrosion potential by perturbing a corroding system with a non-distorted sinusoidal signal of low amplitude. HDA is a low-frequency impedance measurement and, similar to LPR, relies on a steady-state approximation, but requires more mathematical treatment than the LPR technique. The current response to a low frequency voltage sine wave is distorted due to the non-linearities of the charge transfer process. This distortion is analyzed in terms of the higher harmonics, to provide values for corrosion current, the Tafel constants, and hence the Stern–Geary constant B. This technique relates to the EIS in that an alternating potential perturbation is applied to one sensor in a three element probe, with a resultant current response. All kinetic parameters (including the Tafel constants) can be calculated explicitly. No other technique offers this facility. HDA is based on an evolution and improving the performance of the LPR technique. By applying a low frequency sine wave to the measurement current, the resistance of the corrosive solution can be calculated through a harmonic analysis of the resulting signals. With both the polarization resistance and the solution resistance, a more accurate general corrosion rate can be determined. Both EIS and HDA are used for measuring corrosion rates of metals in acids, neutral solutions of electrolytes, and concrete. The advantage of HDA is that mathematical data treatment facilitates direct computation of the Tafel constants and the corrosion rate. The Stern–Geary constant B is obtained by HDA and is stored in the transmitter.

The restrictions of *EIS and HDA*:

1. As in the LPR technique, the assumption of uniform corrosion has to be made.
2. If localized corrosion is taking place, the data are of a qualitative nature, indicating the breakdown of passivity and the possibility of localized attack.

5.5 Monitoring of the Corrosiveness of the Atmosphere

In many cases we must know the corrosiveness of the atmosphere at the plant or in rooms with electronic and electric equipment. Many aggressive variables, such as sulphur oxides, nitrogen oxides, chlorine, ammonia, hydrogen sulphide, ozone, water vapors, salts, and dust, contained in air, influence corrosion of metals in the atmosphere. Failure may occur because of corrosion of electric contacts made of gold, silver, copper, zinc, or their alloys. The WL and ER methods are used for monitoring the corrosiveness of the atmosphere. The WL method allows the classifying of the corrosive level of atmospheres (see Table 3.1). The ER method allows on-line, real-time monitoring and to learn the corrosiveness of the atmosphere. Sometimes the film thickness of corrosion products formed on a copper surface corresponds to

Table 5.4 Classification of the corrosiveness of atmosphere according to the thickness of corrosion products on copper surface (30 days' exposure).

Film thickness of corrosion products on copper, Angströms	Corrosiveness of the atmosphere
<300	Mild
300 to 1,000	Moderate
1,000 to 2,000	Harsh
$>2,000$	Severe

the corrosiveness of the atmosphere (Table 5.4). Electrochemical methods are used for measuring the thickness of film formed on copper surfaces.

A "metal piece test kit" was developed [16], in which several metal strips are placed in the atmosphere for a month. Corrosion products are defined by X-ray fluorescence (XRF), and corrosiveness of the atmosphere is determined according to the presence of sulphur, chloride, and oxygen in corrosion products.

5.6 On-Line, Real-Time Corrosion Monitoring in Industrial Systems

The ER and LPR methods are used for on-line, corrosion monitoring of metallic equipment in plants. The LPR technique entered into use for corrosion monitoring in industry in the 1960s. The problem is that the constants B in equation (5.13) are generally taken to be in the range 26 to 30 mV for most metal-environment systems, and are regarded by most manufactures of LPR instrumentation to be a constant that is configured into an instrument at the factory. The constant B, however, is not constant. Change from 1 to more than 100 mV for various systems, can vary from time to time and even within the same system. Therefore, for accurate corrosion rate measurements, variable constants B should be used. The HDA technique overcomes this limitation and allows the determination of the Stern–Geary B constant. Usually an average constant B is analyzed with time, and then the LPR corrosion rate is post-corrected. Good correlation was shown between the constant B corrected LPR corrosion rate, the WL method, and ultrasonic thickness measurements.

On-line, real-time corrosion monitoring must be integrated with the process control system, i.e., we should monitor technological parameters together with corrosion parameters. Corrosion monitoring allows rapid determination of changes in water or other process stream quality, environmental parameters (temperature, pressure, flow rate, chemical feed rate), and inhibitor performance. The pertinent operating and analytical data are entered into the risk matrix analysis where the data are used to determine the relative risk and the consequence of the corrosion. Corrosion monitoring technology can determine uniform and localized (pitting, crack) corrosion, even when the uniform corrosion rate is low. The CorrDATS system manufactured by Rohrback Cosasco works well for monitoring uniform corrosion, pitting tendency, and heat transfer resistance (criterion of biofouling and other deposits) in water cooling systems.

Three electrochemical methods (LPR, HDA and ENM) are realized by SmartCET Technology (Honeywell International Inc.) and are used for on-line, real-time corrosion monitoring of general and localized corrosion in the oil refining industry, influence microbiological activity in district heating systems, in water injection systems, and suitable optimization of corrosion inhibitors selection, in hot organic streams with small concentrations of water, in multiphase oil-water systems with corrosive gases in liquid and vapor condensing phases, and in natural gas transmission pipelines. Measurements using the three electrochemical techniques only take a few minutes.

The Resistance Corrosion Monitoring (RCM) device was developed for continuous monitoring of pipe-wall thickness in high temperature areas where corrosion is suspected [17]. The RCM operates on the same underlying principal as traditional ER probes except it utilizes the pipe wall as the active sensor element and it provides much more accurate and precise data. The RCM is an array of pins welded directly onto the pipe, covering approximately 1 m^2 area of pipe to be monitored.

The Field Signature Method (FSM) developed by CorrOcean (Norway) is used on subsea pipelines, nuclear power plants, and storage tanks. The FSM method is based on feeding an electric direct current through the selected sections of the structure to be monitored and sensing the pattern of the electrical field by measuring small potential differences set up on the surface of the monitored object. Proper interpretation of these potential differences can lead to a conclusion about the reduction of wall thickness. The FSM provides online information on wall thickness loss, erosion, pitting, or cracking [18].

The *multi-electrode array sensor* probe was developed for monitoring corrosion rates of carbon steels, stainless steels, Ni-Cr, Ni-Cr-Mo, and Cu-Ni alloys in both the liquid and vapor phase till temperature 300°C and pressure 140 atm [19]. This probe allows the measurement of the corrosion rate from 10^{-5} mm/yr (stainless steels in passive conditions) to localized penetration rates for these alloys up to 10 mm/yr. The *multi-electrode array sensor* probe consists of a bundle of metal electrodes, all insulating each other, but connected through a network of resistors. This probe was used for monitoring general and localized corrosion, and coating integrity.

The Concerto MK II system developed by CAPSIS (UK), is used for monitoring localized corrosion in nuclear waste tanks and sour gas pipelines.

We should take into consideration that corrosion monitoring is a very dynamic field in which new companies and devices appear every year.

5.7 Conclusions

Corrosion monitoring is an important part of any plant anti-corrosion management, including corrosion control programs. Comparison of properties and possibilities of

Table 5.5 Comparison of possibilities of corrosion monitoring methods used in industry.

	Corrosion Monitoring Method					
Corrosion phenomenon	Ultrasonic	Coupons (weight loss)	ER probes	LPR	Chemical analytical methods	ENM + HDA
General (uniform) corrosion	Excellent	Excellent	Excellent	Good	Good	Good
Pitting corrosion	Fair	Excellent	NA	NA	NA	Excellent
Galvanic corrosion	NA	Excellent	NA	NA	NA	Good
MIC	Fair	Good	NA	NA	NA	Good
Erosion-corrosion	Good	Excellent	Excellent	NA	NA	Excellent
SCC	Fair	Good	NA	NA	NA	Excellent
Intergranular corrosion	NA	Good	NA	NA	NA	Excellent
Hydrogen-induced corrosion	Fair	Fair	NA	NA	NA[a]	Excellent
Dealloying	Fair	Excellent	NA	NA	Good	Good
Crevice corrosion and under deposit corrosion	NA	Good	NA	NA	NA	Excellent

Notes: ER – Electrical Resistance; LPR – Linear Polarization Resistance; ENM – Electrochemical Noise Measurements; HDA – Harmonic Distortion Analysis; MIC – Microbiologically Induced Corrosion; SCC – Stress Corrosion Cracking; NA – not applicable.
[a]Good for H_2 determination.

some corrosion monitoring methods appropriate for the environmental conditions and for identification of particular corrosion phenomena are shown in Table 5.5. Why are there so many methods of corrosion monitoring? To follow up corrosion is more complicated than to monitor such processing parameters as temperature, pressure, flow rate of streams, and concentrations of reagents because:

(a) There are many corrosion phenomena. One can see from Table 5.5 that not all corrosion monitoring methods are suitable for monitor all corrosion phenomena.
(b) Corrosion may be uniform or localized. Specific methods are suitable for monitoring localized corrosion.

Corrosion monitoring methods must be selected and carried out at any stage of the project of any plant or unit at the plant. It is very important to select the correct corrosion monitoring type and inspection points. Ultrasonic thickness measurements must be taken in the vicinity of the corrosion monitoring points because, even for uniform corrosion the corrosion rate may be significantly changed at different distances. The more places we monitor, the more complete and real corrosion situation may be obtained.

Corrosion probes or their sensing elements should be replaced when they reach 90 to 95% of their nominal life. They must be changed immediately if localized corrosion has occurred. In some cases, the sensing elements need to be cleaned if the data received appears questionable.

Use of the only corrosion monitoring method gives some, but not full information about corrosion situation. The more methods are used, the more we cover a corrosion situation and, as a result, improve production, injection of chemicals, and technological parameters. It is important to use automated multi-technique monitoring capabilities. Corrosion monitoring should be integrated with other plant programs designed to optimize the process conditions, injection of chemicals, and inspection. Only in this way can we successfully manage plant operation.

Recommended Literature for This Chapter

1. Rothwell, N. and Tullmin, M., *The Corrosion Monitoring Handbook*, First Edition, Coxmoor Publishing Company, UK, 2000, 180 pp.
2. Shreir, L.L., Jarman, R.A. and Burstein, G.T., *Corrosion, Vol. 2*, Third Edition, Butterworth Heinemann, UK, 1994, pp. 19:154–19:178.
3. Roberge, P.R., *Handbook of Corrosion Engineering*, McGraw-Hill, New York, 2000, pp. 371–483.
4. Kane, R.D., Selection of Corrosion Monitoring Methods for Refinery Application. In *Corrosion in the Oil Refining Industry*, September 26–27, 1996, Houston, Texas, NACE International, USA, 1996, pp. 14/1–14/17.
5. Chattoraj, M., Fehr, M.J., Hatch, S.R. and Allain, E.J., Online Measurement and Control of Microbiological Activity in Industrial Water Systems, *Materials Performance* **41**(4), 2002, pp. 40–45.
6. Boffardi, B.P., Water Treatments. In: *Corrosion Tests and Standards: Application and Interpretation*, R. Baboian (Ed.), ASTM Manual Series, MNL 20, ASTM, USA, 1995, pp. 697–704.
7. Kane, R.D., Eden, D.C. and Eden, D.A., Real-Time Solutions Integrate Corrosion Monitoring with Process Control, *Materials Performance* **44**(2), 2005, pp. 36–41.
8. Kane, R.D. and Trillo, E., Evaluation of Multiphase Environments for General and Localized Corrosion, CORROSION/2004, Paper No. 04656, NACE International, Houston, TX, USA, 2004.
9. Veazey, M.V., A Paradigm Shift for Process Control? *Materials Performance*, **43**(12), 2004, pp. 16–19.
10. Groysman, A. and Brodsky, N., Corrosion and Quality, Accreditation and Quality Assurance, *Journal for Quality, Comparability and Reliability in Chemical Measurement* **10**(10), 2006, pp. 537–542.

References

1. Annual Book of ASTM Standards, Water and Environmental Technology, Vol.11.01, ASTM, USA, 2008, 1082 pp.
2. Standard Methods for the Examination of Water and Wastewater, 21st Edition, Public Health Association, American Water Works Association & Water Environment Federation, USA, 2005, 1368 pp.
3. NACE Standard RP0192-98, Monitoring Corrosion in Oil & Gas Product with Iron Counts, NACE International, USA, 13 pp.
4. ASTM G51-95(2005), Standard Test Method for Measuring pH of Soil for Use in Corrosion Testing, Annual Book of ASTM Standards, Vol. 03.02, 2006, 3 pp.
5. ASTM G57-06, Standard Test Method for Field Measurement of Soil Resistivity Using the Wenner Four-Electrode Method, Annual Book of ASTM Standards, Vol. 03.02, 2006, 5 pp.

6. ASTM G187-05, Test Method for Measurement of Soil Resistivity Using the Two-Electrode Soil Box Method, Annual Book of ASTM Standards, Vol. 03.02, 2006, 5 pp.
7. Annual Book of ASTM Standards, Pesticides: Environmental Assessment; Hazardous Substances and Oil Spill Responses, Vol. 11.05, ASTM, USA, 2008, 1574 pp.
8. ASTM G148-97(2003), Standard Practice for Evaluation of Hydrogen Uptake, Permeation, and Transport in Metals by Electrochemical Technique, 2003, 10 pp.
9. Groysman, A., Corrosion Monitoring in Industry. In: *Corrosion Science and Technology*, U. Kamachi Mudali and Baldev Raj (Eds.), Narosa Publishing House, New Delhi, India, 2008, pp. 500–550.
10. ASTM G59-97(2003), Standard Test Method for Conducting Potentiodynamic Polarization Resistance Measurements, Annual Book of ASTM Standards, Vol. 03.02, 2006, 4 pp.
11. ASTM G96-90(2008), Standard Guide for On-Line Monitoring of Corrosion in Plant Equipment (Electrical and Electrochemical Methods), Annual Book of ASTM Standards, Vol. 03.02, 2008, 10 pp.
12. ASTM G102-89(2004), Standard Practice for Calculation of Corrosion Rates and Related Information from Electrochemical Measurements, Annual Book of ASTM Standards, Vol. 03.02, 2006, 7 pp.
13. Cottis, R. and Turgoose, S., *Electrochemical Impedance and Noise*, NACE International, USA, 1999, 149 pp.
14. Kearns, J.R., Scully, J.R., Roberge, P.R., Reichert, D.L. and Dawson, J.L. (Eds.), *Electrochemical Noise Measurement for Corrosion Applications*, STP 1277, ASTM, USA, 1996, 476 pp.
15. Scully, J.R., Silverman, D.C. and Kendig, M.W. (Eds.), *Electrochemical Impedance: Analysis and Interpretation*, ASTM, USA, 1993.
16. Iikawa, T., Udo, Y. and Nakajima, E., Assessment of Specific Environments by Metal Piece Test Kits, *Corrosion Science* **35**(1–4), 1993, pp. 735–742.
17. Winslow, M.C., Wilson, M., Lasiuk, B., Allison, P. and Cross, C., Solutions for Processing Opportunity Crudes. In *ERTC (European Refining Technology Conference)*, 10th Annual Meeting, Vienna, Austria, November 2005, 14 pp.
18. Strommen, R.D., Horn, H. and Wold, K.R., New Technique Monitors Pipeline Corrosion, Cracking, *Oil & Gas Journal*, December 27, 1993, pp. 88–92.
19. Yang, L. and Sridhar, N., Coupled Multielectrode Online Corrosion Sensor, *Materials Performance* **42**(9), 2003, pp. 48–52.

Chapter 6
Humanitarian Aspects of Corrosion Science and Technology

People usually perceive corrosion as a nasty phenomenon with which they must cope. The question is how to explain corrosion to everybody? Things that we deal are more interesting and attractive if we look at them from the historical, philosophical, and artistic aspects. Unfortunately, a similar approach to corrosion science and technology has not been available. In the present chapter I try to find the relationship between philosophy, art, and corrosion phenomena. We cannot forget also the historical aspects. Many scientists and engineers contributed to the development of corrosion science and technology, and the younger generation has to know about this. Let me show you the *beauty* of the corrosion world we deal with; to show that we are able to *enjoy* such nasty phenomena as corrosion. We shall discuss the historical, philosophical, and artistic aspects of corrosion.

6.1 History of the Evolution of Knowledge about Corrosion

We have to note that art, science, and technology developed together without any differentiation. This happened only during the last 200 years. The present investigation is based on various historical documents and represents my own view on the history of the evolution of knowledge about corrosion from prehistoric and ancient times to the present. I believe that it is impossible to live and to exist without any analysis of the Past, and without any plans for the Future. Why do we have to know about the history of the evolution of knowledge about corrosion?

(a) Knowledge of the history of the evolution of the studying of corrosion gives an interdisciplinary view of this subject and connects corrosion science and engineering with the humanitarian principles which people have to know.
(b) Predicting new developments. It is obviously understood if we compare achievements in the Past with the carrying out of any research. We have to stop at some stage in every research project and analyze the results obtained. The historical approach helps us to understand what is lacking in the knowledge about cor-

A. Groysman, *Corrosion for Everybody*, DOI 10.1007/978-90-481-3477-9_6,
© Springer Science+Business Media B.V. 2010

rosion. Such an approach helps us to outline the ways for new solutions to the problems concerned with corrosion.

(c) We have to honor and to remember the scientists and engineers who contributed to the evolution of corrosion knowledge.

I have tried to choose the events, subjects, and objects which had a considerable influence on the evolution of our knowledge about corrosion. The more we move away from the events, the more we are able to evaluate and to choose what was important, who was right, and which events had significant influence on the evolution and development of corrosion science and technology. Tolerance in history is wider than in corrosion science, because history is based on subjective descriptions. Subjective approaches exist in corrosion science too. For example, the existence of various explanations of passivation and dezincification say that we do not know the exact mechanism which occurs at the metal-environment border. But ... there is quantitative estimation in corrosion science, which will help to define an exact mechanism in the future. There is no similar quantitative appraisal in history. History is based on the literature and archeological findings. Historical records are not always indubitable. Sometimes scientific and technological records were destroyed, intentionally or non-intentionally, in ancient times. For example, the Roman emperor Diocletian passed a decree about 290 A.D. compelling the destruction of the works on the alchemical arts and on gold and silver throughout the empire, so that the makers of gold and silver should not be able to amass riches which might enable them to organize revolts against the empire. This decree resulted in the disappearance of literature about the early history of the chemical arts.

We shall classify the evolution of corrosion knowledge with time, but it is worth remembering that all things are relative in the world. The history of every subject is divided into several periods. The evolution of art in the history of mankind is often divided into some periods, for example, Ancient, Renaissance, Baroque, Classic, Realism, and Modern periods. I also divided the history of the evolution of the knowledge about corrosion into several periods.

6.1.1 Ancient Times

We shall follow in our description and investigation of the evolution of the knowledge about corrosion according to the definition of *corrosion* as a *phenomenon of destruction of metal due to its reaction with the environment*. An acquaintance with the susceptibility of noble and non-noble metals to react with the environment is as old as the availability of metals to man. The name *metal* came from the Greek word *metallon* which means "mines, pits".

There are above 80 metals among all elements existing in Mendeleev's Periodic Table. Salts, oxides, and hydroxides are found in nature due to the corrosion of metals: reaction with the environment. The pigments from the tomb of Perneb in Egypt (5th Dynasty, about 2400 B.C.) contain a red iron oxide (haematite), a yellow ochre containing clay with iron, a pale blue (copper sulphate carbonate), and

Fig. 6.1 Left – Gold mask of Pharaoh Tutankhamun (14th century B.C.); Right – silver coins (Israel, 69 A.D.). (For a full color version of this figure, see the Color Section)

a green malachite (copper carbonate hydroxide). We have the same question as in the example with the egg and the chicken. Which was the first? We do not know which was first in nature: metal as element or its ions because of metal's corrosion in various media. History shows that the first seven metals known to man were gold ~6000 B.C., copper ~4200 B.C., silver ~4000 B.C., lead ~3500 B.C., tin ~1750 B.C., iron ~1500 B.C., mercury ~750 B.C. This means that the pristine man knew that noble and non-noble metals are prone to corrosion.

Gold and silver ornaments and decorations have been found in royal tombs dating back as far as 4000 B.C. Both gold and silver were used as money by 800 B.C. in all countries between the Indus and the Nile (Figure 6.1).

We may read about these metals in one of the most ancient Jewish songs in the Bible (Numbers, Ch. 31, V. 2, before 1000 B.C.): "… but gold, and silver, and copper, and iron, and lead …". This description in the Bible is the first written proof about the use of five metals in ancient times. Needless to say, we read a lot about these metals in the five books of Moses (Bible): 387 times – about gold, 403 – about silver, 133 – about copper, 87 – about iron, 5 – about tin, and 1 – about lead. Undoubtedly, primitive man observed the changes of the surface of copper, silver, and iron due to a reaction with the environment. Hydrogen sulphide, sulphur oxides, carbon dioxide, oxygen, ozone, water, microorganisms, and other corrosives were always present in sufficient amounts on the Earth. We know that contaminants are divided into natural ones which have always existed and anthropogenic ones which existed only from the period when people began industrial processing. Primitive man encountered corrosion of metals, but was not able to explain this phenomenon because the evolution of knowledge about corrosion was related to the evolution of chemistry and metallurgy. Therefore, we cannot understand the evolution of corrosion science and engineering without following in parallel with the development of our knowledge of chemistry and metallurgy.

The above-mentioned seven metals were familiar as metals of Antiquity in Mesopotamia, Egypt, Greece, and Rome. Ancient Greeks and archeologists divided general history into several periods, and there were metal periods among them: periods of gold, silver, bronze, and iron. We may see that these historical "metallic" epochs changed with the "diminution" of their "nobility". This means that in the historical approach of their evolution people used fewer and fewer noble metals and alloys to those which were more prone to corrosion. Why? The non-noble metals (for example, iron) were present in greater amounts on the Earth.

6.1.1.1 Copper

Copper occurs in native form and was probably the second metal (after gold) discovered and used, but copper was probably the first metal for which man encountered corrosion. Various fashioning tools, utensils, jewelry and weapons were made of copper in the Chalcolithic period lasting from the 5th millenium B.C. in southeastern Europe. We have to mention that there is no clear division of these periods in time. Some sources point out that the Chalcolithic age lasted between 4500–4000 and 1500 B.C. Probably, this depends on the geographical region. The name "chalcolithic" reflects the use of copper, because *chalcos* means copper and *lithos* means stone (translation from the Greek). The use of copper dates from at least 4000 B.C. in Chaldea (part of Babylonia), and perhaps earlier. The native copper was first used as a substitute for stone even by Neolithic man. The Metal Age in Europe started in the early 3rd millennium B.C.

Probably, the first native copper was available near the Earth's surface, easy to find and was used as utensils and ceremonial objects. Some of the proof for such a suggestion includes a copper pendant found in a cave in northeast Iraq and various copper objects found in Egypt (about 3000 B.C.). Another proof of the use of copper was the Qumran manuscripts (scrolls) which were written not only on skin, parchment (oil-paper), and papyrus, but also on copper, and this was in the 8th–3rd century B.C. We may find such proof in the famous poem "De rerum natura" ("On the Nature of Things") by the Latin poet and philosopher Titus Lucretius Carus who wrote in the 1st century B.C.: "after wood, and stone, copper was found". Some pure virgin copper (weight 420 ton) was found not long ago, in 1857, in America in the Great Lakes region. Such virgin copper was usually covered by green or blue salts – *copper corrosion products*. Copper when fractured had a beautiful yellow-red color. Undoubtedly, people who used this native copper wondered why copper was yellow-red inside and outside had a green-blue color.

When native copper was nearly exhausted, attention turned to copper locked up in ores (minerals chalcopyrite, or copper pyrites, that can serve as a source of pure metals). Malachite salts were the source for the production of copper. According to historians, the Egyptians mined copper on the Sinai Peninsula as long ago as 3000 B.C. Copper axes were manufactured by casting in the Balkans in the 4th millennium B.C. The ancient Romans extracted copper ores on Cyprus. This island was given the name of *aes Cyprium* (metal of Cyprus), shortened to *cyprium*, and later corrupted to *cuprum* (from the Latin, copper).

The ancient metallurgists possibly observed that when red-hot copper was plunged quickly into cold water, a bright red copper (I) oxide (Cu_2O) film was formed. This film was otherwise a *copper corrosion product*! If slow heating of a piece of copper occurred in air, a black copper (II) oxide (CuO) film (also a *copper corrosion product*!) was formed.

Man was able to observe the reaction between copper, oxygen, water, and carbon dioxide, forming green malachite salt – a corrosion product of copper:

$$2Cu_{(s)} + O_{2(g)} + H_2O_{(l)} + CO_{2(g)} \rightarrow Cu(OH)_2 \cdot CuCO_{3(s)}. \tag{6.1}$$

If sulphur dioxide (SO_2) was present in the air, the reaction occurred with the formation of blue copper corrosion products:

$$8Cu_{(s)} + 5O_{2(g)} + 6H_2O_{(l)} + 2SO_{2(g)} \rightarrow 2[CuSO_4 \cdot 3Cu(OH)_2]_{(s)}. \tag{6.2}$$

These reactions reflected the corrosion of copper in nature, and everyone could observe copper corrosion products on antique coins, utensils, jewelry, weapons, tools, and then on the roofs, statues, and sculptures.

Mankind's existence was related with animal husbandry and agriculture. Therefore, ammonia was always present in the environment; copper objects encountered it and corroded according to the reaction:

$$2Cu_{(s)} + 8NH_{3(g)} + O_{2(g)} + 2H_2O_{(l)} \rightarrow 2[Cu(NH_3)_4](OH)_{2(aq)}. \tag{6.3}$$

Undoubtedly, microorganisms excreted hydrogen sulphide (H_2S), and man was able to observe the black corrosion products copper sulphide (CuS) and silver sulphide (Ag_2S). Ozone in the air also caused the tarnishing of metallic adornments and other articles used by man because of their corrosion. Corrosion of tools, implements, weapons, and artwork made of copper (and then of bronze and iron) encouraged manufacturing and trade in new metals and alloys. Corrosion of copper helped to solve biofouling on the surfaces of ships. Because of copper's biocidal properties, copper and copper compounds have been used for centuries for the protection of the ships from biofouling. In early wooden ships, a sheathing of metallic copper was used to protect the timbers from being invaded by shipworms. Copper-sheathed ships were protected from the growth of barnacles and algae that increased the frictional resistance of the water against the hull and slowed the speed of ships. Probably, the high speed of Admiral H.V. Nelson's copper-sheathed warships contributed to the British victory over the French and Spanish fleets during the battle at Cape Trafalgar (South Spain).

6.1.1.2 Bronze ("Alloying") Age

A "chalcolithic" man used the pure metals for cooking utensils, weapons, artwork, and jewelry. Pure metals were usually too soft to use for such applications as armour, tools, etc. A "chalcolithic" man discovered (accidentally or experimentally) that the addition of another metal to copper made the material harder. Some historical documents (for example, Leyden Papyrus, the 3rd century A.D.) show that ancient people knew not only bronze (copper alloyed with tin) but other alloys, such as copper with arsenic. Artifacts from 3000 B.C. (Egypt) contain from 1 to 7% arsenic in alloys with copper, sometimes with up to 3% tin. These alloys were used for only a short period as fumes containing arsenic killed many metalsmiths. Only in the 20th century was arsenic added in very small amounts (0.02–0.1%) to copper alloys and fair resistance to dezincification was revealed.

We mentioned that copper was extracted on the Sinai Peninsula about 3000 B.C., although some bronze is dated as early as 3700–3500 B.C. Probably, bronze appeared between 3000 and 2500 B.C., beginning in the Tigris–Euphrates delta, but was rare until extensive trade in tin developed following the discovery of large tin deposits (England and Spain).

An axe consisting of 11% tin and 89% copper dating from 2500 B.C was found in Sumeria. Probably, people sometimes dealt with ores containing tin. Bronze had a melting temperature less than copper and was harder and less brittle than copper–arsenic alloys. The Bronze Age is estimated to be between 2000 to 1000 B.C., until the development of iron in about 1000 B.C. While there may have been some independent development of bronze in varying localities, it is most likely that the bronze culture spread through trade and the migration of peoples from the Middle East to Egypt, Europe, and possibly China. In many civilizations, the production of copper, arsenical copper, and tin bronze continued together for some time. Bronze was the predominant material in 1400 B.C. The "Bronze Age" was marked by the discovery of alloying (the addition of some elements to the main metal with the aim of improving some properties of the base metal) and manufacture of alloys (materials having metallic properties and composed of two or more chemical elements of which at least one is a metal). The development of towns began in this period. There are many archeological findings related to bronze objects in Egypt (2nd millennium B.C.), Israel, Cyprus, Syria, Anatolia, Mesopotamia, Greece, Rome and China (6th century B.C.): ritual vessels, helmets, bells, stamps, statues, pendants, etc.

"Chalcolithic" people who used bronze did not understand and did not see its huge future in industry (gun metal, naval bronze, steam bronze) because of resistance to flowing sea water and biofouling, in art, and in trade. Bell metal, characterized by its sonorous quality when struck, is a bronze with a high tin content of 20–25%. Statuary bronze constitutes tin above 10%. Besides its traditional use in weapons, armour, tools, and other implements, bronze has been widely used in coinage: most "copper" coins are actually bronze, typically with about 4% Sn and 1% Zn.

Pure *tin* does not exist alone in nature (as opposed to gold, silver, or copper). Therefore, the development of metallurgy played the main role in the production of tin, which was used in the manufacture of bronze. Probably, tin played a role in ancient times. We know about Tin (Tinia or Tina) as an Etruscan deity, god of the sky and storm. Today we know about illnesses of tin, "tin plague" or "tin pest", which resulted in the destruction of tin buttons on Napoleon's army's clothes during the cold winter of 1812 in Russia, and the death of Robert Scott's expedition to the

Fig. 6.2 Tin articles.

South Pole in 1912 because of disintegration of tin containers with fuel. It was the chemist Ernst Julius Cohen from Amsterdam who established in the 1890s that an allotropic form of tin is responsible for its strange behavior. Tin has two allotropic forms under usual environmental conditions: β-Tin (white, density 7.31 g/cm^3) exists above 13.2°C, α-Tin (gray, density 5.77 g/cm^3) – under 13.2°C. When the temperature decreases under 13.2°C, white tin transforms into gray tin and losses its metallic properties. The maximum rate of this allotropic transformation occurs at −33°C. The structure of α-Tin is like diamond and it is a gray powder. Because of differences in density of these two allotropic forms, the volume of gray tin increases by 25% and it crumbles. The allotropic transformation of β-Tin into α-Tin is better when there is contact between the two. Therefore "infection" of tin articles may take place and "illness" spreads. The presence of Zn and Al in tin accelerates this transformation. The presence of Bi, Sb, and Pb in tin prevents tin disease.

Tin is neither resistant to alkali or acid solutions, nor to ozone and other oxidizing agents.

Brass, as an alloy of copper and zinc, did not appear in Egypt until about 30 B.C., but after this it was rapidly adopted throughout the Roman world (for example, for currency). The earliest brass, called calamine brass, dates back to Neolithic times. In the Bible, the term brass is often used to denote bronze. The general establishment of a brass industry was one of the important metallurgical contributions made by the Romans. It continued in use through the 14th century in India and the 18th century in Europe, long after metallic zinc was discovered. The ancient Romans used brass primarily in vessels, dress armour, jewelry, and brooches or clasps. From the 13th to the 17th centuries in Europe, monumental brasses were used to commemorate the dead. Ammonia and hydrogen sulphide were always present in the environment. Therefore, there is no doubt that brass corroded, and even the Romans were able to observe dezincification or some other corrosion phenomena with brass.

6.1.1.3 Lead

The first use of lead was probably in 3500 B.C. Specimens made of lead were found in graves in the pre-dynastic times in Ancient Egypt, but lead never found extensive application in Egypt. Vessels and pipes made of lead were later a symbol of the Roman empire. Some books about metallurgy in ancient times state that lead is a resistant metal. This is wrong, because it is not resistant to organic acids (for example, acetic acid was widely used) and to alkali solutions. By 500 B.C., rich lead-bearing silver mines had opened in Greece. Lead was removed from the silver by cupellation (a process of great antiquity in which the alloy was melted in a shallow porous clay or bone-ash receptacle called a cupel). A stream of air over the molten mass preferentially oxidized the lead. Of course, the metallurgists did not understand the chemical essence of this process (oxidizing of lead!), but now we have to emphasize that this was corrosion of lead used for the manufacture of the precious metals (silver, gold) and lead. Lead oxide was removed partially by skimming the molten surface; the remainder was absorbed into the porous cupel. Silver and gold were retained on the cupel. The lead from the skimming and discarded cupels was recovered as metal upon heating with charcoal. Cupellation was also employed to remove from the gold such contaminants as copper, tin, and lead, that is by means of the corrosion of these metals by air.

The Greek physician and pharmacologist Discorides (40–90 A.D.), traveling as a surgeon with the armies of the Roman emperor Nero, described in his "De materia medica" about lead acetate as a chemical drug. We may imagine that lead acetate might be obtained as a result of corrosion (dissolution) of lead in acetic acid. Lead also corroded in the presence of hydrogen sulphide and turned into black sulphides of lead.

6.1.1.4 Iron Age

Iron was known in the ancient times because of meteorites. The latter usually consist of iron and nickel (about 8%). Archeological findings show that iron was known to man of Sumeria and Egypt in 4000 B.C. Tips of spears, daggers and ornaments were fashioned from iron recovered from meteorites. The Egyptians and Babylonians called iron "the metal of heaven" or "the sky stone", which indicates that the first specimens employed were of meteoric origin.

An iron tool was found in the masonry of the Pyramid of Khufu at Giza (built about 2900 B.C.). Iron articles, weapons, dagger blades, etc., were introduced into peoples' lives only in 1200–1000 B.C. It is impossible to mark a sharp division between the *Bronze Age* and the *Iron Age*, but we may find in many historical sources that the *Iron Era* began about 1200 B.C. and lasts up to today. As we do not use pure iron, but an alloy of iron with carbon called *steel*, this historical epoch must be called the *steel era*.

Bronze was succeeded by iron that was less resistant to corrosion. This was not a paradox, because even in the ancient times people took into consideration the economical factors. The substitution of iron for bronze in tools and weapons from about 1000 B.C. was the result of iron's abundance compared to copper and tin rather than any inherent advantages of iron. People did not imagine what would happen with our world because of the appearance of iron corrosion problems.

Iron was present during the *Bronze Age*, albeit scarce. It was used for personal ornaments and small knives, for repairs on bronzes, and for bimetallic items. There is no doubt that people encountered *galvanic (bimetallic) corrosion* in this period because of contacts between dissimilar metals. It was the beginning of the era of *galvanic corrosion* lasting up to today.

Pure iron is softer than bronze. Hardly anyone used iron until the discovery of "Good Iron" (carbon steel) by the Hittites in 1600–1400 B.C. Anatolia was rich in metals, especially iron and silver. In the empire period, the Hittites developed the manufacture of iron, helping to initiate the Iron Age. In early Hittite history iron was 5 times more expensive than gold, 40 times dearer than silver. For centuries, iron must have been the rarest of luxuries. Hittites kept the secret of "Good Iron" for nearly 200 years. During heating in a charcoal furnace, the bloom and iron were exposed to carbon monoxide gas. A reaction between iron and carbon monoxide caused carbon to diffuse into the iron. "Good Iron" is otherwise a carbon steel alloy: a mixture of iron and small amounts of carbon. No one knew that carbon steel was responsible for "Good Iron" until 1773–1786 (3100 years later!). The "Good Iron" was preferable to bronze because of its better mechanical properties. By 1000 B.C., iron was beginning to be known in central Europe. Its use spread slowly westward, and by 500 B.C. it had reached the western borders of Europe. The Romans, who have never been looked upon as innovators but more as organizers, helped to spread the knowledge of iron making so that the output of iron in the Roman world greatly increased. Corrosion of iron objects spread with iron manufacture. We are familiar with many archeological findings of iron (more precisely, carbon steel) all over Europe and Asia: daggers, swords, utensils, etc. Soon people were disappointed with "Good Iron" (carbon steel) because of its rusting, especially in Europe because of the very high humidity of the atmosphere. There can be no doubt that people encountered corrosion of iron, but for lack of knowledge of chemistry, they could not explain corrosion phenomena. Perhaps, metallurgists were the first who observed iron corrosion because of the reaction with oxygen at high temperatures. Iron agricultural implements, tools, weapons, domestic articles, and other instruments made of iron were prone to corrosion because of the presence of water, oxygen and salts. Because of unpleasant rust formed on iron objects, iron was not widely used for decorative, religious, and symbolical purposes. In the 6th–15th century A.D. monasteries were often centers of learning of the arts of metalworking: censers, adornments, statues of saints, bells, etc. Of course, iron articles needed corrosion protection.

Indeed, the Greek philosopher Plato (427–347 B.C.) was the first who described rust as "a soil component separated from the metal". This was the same period when the Greek philosopher Aristotle wrote in his book *On the Heavens*, that the Earth was a round sphere rather than a flat plate.

Many years before Plato's description, the Jews described how to remove corrosion products from the surface of iron and copper: "… Tuval Kain polishes the iron and copper objects …" (Genesis, Ch. 14, V. 22).

A German scholar and scientist known as "the father of mineralogy" Georgius Agricola (1494–1555), 2000 years after Plato, concluded the same in his book *De*

Natura Fossilium: “Rust (*ferrugo* or *rubigo* in Latin) is the iron excretion”. Then Agricola continued: “It is possible to protect iron with coats of red lead, white lead, gypsum, bitumen, or pitch”. All these substances certainly were known from ancient times. Bitumen is used as anti-corrosion material even now. Undoubtedly, with the entry to the *Iron Era*, people tried to paint iron objects, but until the first records by Pliny the Elder in the 1st century A.D. there was no information about the use of paints for the protection of metals from corrosion. The exploitation of linseed oil (a drying oil useful as a carrier) and white pigment zinc oxide in the 18th century brought a rapid expansion of the paint industry in Europe.

While the *Iron Age* peoples of Anatolia and Europe on occasion may have accidentally made *cast iron*, the Chinese were the first to realize its advantages as early as the 6th century B.C. Also brittle and lacking the strength, toughness, and workability of steel, it was useful for making cast bowls and other vessels. Certainly, people who used cast iron equipment had corrosion problems, for example, *graphitic corrosion*. Cast iron was used in the atmosphere, in water, and in soil, but it corroded in the presence of ammonium salts, sulphides, cyanides, stannic, and mercuric salts which were used by alchemists. *Cast iron* was sporadically produced in Europe by the 14th century A.D. The export of knowledge of iron metallurgy and iron objects which began between 1200 and 1000 B.C. changed the face of the world because the utilization of iron for weapons put arms in the hands of the masses and set off a series of large-scale movements of peoples that did not end until 2000 A.D.

There are few monuments of literature left of the early records about metals and all related to their manufacture. One of them, the Egyptian papyrus in Leyden (Leyden Papyrus), comprises about 75 recipes pertaining to the making of alloys, for soldering metals and for coloring the surfaces of metals. This papyrus and the book *Mappae Clavicula: A Little Key to the World of Medieval Techniques* (9th century) tell about corroding copper and lead to make green and white pigments. All this shows that ancient metallurgists and chemists undoubtedly encountered corrosion problems and thought about their protection from corrosion.

6.1.2 Metals and Mysticism

In the early centuries of our era a mysticism developed among people who dealt with metals, due to Egyptian and Chaldean magical ideas. Drawing on a similar astrological heritage, philosophers found correspondences between the metals and planets. The historical and astrologer-writer Olympiodorus in the 5th century A.D. gave the following relation: Gold – the Sun, Silver – the Moon, Mercury – Jupiter, Iron – Mars, Copper – Venus, Tin – Mercury, and Lead – Saturn. Alchemists related the seven metals known them to the Sun and six planets, as they supposed that these seven metals were born as entrails of the Earth under the planet rays’ influence (Figure 6.3).

Fig. 6.3 Miniature from the alchemy book. Seven metals are depicted as gods – planets' symbols sheltered in the cave. (For a full color version of this figure, see the Color Section)

We may suppose (or even emphasize) now that Mars, as a symbol for Iron, was probably chosen not only as a God of War but also because of its red color similar to rust.

Alchemists saw in gold the symbol of regenerated man, resisting temptation, and immune to evil. Another metal, lead, was for alchemists the symbol of sinning, susceptibility to temptation and proneness to evil. Thus, the alchemists saw in some metals (gold – noble) "passivity" and high resistance against *corrosion*, but others (lead – non-noble metal) were prone to *corrosion*.

6.1.3 The Alchemy Ages

During the alchemy period the repertoire of known substances was enlarged. The alchemists played an important role in gaining wide experience in the discovery of new corrosives (salts and acids), they acquired knowledge of corrosion phenomena, but they did not understand them. The alchemists tried to convert any metal into gold, or to find "the elixir of the immortality".

From the earliest records of Indian natural philosophy (5th–3rd centuries B.C.), we know that the six metals known to them (gold, silver, copper, tin, iron, and lead) were "killed" (that is, *corroded*!) to make medicines. The Indians used various solutions (vitriol – copper sulphate, sal ammoniac – ammonium chloride, and saltpetre

– potassium nitrate) for the dissolution of metals. This means that the Indians used corrosion processes for the preparation of medicines.

Hellenistic alchemists also dealt with "killing" the "base" or "sick" metals in order to ennoble them (to gold) or to use as drugs for the rectification of human maladies. Knowledge of "sal ammoniac" (ammonium chloride – NH_4Cl) in the West from the Chinese treatise of the 2nd century A.D., was to be crucial to alchemy, for on sublimation, it dissociates into corrosive chemicals, ammonia, and hydrochloric acid, which readily attacked metals. It is important to mention that even today the formation and subsequent hydrolysis of ammonium chloride results in the formation of acidic solutions and severe *under deposit corrosion* in the overhead of crude oil distillation towers. Then ammonium chloride found extensive use in corrosion and corrosion control: as an electrolyte in dry cells, as a constituent of galvanizing, tinning, and soldering fluxes to remove oxide coatings from metals and thereby improve the adhesion of the solders. Finally, the manipulation of these chemicals was to lead to the discovery of the mineral acids, the history of which began in Europe in the 13th century A.D. The first was probably nitric acid (*strong water – aqua fortis* in Latin), which was used in 1200 A.D. in order to dissolve and purify silver. Many metals are oxidized (corrode!) in nitric acid with the formation of nitrates or oxides (a product's type depends on the concentration of nitric acid). For example, dissolution of silver, copper, and iron in nitric acid is a corrosion process:

$$Ag_{(s)} + 2HNO_{3(aq)} \rightarrow AgNO_{3(aq)} + NO_{2(g)} + H_2O_{(l)}, \tag{6.4}$$

$$3Cu_{(s)} + 8HNO_{3(aq)} \rightarrow 3Cu(NO_3)_{2(aq)} + 2NO_{(g)} + 4H_2O_{(l)}, \tag{6.5}$$

$$8Fe_{(s)} + 30HNO_{3(aq)} \rightarrow 8Fe(NO_3)_{3(aq)} + 3N_2O_{(g)} + 15H_2O_{(l)}. \tag{6.6}$$

Undoubtedly, alchemists had to find material for apparatus that was resistant to corrosion in acids. They tried to use various metals and alloys, ceramics, glass, and wood, and observed various corrosion processes that occurred in the presence of acids. Silver dissolved in hot concentrated sulphuric acid, and this was corrosion. Alchemists used corrosion of copper in sulphuric acid in order to obtain one of the very important salts – copper sulphate:

$$Cu_{(s)} + 2H_2SO_{4(aq)} \rightarrow CuSO_{4(aq)} + SO_{2(g)} + 2H_2O_{(l)}. \tag{6.7}$$

Gold dissolves in "aqua regia" (a mixture of nitric and hydrochloric acids):

$$Au_{(s)} + HNO_{3(aq)} + 4HCl_{(aq)} \rightarrow H[AuCl_4]_{(aq)} + NO_{(g)} + 2H_2O_{(l)}. \tag{6.8}$$

This is a corrosion process and alchemists could observe this reaction many times. The corrosion of gold in a sodium cyanide solution is the basis of the industrial process for obtaining gold from ores:

$$2Au_{(s)} + 4NaCN_{(aq)} + 1/2\ O_{2(g)} + H_2O_{(l)} \rightarrow 2Na[Au(CN)_2]_{(aq)} + 2NaOH_{(aq)}. \tag{6.9}$$

Probably, alchemists observed the different behavior of metals in mineral acids of various concentrations. Such observation must have lead to the discovery of "passivity", but there are no supporting documents. The Russian scientist Michail Lomonosov was the first who wrote about "passivity" in the 18th century. The first chemists were metallurgists. Miners in the Middle Ages knew that if an iron object came into contact with the "vitriol water" ("blue vitriol") of copper mines (an aqueous solution of copper sulphate), the surface of the iron would be covered by a red copper layer. This reaction was mentioned by both alchemists and craftsmen in ancient China (3rd–4th A.D.). Alchemists in the Middle Ages represented iron by Mars, and copper by Venus, and described this reaction in the following way: "The shining armed Mars threw himself into Venus's arms and became red".

$$Fe_{(s)} + Cu^{2+}_{(aq)} \rightarrow Fe^{2+}_{(aq)} + Cu_{(s)}. \tag{6.10}$$

This is corrosion of iron! And miners observed this corrosion process many times but could not explain it.

In the 18th century the German chemist Erxleben described the following corrosion processes which have delighted many generations of chemists to date: "A wonderful tree of silver color grows – Diana's tree or Philosophy's tree". Diana is a goddess identified with the Moon (Silver).

$$Cu_{(s)} + 2Ag^{+}_{(aq)} \rightarrow Cu^{2+}_{(aq)} + 2Ag_{(s)}. \tag{6.11}$$

Everybody who carried out these reactions was able to take pleasure in the growth of beautiful "trees", but nobody knew or understood that they delighted thanks to the fact that corrosion of copper had occurred.

6.1.4 From Alchemy to Chemistry (1500–1791)

Various civilizations existed with seven metals (Au, Cu, Ag, Pb, Sn, Fe and Hg) up to the 16th century. Metallurgists developed various methods for the production, separation, and purification of metals from different contaminants. The majority of these methods include oxidizing - reduction chemical processes. During the 16th century, metallurgical knowledge was recorded and became available. The books *De re metallica* (1556) by the German scholar and scientist Georgius Agricola (the Latin name of Georg Bauer) and *De la Pirotechnia* (1540) by the Italian Vannoccio Biringuccio were especially influential. These books described the etching of steel, the assaying of precious metals, and other aspects of the treatment of metals and alloys. These aspects related to *corrosion* of metals, but authors did not use this concept.

It was the time when the first recorded cast iron pipe was installed in Germany (1562) where it supplied water for a fountain. The first full-scale cast iron pipe system for the distribution of water was installed later (1664) in Versailles, France.

The German chemist, physician, and alchemist Andreas Libavius (Libau) (1540–1616) wrote the book *Alchemy* (1606), which established the tradition for the 17th-century French chemistry textbooks. He discovered the methods for the preparation of ammonium sulphate, hydrochloric acid, and tin tetrachloride which played an important role in the corrosion of metals. The French scientist René Antoine F. de Réaumur published *Memoirs on Steel and Iron* in 1722. When Réaumur, Johann Heinrich Pott, and Wedgwood studied the influence of high temperatures on minerals, they came across corrosion of metals at high temperatures. The American metallurgist Cyril Stanley Smith (1904–1992), one of the scientists who united art, science, and history remarked that the etching of metallic surfaces by application of corrosive substances (*corrosion process*!) to produce decorative effects and a deliberate mode of "weathering" of metals and alloys were practiced during the whole of history of man. People used acid fruit juice and vegetable acids for these purposes. Corrodants based on ferric sulphate [green vitriol, $Fe_2(SO_4)_3$] are also as potent as free acids. They found early use in treating gold and other metals.

Konrad Keyeser, in his *Bellifortis* (1405), mentioned the liquid (HNO_3) made by distilling mixed vitriol and saltpeter, which he called *aqua martis* in clear reference to its use for *etching* iron armor.

When the superior steel of ancient (Damascus) swords was finally duplicated in Europe, its beautiful and purposeful texture brought out by *etching* (corrosion!) led to the discovery of carbon in steel – unknown before 1774. The Swede Tobern Bergman in *A Chemical Essay on the Analysis of Iron* (1781), together with his student Johannes Gadolin demonstrated that the difference between wrought iron, cast iron, and steel were due mainly to the presence of carbon absorbed from the charcoal fuel of the smelter's fire. Another Swedish metallurgist Sven Rinman qualitatively identified as carbonaceous the residues left by *etching* in 1774. In Europe, distilled hydrochloric acid (HCl) was used to *etch* (corrode!) designs in the mid 16th century. Iron is the most easily corroded metal known since ancient times. Etched surfaces appeared on iron swords as early as the La Téne period associated with Celtic culture (about 400 B.C.). Bronze was etched in China at about the same time. Etching (corrosion!) was also used for preparing of an iron surface for *gilding*. The etching was carried out with a corrosive mixture of salts, variously incorporating ferric sulphate, alum, sal ammoniac (ammonium chloride), saltpeter (potassium nitrate), mercuric chloride, and vinegar. The etching was a selective chemical attack – a desirable use of corrosion. The Indians of pre-Columbian Central and South America (1000–1500) had great skill in fashioning gold and its alloys (with copper or silver) "owing to corrosion" of less (based) noble metal and leaving a thin surface of pure gold. Innumerable "books of secrets" with recipes for etching of metals were published in Germany in the 16th century.

6.1.5 Discovery of New Metals

Platinum (Pt) was brought to Europe by Spaniards from Mexico in about 1500. This is the only natural metal first discovered in America. Alchemists observed the dissolution of platinum in "aqua regia" when heated:

$$3Pt_{(s)} + 4HNO_{3(l)} + 18HCl_{(aq)} \rightarrow 3H_2[PtCl_6]_{(aq)} + 4NO_{(g)} + 8H_2O_{(l)}. \quad (6.12)$$

This is a corrosion process! Platinum in contact with the alkalies, cyanides, and sulphides of alkali metals corrodes in the presence of oxygen and other oxidizers at high temperature. In any case, platinum is very resistant to many environments. Therefore, it found application as an inert electrode in the standard hydrogen electrode, as a counter electrode in electrochemical measurements, as the standard metre bar kept in Paris, and as a catalyst in many chemical processes.

Zinc. In spite of the use of alloys of zinc and copper in ancient Egypt, Greece, and Rome, pure zinc was not obtained until the 13th–14th century in China and India by reducing calamine $\{Zn(OH)_2SiO_3\}$ with wool or charcoal. The Greeks knew of the existence of zinc and called it *pseudargyras* (false silver), but they had no method of producing it in quantity. Pure zinc was imported into Europe up to the second half of the 18th century. Wide use of zinc began in the middle of the 18th century, after the introduction in 1743 into England of the process of producing pure zinc that had been known in India from the 14th century. Probably, chemists observed the fast corrosion of zinc in acids and strong alkali solutions during this period. Then in the 1830s, the Swiss physicist Auguste-Arthur de la Rive carried out experiments on zinc dissolution in acids and suggested a hypothesis about electrochemical reactions between zinc and impurities contained in zinc.

Nowadays, the most important use of zinc, approaching 50%, is in the corrosion protection of iron and steel; about 15 to 20% is used in brass and other alloys; and about 12% is used in zinc dust and other chemicals.

The development of the chemical analysis methods resulted in the discovery of 12 metals in the 18th century: Cr, Be, Y, Ti, Zr, U, W, Te, Mo, Mn, Ni, and Co. Before 1800, 11 metals were used: Au, Ag, Cu, Pb, Hg, Fe, Sn, Pt, Sb, Bi, and Zn. The quantity of metals known to people doubled during the 18th century. Possibilities of the *corrosion phenomena* occurring increased!

6.1.6 Definition of "Corrosion"

There is no clear record of where and by whom the word corrosion was used for first time. There is no doubt that the concept corrosion, which means "gnaw, eat away", translated from the Latin *corrodere*, was used in England in 1667 in the magazine *Philosophical Transactions*. The Anglo-Irish chemist and natural philosopher Robert Boyle, known as the first physical chemist, studied the calcination (the conversion of metals into their oxides as a result of heating to a high temperature) of metals, and discovered (1674) that when acid interacts with certain metals (corrosion!) a flammable gas is produced (known as hydrogen).

Robert Boyle (1627–1691)

Boyle, in his *The Works of the Honorable Epitomized* (1675), used the word *corrosion* for the description of the process between a metal and the environment:

$$\text{Metal} + \text{Environment} \rightarrow \text{CORROSION.} \tag{6.13}$$

He stands between alchemy and chemistry, and opened the door to chemistry.

The German physician and chemist Georg Ernst Stahl introduced in 1697 the idea of phlogiston as an agent of combustion and *rusting*. The word *corrosion* appeared in Germany only in 1836 after translation of Sir Humphry Davy's works.

The strict definition of *corrosion* as a phenomenon or process was not clear up to the 1960s, and numerous discussions took place. Only in the 1970s was *corrosion* strictly defined as "the chemical or electrochemical reaction between a material, usually a metal, and its environment that produces a deterioration of the properties of the materials and its environments".

The Swedish chemist and apothecary Carl Wilhelm Scheele and the English clergyman, political theorist, and physical scientist Joseph Priestley discovered (1772–1774) oxygen by heating nitric acid, potassium nitrate, manganese dioxide, and mercuric oxide (II).

There is a note that the Polish alchemist Michael Sendivogius discovered how to make oxygen from saltpeter (potassium nitrate) in around 1600, and the Dutch inventor and alchemist Cornelius Drebbel used it in a practical way in 1620. Among Scheele's other discoveries were glycerine, the pigment called Scheele's green (copper arsenite), and the toxic and corrosive gases hydrogen sulphide, hydrogen cyanide, and hydrogen fluoride. Joseph Priestley also discovered that graphite conducts electricity and noted the relationship between electricity and chemical change, discovered the gases ammonia, sulphur dioxide, nitrogen, carbon monoxide, and observed that oxides are changed to the metallic state (reduction – the reaction opposite to corrosion) when heated in hydrogen.

Many investigations into corrosion gave impulse to research in chemistry, for example, in chemical kinetics. The chemist Carl Wenzel noted in 1777 that the rate

Left: Joseph Priestley (1733–1804); Right Carl Wilhelm Scheele (1742–1786)

of dissolution of zinc and copper in an acid solution depended on the concentration of acid.

The French chemist, the "Father" of Chemistry, Antoine-Laurent Lavoisier, first recognized oxygen as an element, coined its name (1789), and explained the reaction with oxygen as a *chemical reaction*. Lavoisier destroyed the myth about phlogiston, and the first investigation he made into this was the burning of tin. He understood that combustion of tin was combining with oxygen (oxidation = corrosion). Thus, Lavoisier defined that metals react during oxidation with the active part of air – oxygen. We may be surprised why Lavoisier did not see the *corrosion phenomenon* in this reaction. But corrosion of tin and other metals in air during combustion helped Lavoisier to discover the law "conservation of matter" in chemical reactions which became a fundamental principle in science.

Antoine-Laurent Lavoisier (1743–1794)

Thus, the discovery of oxygen and evolution of understanding of oxidation – reduction reactions (combining of metals with oxygen) – stimulated evolution and understanding of *corrosion as a chemical process*. But this was to happen later.

Antoine Lavoisier classified substances into four groups: one of them being metals that are capable of being oxidized (corrosion!). Lavoisier laid out the new principles of chemistry which helped corrosion to be recognized as a particular chemical process.

Michail Vasilievich Lomonosov (1711–1765)

Forty years before, in 1750, the Russian scientist Michail Lomonosov wrote in his investigation "Dissolution of metals in acids", that "the metals dissolve in acidic alcohols in a different manner than salts in water". Later, after long investigation during 1748–1756, Lomonosov suggested the first theory of a metal's oxidation: "The metals reacted with air at high temperature, and scale formed on the metal's surface". Unfortunately, these works of Lomonosov were unknown in Europe. Evolution of chemistry resulted in the beginning of research into *corrosion science*.

6.1.7 "Renaissance" Era (1791–1890)

The end of the 18th century was marked by a revolution in politics (France, America), in technology, and ... in *corrosion science*. The Italian physician and physicist Luigi Galvani began experiments with frogs and various metals in 1780.

Galvani discovered that contact of two different metals with the muscles of a dissected frog resulted in their contraction. He published his observations of frog muscle twitching in 1791 in his book *Animal Electricity*. Galvani's work created the beginnings of two fields of scientific inquiry, which is why it indeed "galvanized"

Luigi Galvani (1737–1798)

the scientific community of his day. The first field was that of *electrochemistry*, which largely followed on from Galvani's observation that various metals can cause frogs' legs to twitch. Based on these experiments, Alessandro Volta developed in 1797 a useful device, the battery, to order the metals in an electromotive sequence. Thus, 1791 is the year of discovery of electrochemistry as the field for study of the metal-solution interface. The second field Galvani discovered was *electrophysiology* – the study of ionic electricity as it occurs in the nervous system. Because of an incorrect conclusion of the Italian physicist Alessandro Volta and the misinterpretation of Galvani's discoveries, for nearly 200 years many scientists did not believe in "animal" electricity that runs entirely without metals. Galvani's experiments played a huge role in the evolution and development of many branches of science and technology: chemistry, physics, medicine, and, of course, corrosion science. This great man gave the name to corrosion failure – *galvanic corrosion*, and to a device, the *galvanometer*, for measuring electric currents.

A second scientist who played a distinguished role in the evolution of corrosion science and many other scientific and technological fields was the Italian physicist Alessandro Jiuseppe Antonio Anastasio Volta.

After publication of Volta's friend Galvani's book in 1791, Volta began experimenting in 1794 with metals alone and found that animal tissue was not needed to produce an electric current. Volta (1797, Pavia, Italy) constructed the first battery from alternating disks of zinc and copper (or silver) separated by wet, absorbent paper. Luigi Galvani thought that "animal electricity" existed in any organism without relation to dissimilar metals. Alessandro Volta did not believe in the "animal electricity" and explained that the electric current appeared because of contact of two dissimilar metals in any electrolyte. However, Volta did not understand that the electric current took place in the battery because of the appearance of different electric potentials on electrode surfaces, namely, because of corrosion – the electrochemical

Alessandro Jiuseppe Antonio Anastasio Volta (1745–1827)

reaction of one less noble metal which contacted with another more noble metal in general electrolyte.

The *voltaic pile* (or *voltaic cell*) was the first convenient source of electric current. In 1801 in Paris, Volta gave a demonstration of his battery's generation of electric current before the French emperor Napoleon Bonaparte, who made Volta a count and senator of the Kingdom of Lombardy. The English Romantic poet Percy Bysshe Shelley (1792–1822) was fascinated by the voltaic pile and had dreams of a world transformed by electricity. Volta was responsible for the phrases "electromotive force" and "electric current" now used by all corrosion specialists, physicists, and chemists. The *volt*, a unit of the electromotive force that drives electric current, was named in his honor in 1881. Contemporaries of Volta told that he had a great sense of humor.

The German physicist Johann Wilhelm Ritter (1797–1799) and the English scientist William Hyde Wollaston (1801) suggested that galvanic phenomena are the result of chemical reactions of metals with solutions. The former observed the similarity between the "galvanic activity of metals" and their affinity to oxygen. This point of view was supported by Sir Humphry Davy, Michael Faraday, the Swedish chemist Jans Jacob Berzelius, and the Russian physico-chemist Nikolai Nikolayevich Beketov. The "galvanic phenomena" roused much interest among many scientists and governments in Europe, especially in France, England, Italy and Russia. Napoleon Bonaparte (1769–1821) personally contacted Volta, coined a medal in his honor, and provided him with financial assistance for many years. The Russian ambassadors in Paris and London had to report about new experiments with the "galvanic phenomena". Probably these reports stimulated the study of galvanic phenomena by V.P. Petrov, Lenz, B.S. Yacobi, Savelyev, and Kristian Iogann Ditrih Grotgus in Russia. Thus, European governments encouraged investigations into electrochemistry and thus into *corrosion science*.

The English chemists William Nicholson (who discovered the electrolysis of water) and Carlisle (1800) constructed similar "voltaic piles" in England, and new discoveries of Sir Humphry Davy, Michael Faraday and other scientists appeared soon after.

Iron structures began to enter into wide use at the border of the 18th–19th centuries. The first bridge across the Severn (England) was made of cast iron in 1779. The first ship "Vulcan" was made of carbon steel in 1819. The first railway was built in 1825 in England. Intensive need for iron caused the development of metallurgy in the 19th century. Corrosion processes related to iron and steel followed shortly afterwards. Engineers and scientists began to study this nasty phenomenon at the end of the 19th–beginning of the 20th century.

The next scientist who contributed much to corrosion science was the English chemist Sir Humphry Davy.

Sir Humphry Davy (1778–1829)

Using the voltaic pile, Davy through electrolysis discovered and isolated six metals: sodium (1807), potassium (1807), magnesium (1808), calcium (1808), strontium (1808), and barium (1808), and the element boron. He is one of the "Fathers" of cathodic protection. Davy (1824) designed a method whereby copper-clad ships could be protected by having zinc or cast iron plates connected to them. The warship "Sammarang" was the first ship with cathodic protection of the copper sheathing. This ship made a voyage to Canada in 1824–1825 with cast iron sacrificial anodes covering about 1.2% of the ship's surface. Probably, Davy was not the first who invented *cathodic protection*. In 1936, in Khuyut Rabuach, near Baghdad, clay jugs with copper cylinders were found containing an iron kernel. Archeologists suggested that it was a *battery* working on the *sacrificial* iron anode principle. In Imperial Rome (1st century B.C. to 4th century A.D.) *batteries* were used for the electrolysis and formation of gold coatings on jewelry. Davy showed that a voltaic pile produced no current when the discs were wetted with pure water (poor electrolyte). Corrosion of metals may occur in the presence of electrolytes!

Jans Jacob Berzelius (1779–1848)

The Swedish chemist Jans Jacob Berzelius isolated silicon (1823), zirconium (1824), and titanium (1825), elements which were to find much use in corrosion control.

Important observations and steps were made during this period on the long road towards the definition of the electrochemical corrosion mechanism. Austin (1788) pointed out that neutral water in contact with iron became alkaline. But only in the beginning of the 20th century could scientists show that the cathodic reduction of dissolved oxygen in water on a metal surface was responsible for the increased alkalinity:

$$O_{2(g)} + 2H_2O_{(l)} + 4e^- \rightarrow 4OH^-_{(aq)}. \tag{6.14}$$

The French chemist Louis-Jacques Thénard suggested in 1819 that "corrosion is an electrochemical process".

Louis-Jacques Thénard (1777–1857)

The Swiss physicist Auguste-Arthur de la Rive (1801–1873) explained the faster reaction of non-pure zinc compared with pure zinc in acid because of an electrical reaction between zinc and the impurities contained in it (1830). In 1840 he invented the process of *electroplating* (process of coating with metal by means of an electric current) gold onto silver and brass, discovered that ozone is created when electrical sparks pass through oxygen, and shared the view of Michael Faraday, Sir Humphry Davy, Jans Jacob Berzelius, Johann Wilhelm Ritter, and William Hyde Wollaston that voltaic electricity was caused by chemical action.

Modern *electroplating* started in 1800 with Volta's discovery of the voltaic pile, or battery. The battery was employed to deposit lead, copper, silver, and zinc on a variety of basis metals, such as gold and iron. Electroplating on a commercial scale began in about 1840 and was accelerated by the discovery of cyanide solutions for plating silver, gold, copper, and brass.

A German chemist Friedrich Stromeyer discovered cadmium in 1817, which found a large use in electroplating onto iron, steel, and other alloys to protect them from corrosion. Unfortunately, he did not know about the high toxicity of cadmium.

The next scientist (after Galvani, Volta, and Davy) who contributed much to corrosion science as well in electrochemistry was Michael Faraday, one of the greatest experimenters ever.

Michael Faraday (1791–1867)

He defined the relationship between electric current and chemical reaction (1831–1840), in spite of the fact that he had no grasp of mathematics. The Laws of Faraday, 100 years after his discovery, were used in the definition of the corrosion of metals in electrolytes as an electrochemical process. Today all electrochemical methods of corrosion monitoring allow the determination of corrosion rate data based on the principles determined by Faraday in the 1830s.

Michael Faraday and James Stodard tried to "invent" *stainless steel* in 1819–1822 adding chromium to iron. They added less than 12% chromium and noted improvement in resistance to the atmosphere, but found only marginal improvement in corrosion resistance to hot sulphuric acid. They also attempted to duplicate wootz (a steel alloy having a pattern of bands or sheets of micro carbides within a tempered martensite or perlite matrix; developed in India about 300 A.D.), and watered Damascus steel.

After consulting with the English philosopher and historian William Whewell (1794–1866) who coined the word "scientist" in 1834, Faraday adopted the words *electrode, anode, cathode, ion, anion, cation, electrolyte, electrolysis* (*electrolyse*). But, Faraday consistently did not believe in existence of atoms.

This period was also useful for research into *passivity* – a special condition of metal in which corrosion decreases drastically. The German chemist Christian Friedrich Schönbein named, in the famous correspondence with Faraday (1836), "*passive* iron" after its immersion in concentrated nitric acid.

Christian Friedrich Schönbein (1799–1868)

Schönbein is famous for the discovery of ozone (1839), nitrocellulose, collodium, and is nowdays assumed to be "the father of atmospheric chemistry".

Long before, Mikhail Lomonosov wrote in his thesis "General Activity of Solvents" (1743) that "When nitric acid is used for dissolution of metals, dissolution is stopped quickly, because the acid ceased its action".

Later James Keir (1790) noted:

> There are no changes in the properties of iron being in contact with concentrated nitric acid. Now the iron is resistant to the relatively diluted nitric acid. If the iron was not previously in contact with concentrated nitric acid, it was attacked quickly and severely by the dilute nitric acid.

Faraday attributed the resistance of iron in concentrated nitric acid to a protective iron oxide film, but only the English scientist Uhlick R. Evans provided direct evidence of its existence in 1930 by means of the electrochemical technique of peeling the film from a passive surface and viewing it under a microscope.

The discovery and preparation of aluminum (1825) by the Danish physicist and chemist Hans Christian Oersted and development of the industrial process of the manufacture of aluminum in the 1850s by the French scientist Henry Sainte-Claire Deville opened the way for corrosion problems connected with the use of this "luster metal somewhat resembling tin" as Hans Christian Oersted wrote.

The German chemist and an outstanding teacher Friedrich Wöhler (1800–1882), who was a great friend of Berzelius, first determined (1828) that aluminum had high stability (corrosion resistance!) in air.

The English physicist James Prescott Joule (1818–1889) had used *dissolution* (corrosion!) of metals in acids and in aqueous solutions of salts during 1840–1843 in order to compare the electrical energy of "galvanic elements" with the heat evolved in a calorimeter. Thus, corrosion of metals helped Joule to suggest that electrical energy is equivalent to chemical energy, and to define the first law of thermodynamics. The British inventor and engineer Sir Henry Bessemer (1813–1898) patented (1856) his converter process for blowing air through molten pig iron, and for 30 years millions tons of steel poured into industry, giving rise to many corrosion problems.

Iron constitutes 5% by weight of the Earth's crust and it is the fourth most abundant element after oxygen, silicon, and aluminum, but it soon became the centre of all corrosion problems in industry and society.

Calvert and J. Johnson reported in 1866 about dealloying brass. Later this corrosion phenomenon received the name *dezincification*.

The American scientist Josiah Willard Gibbs, in his most famous paper "On the Equilibrium of Heterogeneous Substances" (1876), and then the German scientist Hermann Ludwig Ferdinand von Helmholtz, who was completely unaware of Gibbs' work, discovered in 1882 the very important thermodynamic equation:

$$\Delta G_T^{\circ} = \Delta H_T^{\circ} - T \cdot S_T^{\circ} \tag{6.15}$$

which allowed the Belgian scientist Marcel Pourbaix (in 1938) to create his famous "potential–pH" diagrams for the equilibrium "metal–water".

Left: Josiah Willard Gibbs (1839–1903); Right; Hermann Ludwig Ferdinand von Helmholtz (1821–1894)

We have also to mention two contributions of Gibbs to corrosion science. One of them, the *Gibbs phase rule* (1878) which plays a very important role in phase diagrams in metallurgy (development of new alloys), and the other, interpretation of entropy as "randomness" (1901) which plays a very important role in the understanding of corrosion processes.

The first evidence that a metal with internal stresses would crack in the presence of certain aggressive environments was reported by the English metallurgist William Chandler Roberts-Austen (1843–1902) in 1886 (his name was given to the γ-iron phase austenite). He showed that 13-carat gold with internal stresses developed cracks when put into contact with a ferric chloride ($FeCl_3$) solution.

6.1.8 "Baroque" Era (1890–1935)

The German scientist Walther Hermann Nernst, who was one of the founders of modern physical chemistry, suggested (1889) that the oxidation of metals in electrolytes occurred as the reaction

$$M \rightarrow M^{n+} + ne^{-}. \tag{6.16}$$

Walther Hermann Nernst (1864–1941)

This means that the electric potential of metals is formed as a result of red-ox reaction between a metal and metallic cations in a solution. Later the Russian scientists L.V. Pissarzevski (1914), A.I. Brodsky and N.A. Izgarishev (1926) suggested that water molecules take part in the dissolution of metals:

$$M_{(s)} + mH_2O_{(l)} \rightarrow M^{n+} \cdot mH_2O_{(l)} + ne^{-}. \tag{6.17}$$

Walther Nernst, when conducting important research into the theory of galvanic cells and the thermodynamics of chemical equilibrium, deduced a formula for the influence of the concentration (activity) of reagents upon voltage (1900):

$$E = E^{\circ} + \frac{RT}{nF} \ln \frac{a_{\text{ox}}}{a_{\text{red}}}. \tag{6.18}$$

where E is an electromotive force; E° is a standard electromotive force ($a_{ox} = a_{red} = 1$); R is a universal gas constant; T is an absolute temperature; n is the number of electrons taking part in the reduction-oxidation (redox) process; F is a Faraday constant; a_{ox} and a_{red} are activities of oxidized and reduced species, respectively.

In 1899 it was reported about inhibitive properties of sodium nitrite ($NaNO_2$). For about 40 years sodium nitrite remained effectively unnoticed until the 1940s (owing to the works of chemists Aaron Wachter and S.S. Smith). It is now one of the most widely employed corrosion inhibitors for carbon steel in closed water-circulating systems. The German scientist Julius Tafel (1862–1918) in 1905 experimentally confirmed the theoretically later defined equation by the British physical chemist John Alfred Valentine Butler (1899–1977) and the German chemist Max Volmer (1885–1965) which was the main equation for electrochemical kinetics, connecting the overpotential with the corrosion rate of metal:

$$\eta_a = \pm\beta \log(i/i_{\circ}), \tag{6.19}$$

where η_a is an overpotential, β is a constant, i is the rate of oxidation or reduction in terms of current density, $i_{\circ}$ is the exchange current density. Tafel extrapolation is now widely used for the determination of corrosion rates.

The "*season cracking*" of brass was described by P.R. Sperry, and the effectiveness of phosphoric acid as a corrosion inhibitor was discovered by the Englishman Coslett in 1906. The German physical chemist Fritz Haber invented (1906) a device for measuring current density and soil resistance.

Fritz Haber (1868–1934)

The American chemist Leo Hendrik Baekeland invented the first man-made polymer bakelite in 1907. Fritz Haber made some important inventions for which all corrosion specialists and electrochemists remember him. He improved (1900) the probe of his friend Hans Luggin (1863–1899), and we are familiar with the Luggin–Haber probe (capillary) for the diminishing IR-drop. Then he invented (1908) the non-polarized zinc-zinc sulphate reference electrode. D.M. Buck reviewed investigations dating from 1908 that showed the beneficial effect of copper as an alloying element in steels for atmospheric exposure.

The American inventor Elwood Haynes (1857–1925) created the high wear-resistant *stellite* alloy (cobalt-chromium) in 1910.

It was proved during this period (1900–1934) that the corrosion of metals in electrolytes occurs according to an electrochemical mechanism. The Swedish scientists Ericson-Auren and Palmaer (1901) defined the existence of "local elements" (macro-elements) on a metal's surface immersed in the electrolyte. The American scientist W.R. Whitney (1903) determined that the corrosion of iron in aqueous solutions is related to an electric current.

Intergranular stress corrosion cracking was known in 1919.

The British scientists Whitman and Russel (1924) and the Russian scientists Akimov and Clark (1935) showed that corrosion was intensified when a small anode is connected to a large cathode.

The British scientist Ulick Richardson Evans defined (1923–1925) the existence of differential aeration currents (cells). The first book of U.R. Evans, *The Corrosion of Metals*, was published in 1923.

Ulick Richardson Evans (1889–1980). From the collection of Antoine Pourboix.

Evans, Hoar, Bannister, Thornhill, and Agar in England, and Mears with Brown in the United States (1931–1939) showed by means of Faraday's law that the quantity of corroded metal in an electrolyte equals an electric current. Thus, the electrochemical mechanism of corrosion was proved and "opened the door" to the use of electrochemical investigation methods of corrosion. The electrochemical approach to corrosion phenomena allowed Evans to suggest in 1929 the "Evans diagram" (relationship between electric potential and current). The twenties are marked by flourishing of research and study in corrosion science. The English metallurgist Sir

Robert Hadfield (1922) estimated the economic loss from corrosion – £700,000,000 for the annual cost of wastage due to rusting of the world's iron and steel.

N.B. Pilling and R.E. Bedworth (1923) began their experiments on high-temperature oxidation of metals and alloys. The introduction of chromium plating in 1925 stimulated repercussions all through the plating industry. W.H.J. Vernon, J.C. Hudson, and Patterson (the 1920s, England) showed that atmospheric corrosion required both wetness and pollutant, and defined the existence of a critical humidity for atmospheric corrosion to occur.

Corrosion education began in the 1920s. The first corrosion course for students was held in Cambridge University in 1924. Today students may study corrosion science and engineering in nearly 300 universities around the world.

Cathodic protection received a theoretical explanation long after its first use in 1824. One hundred years later, the electrical engineer Robert J. Kuhn (1928–1933, USA) experimentally defined the criteria for the cathodic protection for steel gas piping: the electric potential must be less than −0.85 Volt with reference to a copper-copper sulphate electrode. The latter was invented by McCollum in 1910. K. Cohen (1902) was the first who used impressed current cathodic protection of metallic structures in liquid electrolytes. E. Cumberland (1905, USA) used impressed current for the protection of boilers. The term *electrochemical protection* for a method using sacrificial anodes was defined in 1913 at a conference in Switzerland.

This period was marked by the discovery (1910–1915) of martensitic stainless steel by Harry Brearley in England (the steel containing 12.8 wt% Cr for producing cutlery), ferritic stainless steels (14–16 wt% Cr) by Frederick Becket and Christian Dantsizen (1911) in the United States, and austenitic stainless steels (18 wt% Cr-8 wt% Ni) by Eduard Maurer and Benno Strauss (1912) in Germany, which opened a new era of knowledge, problems, and investigations in corrosion science and engineering. The latter material was exploited by the chemical industry from 1929 onward and became known as 18/8 austenitic stainless steel. Harry Brearley observed during metallographic work that an Fe-Cr alloy resisted attack by etchants (nitric acid in alcohol), vinegar, and lemon juice, and called his invention *rustless steel.* Later Ernest Stuart, a cutlery manager, gave the name *stainless steel* to corrosion resisting knives. This name was then also applied to the austenitic Fe-Cr-Ni steels. It is interesting to mention that in 1892 the British metallurgist Sir Robert Abbott Hadfield, investigating the alloy contained 9 wt% Cr, concluded that chromium did not influence the corrosion resistance of steel. However, Philip Monnartz (1908–1911) in Germany discovered the real virtues of chromium additions with respect to corrosion in oxidizing media, and recognized that the improved corrosion resistance was due to a passive surface layer (*inoxydable* in French or *rostfrei* in German).

Swiss engineer Max Ulrich Schoop (1913) discovered flame spraying (thermal spraying, metalizing, or metal spraying) coatings.

A Danish manufacturer, J.C. Hempel, invented the first antifouling marine paint in 1915.

Alkyd materials, polyvynilchloride and urea-formaldehyde began to be produced in 1920s.

E.C. Bain and W.E. Griffiths presented the first duplex stainless steel phase diagrams in 1927.

The 1930s were marked by the first impetus in the investigation of microbiologically induced corrosion (MIC): Von Wolzogen Kühr and van der Vlugt (1934, Denmark) showed that microorganisms in soil influence metallic corrosion. The genus *Desulphovibrio*, discovered by M.W. Bezerinck in 1895, was found to be responsible for anaerobic corrosion of cast iron in soil.

Stabilized stainless steels (SS 321 and SS 347) resistant to intergranular corrosion, were introduced during the early 1930s. Ni-Cr-Mo alloys were introduced in 1932. "Weathering steel" was developed in the USA in 1933 as CORTEN, a high-strength, low-alloy structural steel with high corrosion resistance in the atmosphere. This steel later found wide use in architectural design and statues (see Figure 3.7). The book *Corrosion, Causes and Preventing* (1933, USA) was written by Frank Newman Speller.

6.1.9 "Classical" Period (1935–1960)

The thermodynamic and kinetic approach to corrosion processes became established, new materials were produced for corrosion control, and new devices, instruments, and techniques were developed during this period. The year 1938 was marked by two events in the thermodynamics and kinetics of corrosion. The Belgian physical chemist Marcel Pourbaix carried out thermodynamic calculations based on the Nernst and Gibbs–Helmholtz equations for the iron-water equilibrium at 25°C. This approach allowed him to invent the *electrode potential-pH diagrams* and thus to define the conditions of corrosion, passivity, and immunity for iron in water (see Figure 4.4). Then Pourbaix created such diagrams for all metals in water, and in the presence of some contaminants, and published the *Atlas of Electrochemical Equilibria*.

Pourbaix initiated the formation of the Commission of Electrochemistry of the International Union of Pure and Applied Chemistry (IUPAC), the commission which received sign conventions for equilibrium electrode potentials (1953), and the period of misunderstanding in signs of electrode potentials came to an end.

Marcel Pourbaix (1904–1998)
(From the collection of his son Antoine Pourbaix)

The second event was attributed to C. Wagner and W. Traud (1938, Germany) who developed the *mixed-potential theory* based on the interpretation of corrosion processes as a superposition of the electrochemical partial reactions, which provided the theoretical foundation for the application of electrochemical corrosion testing techniques. Another important contribution by Wagner and Traud was the discovery that the slope of the total potential-current curve at the mixed potential (E_{mix}) can be used to calculate the rate of the corresponding electrochemical reaction, where the corrosion current is zero.

Victor Nightingall (1938, Australia) invented inorganic zinc silicate paints.

The Americans J.C. Hodge and J.L. Miller in 1940 identified stress corrosion cracking (SCC) of austenitic stainless steels associated with environments containing chlorides.

The shortage of nickel during World War II resulted in the development of high-manganese austenitic stainless steels, in which some or all of the nickel was replaced with manganese.

The American chemist Aaron Wachter in 1941–43 suggested using the organic compound *dicyclohexylamine nitrite* which, in the form of powder, was put inside a cannon, and sublimed, thus protecting the inner surface against atmospheric corrosion. Thus he invented volatile phase inhibitors.

A. Hickling (1942) described an electrical device for potential control in polarography which he called a *potentiostat*. The National Association of Corrosion Engineers was founded by 11 corrosion engineers in the pipeline industry in the USA in 1943. Low carbon stainless steels 304L and 316L, resistant to intergranular corrosion, were introduced about 1947. Precipitation hardening stainless steels (17-4 PH, 17% Cr and 4% Ni) were developed in the USA in the 1940s. H.H. Uhlig estimated the cost of corrosion to the economy in the USA in 1949.

High-performance austenitic alloy Carpenter 20, nickel alloys 200, 400, 600, and Ni base alloy "B", were developed during the pre-1950s period. High-performance austenitic alloy 20Cb, nickel alloys 800, 825, and Ni-Cr-Mo alloy "F", were also developed in the 1950s. M.H. Roberts (1954, England) indicated some drawbacks to Hickling's potentiostat, particularly the fact that the output current was unidirectional, and he described the design of a circuit with the output current reversible through zero and with a higher current capacity.

C. Edeleanu (1954, USA) suggested *anodic protection* for metals with active-passive transitions such as iron, chromium, nickel, titanium, and their alloys. J.L. Rosenfeld (1950s–1970s, Russia) carried out fundamental research into atmospheric corrosion and the protection mechanism of volatile phase inhibitors (VPI). N.D. Tomashov (1950s, Russia) developed the method of noble-metal alloying. The term *polarization resistance* was first used in 1951 by K.F. Bonhoeffer and W. Jena.

Milton Stern and A.L. Geary (1957) developed the *linear polarization resistance technique* for fast electrochemical corrosion rate measurement in liquid electrolytes. M. Romanoff (1957, USA) summarized the studies of underground corrosion conducted by the National Bureau of Standards in the USA from 1910 to 1955.

Duplex stainless steels (alloys with a microstructure consisting of about half ferrite and half austenite) were put into practice in the 1950s to meet the needs of the chemical industry for high strength linked to corrosion-resistance and wear-resistance.

The corrosion world saw the brilliant books on corrosion science and technology of U.R. Evans (1948, UK), H.H. Uhlig (1948, USA), J.L. Rosenfeld (1953, USSR), Todt Fritz (1958, Germany), and N.D. Tomashov (1959, USSR) in this period.

6.1.10 "Modern" Period (after 1960)

This period may be described as a period of corrosion monitoring, introduction of new materials, new methods of corrosion control, and intensive dissemination of corrosion knowledge. The First International Corrosion Congress was held in London in 1961. *Electrical resistance* and *linear polarization resistance* probes (corrosometers) were introduced in industry for on-line corrosion monitoring.

I. Epelboin (France) developed the *electrochemical impedance spectroscopy* (EIS) in the 1960s as a corrosion mechanism analytical tool.

Nickel alloy 800H, and Ni-Cr-Mo alloys G, 625, N and C-276 were developed in the 1960s.

Fusion-bonded epoxy (FBE) and three-layer (liquid or powder epoxy primer, co-terpolymer adhesive plus polyethylene or polypropylene) pipeline coatings were also introduced in the 1960s.

Austenitic steels 317L, 904L and 28, super-austenitic steels AL6X and 254 SMO, nickel alloy 800HT, Ni base alloy B-2, Ni-Cr-Mo alloys G-3, C-4, and high temperature alloys 188, 617 and 214 were developed in the 1970s. Austenitic steel 317LMN, super-austenitic steels AL6XN/25Mo6, 1925hMo/31, and 654 SMO, Ni-Cr-Mo alloys G-30, C-22/622, high temperature alloys 230 and HR160 were developed in the 1980s. Nickel base alloys B-3, B-4, B-10, Ni-Cr-Mo alloys 59, MAT 21, 686, C-2000, alloy 33, high temperature alloys 45TM, 602CA, 603 GT, 2100GT, and 626Si were developed in the 1990s.

Microscopic and spectroscopic methods in corrosion science were introduced in the 1970s–1980s. The use of *ellipsometry*, *Auger electron spectroscopy* (AES), *surface-enhanced Raman spectroscopy* (SERS), *electron spectroscopy for chemical analysis* (ESCA) or *X-ray-induced photo-electron spectroscopy* (XPS), *ion microprobe analysis* (SIMS), *scanning electron microscopy* (SEM) and *energy-dispersive spectroscopy* (EDS), *atomic force microscopy* (AFM), *transmission electron microscopy* (TEM), *Mössbauer spectroscopy*, *X-ray diffraction*, and *X-ray fluorescence*, *Fourier Transform Infrared Spectroscopy* (FTIR) were introduced for the study of corrosion products and films formed on metal surfaces during corrosion processes.

The first *powder coatings* appeared in Australia in 1967. The first publication about *electrochemical noise* (voltage and current transients, or fluctuations in corroding metals) was presented by Warren P. Iverson in 1968, but measurements based on electrochemical noise (EN) have been widely used to study corrosion processes, especially localized corrosion, since only the early 1980s. Today, on-line real-time monitoring of localized corrosion in industrial systems is carried out by means of EN measurements.

Surveys of corrosion costs were carried out in the United States, the United Kingdom, the former USSR, Germany, Sweden, Finland, India, and Japan in the 1960s–1970s. Studies in different countries identified corrosion costs as from about 2 to 4% of the gross national product, of which about 25% was identified as avoidable if the best anticorrosion measures and education were used. Similar studies followed in Australia (1983) and Kuwait (1995). Corrosion cost analysis was carried out in the USA in 1998. Many books on corrosion were published in the USA, UK, USSR, Germany, and Sweden after 1960. The encyclopedia *Corrosion* (Vol. 13) was published in 1987 (USA).

The events of the last 50 years need careful, further investigation and analysis because corrosion science and engineering encompasses many fields such as chemistry, metallurgy, materials science, microbiology, physical chemistry (especially, electrochemistry and surface chemistry), physics of solids, electronics, environmental chemistry, and ecology. Corrosion has become a wide interdisciplinary subject.

6.1.11 Conclusion

> What will history say?
> History will lie as always.
> George Bernard Shaw (1856–1950), the Irish dramatist, literary critic, and playwriter

6.1.11.1 A View to the Future

We are wary of the future, have to think, dream, and predict our future.

1. Who dealt with corrosion in the 19–20th centuries? Chemists and metallurgists. Electrical engineers dealt with cathodic protection. Mechanical engineers were responsible for the decision of corrosion problems in industry. Today the corrosion specialist must be a well and broadly educated person and know chemistry, physics, physical chemistry, materials, microbiology, and psychology. Integration of different sciences into corrosion on the one hand, and division of corrosion into various parts (according to the demands of industry) on the other hand, will be necessary. This integration will give the young generation a challenge in dealing with corrosion.
2. The *People Factor* is very important in corrosion failures. In 65 to 85% of corrosion failures, people are responsible. Our task: to teach and to learn, to find information quickly, to think integrally and interdisciplinarily, to connect knowledge in various disciplines in order to solve corrosion problems.
3. Corrosion science was determined as a science for metals, and was well studied in the 20th century. Today, we deal with corrosion of all materials: *polymers,*

ceramics and glasses, composites, semi-conductors, and *biomaterials*. Corrosion will be divided into corrosion of metals and corrosion of non-metallic materials.
4. Stephen Hawking: "If you could see what is going to happen in the future, you could change it". It is impossible to foresee the chemical behavior of materials in many cases, because there are many influencing and non-controlled factors. Only experiments are capable in the majority of cases of answering the main questions: will a material withstand given conditions? How long? What to do in order to prolong the life of the equipment?

Corrosion today is at the same stage as *alchemy* in the Middle Ages. We have to use the methods of nano-chemistry more and more, scanning microscopy in-situ, quantum mechanics in order to study the corrosion mechanism on the atomic and molecular level, and to predict the chemical behavior of materials under various conditions. I am not sure that we will be able to create simple models for all corrosion processes and to predict them because of the uncertrain behavior of atoms under particular corrosion conditions.

Recommended Literature

1. Evans, U.R., *An Introduction to Metallic Corrosion*, Edward Arnold Ltd., London, UK, 1963, pp. 217–227.
2. von Baeckman, W.G., Schwenk, W. and Prinz, W., *Handbook of Cathodic Corrosion Protection*, Gulf Publishing, USA, 1997, 520 pp.
3. Smith, C.S., *From Art to Science*, MIT Press, USA, 1982, 118 pp.
4. Smith, C.S., *Search for Structure: Selected Essays on Science, Art and History*, MIT Press, Cambridge, MA, USA, UK, 1981, 410 pp.
5. Revie, R.W., *Uhlig's Corrosion Handbook*, Second Edition, Wiley-Interscience, 2006, 1302 pp.
6. Frankel, G.S., Corrosion Science in the 21st Century, *J. Corr. Sci. and Eng.* **6**, Paper 028, Proceeding of "Corrosion Science in the 21st Century", UMIST, Manchester, England, 2003, 15 pp.

6.2 Corrosion and Philosophy

In many cases we do not know what will happen with metal, alloy, coating, inhibitor, polymer, or ceramic materials in contact with the environment if we do not carry out experiments.

The word *science* itself is the Latin word *scientia* meaning *knowledge*. Until the 1840s, that which we now call *science* was known as *natural philosophy*. It is not necessary to prove that many scientists are philosophers and vice versa. The absence of differences between *philosophy* and *science* in the past is reflected in society. All students receive the third degree PhD – Philosophical Doctor (Doctor of Philosophy). Therefore, the relationship between *philosophy* and *corrosion* as a part of science is a legitimate subject.

First of all, we have to define *philosophy* and subjects dealt with. Philosophy, in its meaning as *love of wisdom* (translation from Greek) has existed since 5 B.C. and reflects man's searching to understand the Universe and study the fundamental relationships between science, art, and life.

Philosophy encourages people to think critically, deeply, and rigorously. Therefore, any attempts to use philosophical methods are useful for a deeper understanding of corrosion phenomena.

6.2.1 Duality and Uncertainty in Corrosion

We are familiar with duality and uncertainty both in art and science. For example, the famous picture of Salvador Dali "Slave Market with the Disappearing Bust of Voltaire". We may see a bust of Voltaire or two nuns. This depends on our perception and concentration on the picture. We are familiar with the wave-particle duality in physics. The electron has a *unity*, while its wave or particle "faces" are the results of our experiments.

Duality and uncertainty exist in corrosion science. Red colored corrosion products show the presence of iron oxides (Fe_2O_3) or iron hydroxides (FeOOH). Black colored corrosion products show the potential presence of various sulphides (FeS, CuS, Ag_2S) or oxides (CuO, Fe_3O_4). Only additional examination will determine what they are.

Here are some other examples of duality and uncertainty in corrosion science:

(a) Pits may be formed as a result of various corrosion processes: galvanic corrosion, differential aeration cells, chloride attack, MIC, crevice corrosion, erosion, and cavitation. This means that some uncertainty exists in determining the corrosion process, its mechanism, and causes. We need to use additional methods in order to define the real mechanism and its causes. Sometimes this is impossible. For instance, MIC exists as a rule together with differential aeration cells and crevice corrosion. Therefore, uncertainty remains.

(b) Galvanic corrosion occurs if two metals contact each other in an electrolyte. Sometimes galvanic corrosion occurs if two metals do not contact each other. For instance, copper cations in water may result in galvanic corrosion of aluminum and carbon steel.

(c) Stress corrosion cracking (SCC). Cracks may occur as a result of the presence of chlorides, caustic, etc. It depends on the metal type and temperature. Even if we know the metal type, uncertainty remains in the definition of the chemical cause of SCC because several compounds may be present simultaneously and cause SCC.

(d) Several variables may together take part in cathodic corrosion processes. When corrosion of iron occurs in acidic solution, three components, cations H_3O^+, dissolved oxygen (O_2), and cations Fe^{3+}, may take part in cathodic reactions.

That is, there is uncertainty in the definition of components and their role in the cathodic processes.

Given knowledge of a cause, we can usually predict what the *effect* will be. The problem is that the relation between *cause* and *effect* is not symmetrical. Given a *cause*, there will be one effect. But given an effect, there could have been many causes able to produce the same effect. This is the main problem in the failure analysis in corrosion. We have to define the causes of failure if we know a result. A corrosion hole may be the result of pit or crack propagation. As in a philosophy, the causes and mechanism of formation of a hole in metal may be different. The study and definition of real corrosion mechanisms resembles the work of a philosopher based on some philosophical principles and laws which exist in science, in life, and in art.

6.2.2 Dialectics of Corrosion

Dialectics is a philosophical concept of evolution applied to diverse fields including nature, science, and technology. It is also a tool to understand the way things are and the way things change.

The first law of dialectics, *unity of opposites*, points out that every object and process in nature, life and art is made of opposing forces (sides). In mathematics: plus and minus, integral and differential. An atom is a unity of two opposites: positive nucleus and negative electrons. Many chemical reactions occur in two opposite directions at the same time. *Corrosion is the unity of two reactions, anodic and cathodic, occurring at the same time*. There is no corrosion if there is no participation of anodic or cathodic process. Many corrosion processes are described by a curve with a maximum: dependence of corrosion rate of carbon steel from the electrolyte concentration (see Figures 1.21, 1.22, 4.9 and D.1), influence of temperature on the corrosion rate of carbon steel in water in an open system (see Figure 1.25), dependence between electric potential and electric current for passivated metals and alloys (see Appendix D). These maximums reflect two opposites, more precisely, two factors or two tendencies that influence corrosion in opposite directions. In the case of the influence of NaCl concentration, these two factors are the concentration of chloride anions (conductivity) and dissolved oxygen concentration. The first factor prevails up to some added NaCl concentration. Added salt causes salting out of dissolved oxygen from a solution. Reduced oxygen (cathodic participant) concentration results in the decreasing of corrosion (see Figure 1.22). If we add peroxo-compounds in water, corrosion increases because of intensification of anodic process, but then corrosion decreases because of passivation by the same peroxo-compounds (when their concentration is enough large) (see Figure 4.9). If we increase the temperature of water, corrosion of iron increases because of temperature acceleration of the electrochemical anodic and cathodic reactions. But oxygen concentration in water decreases with an increase of temperature, and this factor prevails at temperatures above 80°C (in open water system) (see Figure 1.25). The presence of “maxima” always shows the existence of dialectical law *the unity of opposites*.

Here are some examples of the philosophical approach to corrosion:

1. Stainless steel (SS) is not recommended to use in sea water under stagnant conditions. However, SS is used in the same water when circulated.
2. Sodium carbonate is dangerous for aluminum and its alloys. But the same sodium carbonate is a corrosion inhibitor for iron alloys at ambient temperatures, and the same salt may cause stress corrosion cracking of iron alloys (caustic embrittlement) at high concentrations and at high temperatures.
3. Dissolved oxygen may cause and stimulate corrosion of iron alloys. A high oxygen concentration in water may cause passivation of iron alloys.
4. Small concentrations of anodic inhibitors may cause pitting corrosion. Large concentrations of the same inhibitors result in good protection.

“Philosophical” principles exist in many chemical phenomena: adsorption – desorption, dissolution of crystals – deposition of crystals. Corrosion and corrosion control is the *unity* of two reactions occurring in opposite directions:

Corrosion →

$$Fe_{(s)} \leftrightarrow Fe^{2+}_{(aq)} + 2e^{-}$$

← Protection

The second law of dialectics, *quantitative change becomes qualitative*, shows that gradual changes lead to turning points. The German philosopher Georg Wilhelm Friedrich Hegel said that “metal oxides are formed at certain quantitative points of oxidation”. He did not know that this example related to corrosion! Different quantitative ratios of iron and oxygen result in various quality of iron oxides:

$$2Fe_{(s)} + O_{2(g)} \rightarrow 2FeO_{(s)}, \tag{6.20}$$

$$3Fe_{(s)} + 2O_{2(g)} \rightarrow Fe_3O_{4(s)}, \tag{6.21}$$

$$4Fe_{(s)} + 3O_{2(g)} \rightarrow 2Fe_2O_{3(s)}. \tag{6.22}$$

We shall give two examples of this law. The Pourbaix diagram shows that change of electric potential and/or pH causes drastic change to a metal’s resistance: from full dissolution through passivity to immunity (see Figure 4.4). The second example concerns the stainless steel formation. If we add chromium to iron we receive stainless steel at 12% of chromium.

The third law of dialectics, *negation of negation*, shows that change moves in spirals, not circles. Many changes are cyclical. First one side dominates, then the other: day – night, autumn – winter – spring – summer, etc. Dialectics suggests that these cycles do not come back exactly to where they started. They do not make a perfect circle. Instead, change is evolutionary, moving in spiral. Hegel described the following example: "The bud disappears when the blossom breaks through ... the former is refuted by the latter ... the fruit appears ... in place of the blossom". The periodic table of the chemical elements of Mendeleev consists of periods of elements with properties changing according to atomic number. Every period returns to the "same and not the same" electron orbitals: lithium – sodium – potassium – rubidium, etc. Here are some examples of the "negation of negation" in corrosion.

Ore – iron is manufactured from the ore – equipment is produced from the iron – corrosion of equipment results in formation of the corrosion products (ore). Every stage is negated by the latter. Here is a scenario of MIC in the interpretation of the "negation of negation": formation of aerobic bacteria settlement – anaerobic bacteria settlement underneath – injection of biocides kills some bacteria, but not all; dead bacteria are the food for further development of new bacteria-induced corrosion. It should be noted that the law of "negation of negation" in corrosion in many cases is developed in the "negative" spirals.

6.2.3 Time in Corrosion, Philosophy and Art

Time is a very important category in philosophy. Time is a part of our existence. Time has a different meaning in science. For example, how do we relate to time in reversible and irreversible processes? Time is not important in chemical thermodynamics. Simply, there is no time in thermodynamics. Only entropy is related to time, because irreversible processes give birth to entropy. Increase of entropy distinguishes the future from the past, therefore the time arrow exists. In irreversible processes when heat "flows" from the higher to lower temperature, a marked direction of time appears. Time is very important in corrosion science and technology. We have to relate philosophically to time, when we use it in corrosion. It is very important to define the corrosion test duration. That is, how long must a certain metal be immersed in a particular media in order to define the corrosion rate? For example, the immersion time would be different in the case of iron in hydrochloric acid, in water, and in gasoline (see Section 1.4). There is no physical meaning of corrosion rate at the first moment: we cannot divide by zero ($t = 0$).

We cannot correctly define pitting, crevice, galvanic, and some other corrosion phenomena, if the duration time is less than an initiation time. Blisters on the coating surface do not appear at once. What is the lifetime of metallic construction in the case of MIC? How long have we to examine a titanium specimen in some medium? Why should we construct kinetic curves? How to define an optimal time for a corrosion test of alloys, coatings, and inhibitors in various media? What do these kinetic curves mean? The replies to these questions are connected with *time*. If we do not choose the time correctly in defining corrosion, we may fail tragically.

Time is both an "abstract" and a "real" term. One second or one million years – neither have a real meaning for a person. Time has real meaning for people when the period is optimal: a day, a week, a month, a year. The concept of time is relative, as described by Albert Einstein. It is recognized that time has only one direction.

However, the English cosmologist Stephen Hawking suggested that time had "real" and "imaginable" (unreal) values for the description of the Big Bang and the Big Crunch. In spite of the measuring of time very precisely, we feel its abstract essence.

We should mention that time is not only a physical category, but also a philosophical and poetic one. Time has many faces and destinations in art. The poet Joseph Brodsky (1940–1996) wrote: "Time is neutral ... Water is an appearance of time". This is correct, because water is neutral in the physico-chemical sense. That time is neutral, is correct for thermodynamics but not for kinetics. Sometimes time is stopped. The Israeli painter Alexander Okun wrote:

> It is wrong that *time passes* together with the page removed from the calendar. The time is settled in gardens and parks, in the fascinating old halls ... Time continues to exist tens, hundreds of years, after the calendar leaf decays and turns into nothing.

This means that we can stop time in art and in history, to enjoy by fixation of senses and events in the works of art. What about time in corrosion? Comparing with the "freezing time" in art and in history, we can say the same about the fixation time in corrosion phenomena. Pits, holes, cracks, corrosion products of different color tell about the time which was stopped for metals. Sometimes we may hear that "metal itself tells all about its history".

6.2.4 Corrosion and Entropy

The second law of thermodynamics states that *the entropy of the universe as an isolated system always increases*. Entropy is a property describing chaos and disorder. Corrosion is one of the examples of this law in reality showing an increase of entropy that is the increase of disorder and chaos. When we look at corrosion products, rust – substances of degradation – we see the increase of "chaos and disorder", increase of entropy on the "macro" scale. But when we look at the same corrosion products – substances of degradation – by means of a scanning electron microscope (SEM), we see some order in the "micro" scale (see Figure 3.12b). We may say that "micro-order" is mixed with "macro-chaos". The local decline of entropy (the beautiful harmonic shape, orderly arranged form of crystals) is mixed with the general increase of the entropy of the corroded system. We may observe here a great role of disorder and chaos. Disorder provides the freedom to design local order/environment.

The English modern cosmologist Stephen Hawking, when he introduced "imaginary" and "real" time in searching for the beginning and the end of the Universe, explained this in the following way:

> The curvature of space-time caused by the matter in the universe can lead to the three space directions and the imaginary time direction meeting up around the back. They would form a closed surface, like the surface of the Earth, without boundaries or edges. It wouldn't have any point that could be called a beginning or end, any more than the surface of the Earth has a beginning or end.

Fig. 6.4 Sasha Okun "Woman". (Reprinted with permission) (For a full color version of this figure, see the Color Section)

The same eternal principle exists in corrosion. There is no beginning or end for corrosion. Similar laws probably exist in art, science, engineering, and in society. The harmony of *beauty* exists together with *disharmony* in art. For example, the picture of modern Israeli painter Sasha Okun (Figure 6.4).

Such distortions are present in every person, object, or phenomenon. Therefore, we may say that *corrosion is a disharmony in technology, in engineering, and in life*.

It is well known that color, music, and words as forms of art may relax or excite and disturb, that is, they may create harmony or chaos. A similar thing exists in corrosion. Corrosion results in destruction, disorder, and chaos. But corrosion is not always an undesirable phenomenon.

6.2.5 Beneficial Applications of Corrosion Phenomena

The existence of "desirable" corrosion phenomena is the best proof of the philosophical approach to corrosion.

(a) *Use of galvanic corrosion.*

1.1. *Batteries* consist of two electrodes, anode and cathode, in electrolyte. The anode corrodes and electrical current is formed which is used in various devices for peoples' benefit.
1.2. *Cathodic protection*: sacrificial anodes. Even in the desirable corrosion phenomenon of the cathodic protection, we may consider the element of philosophy, as cathodic protection can cause the following problems: stray current corrosion; hydrogen embrittlement and coating debonding.

1.3. *Cleaning silver.* Many of the stains on silverware are due to silver sulphide. A simple electrochemical cleaning method consists of placing the silver in an aluminum pan containing water and baking soda. The electric current generated by the contact between silver and aluminum causes the silver sulphide to be reduced back to silver and corrosion of aluminum.

(b) *Control MIC* by means of intentional corrosion of copper in water. Corrosion of silver and, as a result, the presence of silver ions in water disinfect water. The army of Alexander the Great (Macedonian) was saved because the soldiers drank water from the river with helmets and vessels made of silver.

2. *Protection by means of corrosion products.*

2.1. *Patina* is the corrosion product formed on copper and bronze roofs, statues, and other artifacts, which is protective and provides a decorative appearance on the copper surface.

2.2. If iron is heated in dry air above 200°C, a thin black protective layer of Fe_3O_4 is formed on an iron surface. This method is called *blueing* and was widely used for protection of guns.

2.3. *Anodizing* of aluminum is a corrosion process which results in the formation of protective oxide layers of Al_2O_3 of good appearance.

3. Owing to corrosion of iron by water vapors at high temperatures

$$3Fe_{(s)} + 4H_2O_{(g)} \xrightleftharpoons{T<570^\circ C} Fe_3O_{4(s)} + 4H_{2(g)}, \tag{6.23}$$

$$Fe_{(s)} + H_2O_{(g)} \xleftrightarrow{T>570^\circ C} FeO_{(s)} + H_{2(g)} \tag{6.24}$$

hydrogen gas was firstly obtained in industry, and Antoine Lavoisier determined that water was composed of two elements.

4. Cations Fe^{2+}, Co^{2+}, and Ni^{2+} formed as a result of corrosion were included in ancient glasses for coloring.
5. *Electropolishing* is a "corrosion" technique used for the preparation of metallographic specimens, in which a high polish is produced by making the specimen the anode in an electrolytic cell, where preferential dissolution at high points smoothes the surface.
6. Corrosion of iron and steel led the English geologist Henry Clifton Sorby in 1863 to polish and etch polycrystalline alloys and to see grain boundaries by means of an optical microscope. Thus, he discovered the microstructure of steel owing to corrosion.
7. Catalysts with large surface areas are prepared by means of the use of the corrosion phenomenon *selective leaching*. *Raney nickel* is a solid catalyst produced from a *nickel-aluminium* alloy by treating with concentrated sodium hydroxide. Aluminum corrodes and is leached of the alloy, and porous structure left behind has a large surface area, which gives high catalytic activity and is used in many chemical industrial processes.
8. The operation of ER (Electrical Resistance)-probes, LPR (Linear Polarization Resistance)-corrosometers, galvanic probes and ENM (Electrochemical Noise

Measurements) instruments used widely in corrosion monitoring, is based on the corrosion of sensitive metallic elements.

9. Manufacture of salts and other chemicals by reaction of metals with different reagents. Copper sulphate is produced by the dissolution (corrosion!) of copper in sulphuric acid.
10. Assaying includes cupellation and parting. Cupellation is the separation of gold or silver from impurities by melting the impure metal in a cupel and directing a blast of hot air onto it. The impurities, including lead, copper, tin, and other unwanted metals, are oxidized (corroded!). Parting (the separation of gold and silver) is accomplished by boiling gold and silver in nitric acid. The silver dissolves (corrodes!) forming silver nitrate, leaving a residue of gold.
11. Without corrosion of iron, life would not exist. The cations Fe^{2+}, the corrosion product of iron, are present in hemoglobin. Lack of iron in the human organism causes anaemia. Recipes of many iron medicines have been known since ancient times. Chlorophyll contains the corrosion product of magnesium (Mg^{2+}). Vitamin B12 is a porphyrin containing a corrosion product of cobalt (Co^{2+}) in the center of a large organic molecule.
12. Corrosion of weapons weakens or strengthens an army. Corrosion of our weapons is a detrimental phenomenon for us, but is beneficial when the enemy's weapons are affected.
13. Bacteria *thiobacillus thiooxidans* produce sulphuric acid which is used for cleaning concrete surfaces from radioactive contaminants, in spite of the same bacteria participate in MIC.
14. Cavitation is a detrimental corrosion phenomenon for impellers, blades, propellers, etc. But cavitation may be used for MIC control, namely, for killing microorganisms in water, and for environmental protection dispersing spilled oil.
15. Corrosion gives "employment" and "food" for many people: those who manufacture metals from ores to make new equipment instead of corroded ones, and of course, to those of us who deal with corrosion and "feed" our families. There is a little corrosion humor there. Corrosionists are funny guys, really?
16. Anodic polarization usually causes intensive corrosion of metals. But anodic polarization (controlled oxidation – corrosion!) of some metals and alloys under particular conditions causes passivation.

I would like to mention a joke. Moses said: "All is there", and showed the Sky (God). King Shlomo said: "All is here", and showed the Head. Jesus said: "All is here". And put his Hands on the Heart. Karl Marx proclaimed: "All is here", and knocked the stomach. Sigmund Freud said: "All is here" ... and showed something below the stomach. Albert Einstein came and said: "All is relative!"

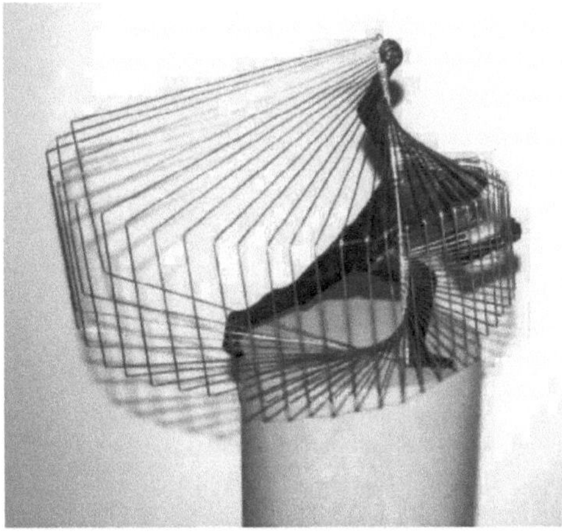

Fig. 6.5 Noam Ben-Jacov "Motion". (Reprinted with permission) (For a full color version of this figure, see the Color Section)

6.3 Corrosion and Art

We want to connect two different topics, corrosion and art, both playing an important role in life. Imagination and the inclusion of all our senses help to describe corrosion phenomena as well as anticorrosion measures through the medium of art, and in such manner as to see the beauty of the corrosion world.

There is much in common between *science* and *art*. For example, the mutual influence on society and its development. The French writer Jules Verne (1828–1905) inspired several generations of scientists and engineers. Some scientists published their ideas in fiction.

We may mention the Latin poet and philosopher Titus Lucretius Carus who in the 1st century B.C. wrote a poem "On the Nature of Things" in which the physical theory of the Greek philosopher Epicurus (341–270 B.C.) was described. Or more recently, the Swiss-French philosopher Jean-Jacques Rousseau (1712–1778) propagated his theory about education of the new generation in his literary productions "Émile", "Equal Rights", and others.

Certainly, there are not only similarities between art and science. There are many differences. For example, when Charlie Chaplin invited Albert Einstein to an evening party, and saw how people were delighted with him, he said: "They admire me, because I do something that all understand. And they are fascinated with you, because you do something that nobody understands".

The Israeli sculptor Noam Ben-Jacov deals with different corrosion resistant alloys and fuses his sculptures with dancing (Figure 6.5). Many artists connect various fields of art and combine them with materials engineering.

A similar situation exists in science. People connect various, sometimes contrary, forms of science. But in contrast to art, the combination process in science goes naturally. For example, we know the usefulness of physical chemistry and ma-

Fig. 6.6 Beauty of corroded Raschig rings in an oil refinery. (For a full color version of this figure, see the Color Section)

terial science in corrosion and in medicine. Corrosion science and engineering include chemistry (especially physical chemistry, electrochemistry, surface phenomena), metallurgy, materials science, physics, electricity, microbiology, economics, and even psychology.

Photographers, with the help of the camera's lens, derive pleasure from art works combining industrial elements. Here is an example of how photographers see art in industry (Figure 6.6).

I write about this in order to support my investigation and demonstrate the relationship between corrosion and art. Corrosion is divided into corrosion science and corrosion engineering. Synergism exists between these two fields. They complement each other. Corrosion influences, directly or indirectly, all spheres of our life.

Corrosion science and corrosion engineering deal with deterioration of materials and preventive protection methods. Art deals with creation or expression of something beautiful, especially in visual, auditory music, or verbal forms. The aim of this section is to find a relationship between corrosion and various forms of art.

One interesting unusual structure attracted attention at the World Exhibition in Brussels in 1958 (Figure 6.7a). This is a model of an elementary cell of α-iron with the magnification of 165 billion. This structure is very important in corrosion science. Many metals have such a structure: vanadium, tantalum, tungsten, α-chromium, molybdenum, niobium, barium, and alkali metals.

Statues and sculptures have been made of various metals and alloys over hundreds of years. They demanded preventive anticorrosion measures. Stainless steel with a decorative and protective titanium nitride (TiN) coating was used instead of gilded iron for the roof of the Cathedral of Christ the Savior in Moscow (Figure 6.8).

The sculptor Natasha Carsberg from England creates sculptures from carbon steel treated with chemicals and natural pigments (Figure 6.9). We take pleasure from this corroded industrial material which resembles organic and ephemeral forms. The people of art see science in art, we see art in science.

(a) **(b)**

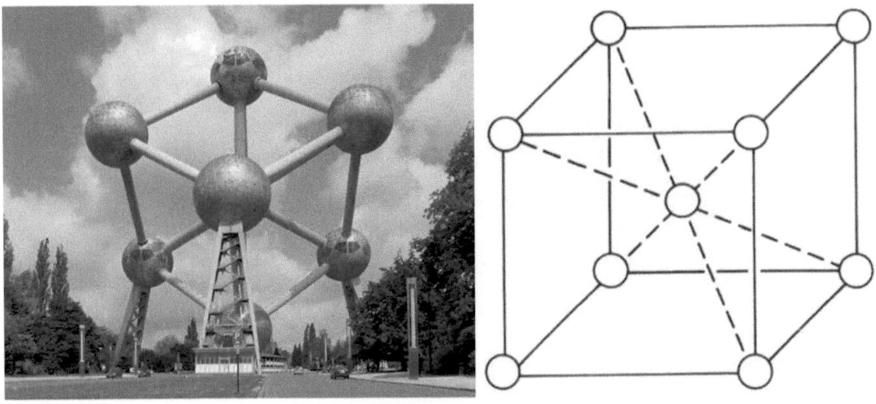

Fig. 6.7 (a) Atomium, World Industrial Exposition, 1958, Brussels, Belgium and (b) structure of α-iron. Photo: Henk Binnendijk (Reprinted with permission)

Fig. 6.8 Cathedral of Christ the Savior (Moscow), stainless steel with decorative and protective titanium nitride (TiN) coating. (For a full color version of this figure, see the Color Section)

In the open air museum in Tefen, Israel, there are many statues and sculptures, which bring thoughts about relationships between corrosion and art. One of them is a statue made of bronze "Come on! Let us fight against corrosion!" (Figure 6.10). One sees the good resistance of bronze to the atmosphere of northern Israel.

We may find in some works of art an analogy of corrosion phenomena. The weightless statue "Tall Figure" of the Swiss sculptor and painter Alberto Giacometti resembles *uniform corrosion* (Figure 6.11).

If the work of Giacometti has been compared to that of the existentialists in literature, uniform corrosion of equipment would be represented as "corrosion exist-

Fig. 6.9 "Bone structure V", carbon steel, treated with vinegar and natural pigments, Natasha Carsberg, England. (Reprinted with permission) (For a full color version of this figure, see the Color Section)

Fig. 6.10 "Come on! Let us fight against corrosion!" Bronze, Israel. (For a full color version of this figure, see the Color Section)

entialism". As existentialists interpreted human existence in the world that stresses its concreteness, subjectivity, individual freedom, and choice, metals' existence is also stressed with variables in the environment. The "Pit" of the American painter Philip Guston (1913–1980) resembles pitting corrosion in the form of shallow pits (Figure 6.12).

The sculpture of the Italian futurist painter and sculptor Umberto Boccioni can be seen as a Neitzschian superman dynamically striding through time and space (Figure 6.13). We see in Boccioni's sculpture the erosion phenomenon which coincides with speed, according to the French philosopher Henri Bergson (1859–1941),

Fig. 6.11 Alberto Giacometti (1901–1966) "Tall Figure" (1949, painted bronze), resembles uniform corrosion. © Photo Scala, Florence, The Museum of Modern Art, New York/Scala Florence, 2009. (For a full color version of this figure, see the Color Section)

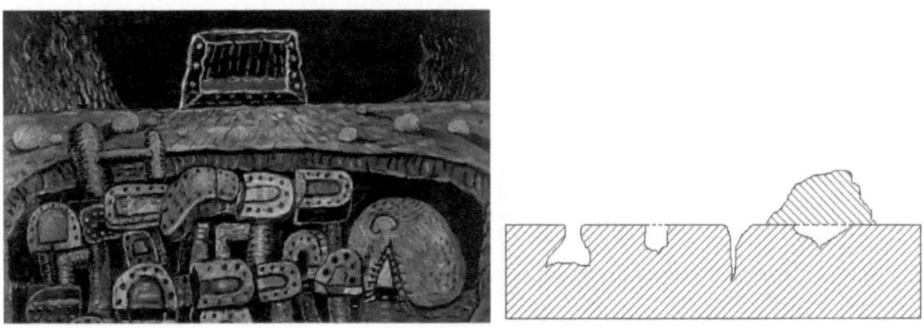

Fig. 6.12 Philip Guston "Pit" (1976) National Gallery of Australia, Canberra. © The Estate of Philip Guston. Pitting corrosion – severe local corrosion. Reprinted with permission. (For a full color version of this figure, see the Color Section)

who saw in art "the material world melted back into a single flux and a continuity of flowing". We may look at this sculpture as a "kinetic sculpture". Exactly, erosion is a kinetic phenomenon.

a b c

Fig. 6.13 Umberto Boccioni (1882–1916) "Unique Forms of Continuity in Space" (1913) or "Erosion". (a, c) Erosion inside brass tube; (b) the sculpture of Umberto Boccioni. © Photo Scala, Florence, The Museum of Modern Art, New York/Scala, Florence, 2009. (For a full color version of this figure, see the Color Section)

Hieronymus Bosch (1450–1516) with his brilliant abstract and surrealistic pictures contributed a lot to the "description" of corrosion processes. One may imagine and see in his triptych "The Garden of Earthly Delights" three periods of the metallic life (Figure 6.14).

This work is filled with symbolic detail: the left panel "Paradise" represents the creation of Adam and Eve (imagine new metallic structure without defects); the central panel "Haywain" shows life on earth as a continuous process of sinning (imagine beginning of the corrosion), and the right "Hell" depicts an eternal hell (imagine the jubilation of corrosion – destruction). Here we may compare with the ancient Chinese Taoism philosophy: all things natural are continually in a process of appearing and disappearing. All processes in nature and industry resemble the swing of the pendulum between being and nothing, growth and decay, *activity* and *passivity*. Therefore, corrosion is a brilliant example of such a philosophical approach to the eternal processes. This means that it is impossible to completely prevent degradation of metallic equipment as it is impossible to prevent flowing time. And we can only delay and control the corrosion process.

There are many examples how anti-corrosion coatings combine with aesthetics. A fuel storage tank farm was close to Tel-Aviv (Israel). This is a general situation in many countries. Local painters created good image, and tanks were painted according to requirements of anticorrosion protection, landscape, beauty, and aesthetics (Figure 6.15).

Rust and rusting usually have a bad name, even negative, in literature. The Russian writer Anton Chekhov, in his story "My Life", said: "A plant-louse eats grass, rust eats iron, and a falsehood eats a soul". But we know that rust possesses protective properties in particular conditions, for examples, in the atmosphere. One of the applications of protective corrosion product films is in weathering steels *CORTEN* (see Figure 3.7).

Fig. 6.14 Hieronymus Bosch "The Garden of Earthly Delights" (1510–1515) and three periods of car's life: new car, beginning of corrosion, and destruction. (Reprinted with permission, © Museo Nacional del Prado, Madrid, Spain.) (For a full color version of this figure, see the Color Section)

Fig. 6.15 Anti-corrosion protection as well as decoration of a storage tank (Israel). (For a full color version of this figure, see the Color Section)

We can read in many books where writers mention *rust* or compare people with metals. For example, we may read: "His face was *bronze*", "His body was like *iron*", "He touched a door-handle coated with *chromium* and it had *rusted* spots", "The *lead* clouds", etc.

Fig. 6.16 Russian version of the "The Wizard of Oz" (1939): "Iron Woodman". Dorothy brings an oil-can against atmospheric corrosion. (For a full color version of this figure, see the Color Section)

The American fairy tale "The Wizard of Oz" written by Lyman Frank Baum (1856–1919) in 1900, is the first literature story which describes the corrosion resistance of metallic implants made of *tin* under atmospheric conditions and preventive anticorrosion measures. Every time the woodchopper cut his legs, arms, body, and head, a tinsmith made him new ones of tin, and his name became *Tin* Woodman. Nowdays, one hundred years later we know that implanted organs are made of titanium alloys, stainless steel, and platinum, because their corrosion resistance is higher than that of tin. The Russian writer Alexander Volkov (1891–1977) when translating "made" the woodchopper from iron (Figure 6.16). The anticorrosion measure against atmospheric corrosion was described in this fairy tale. Oil! The girl Dorothy had to bring an oil-can and to oil in order to save "Iron" woodman from corrosion. Thus, this tale teaches children correct anti-corrosion treatment of iron objects.

The world of fairy tales is a source in study of the behavior of metals. The German authors Brothers Grimm wrote "The *Golden* Key", "The *Golden* Duck (Goose)", "The *Iron* Hans", "The Frog King and *Iron* Heinrich", and "The *Iron* Stove". Children may understand from these tales that iron rusts in contact with water, but *gold* is a resistant material in aqueous solutions, atmosphere and soil. "The *Golden* Key" is an example of the description in literature of prevention of galvanic corrosion. The *gold* key was stored separately from the small *iron* chest. One should understand from this tale that two dissimilar metals (gold and iron) must not be in physical contact in the soil (electrolyte). Otherwise, galvanic corrosion will occur.

When we read the tale "The Brave Tin Soldier" by Hans Christian Andersen we learn that a soldier made of tin is resistant to water and organic liquids contained in a fish's stomach. But the melting point of tin is low, therefore "the brave tin soldier" disappeared after being thrown into a furnace.

Rusted equipment does not give a good feeling or pleasant sense. But when we look at *rust* through "art eyes" or the scanning electron microscope (SEM) we can imagine various interesting forms competitive with art. Here are several examples of such pictures of rusted surfaces by means of SEM and even naked eyes.

a b

Fig. 6.17 (a) "The Old Trees and Waterfall" (China, 1549, National Palace Museum, Taiwan, Republic of China); (b) carbon steel strips after immersion in sodium peroxocarbonate aqueous solutions. (For a full color version of this figure, see the Color Section)

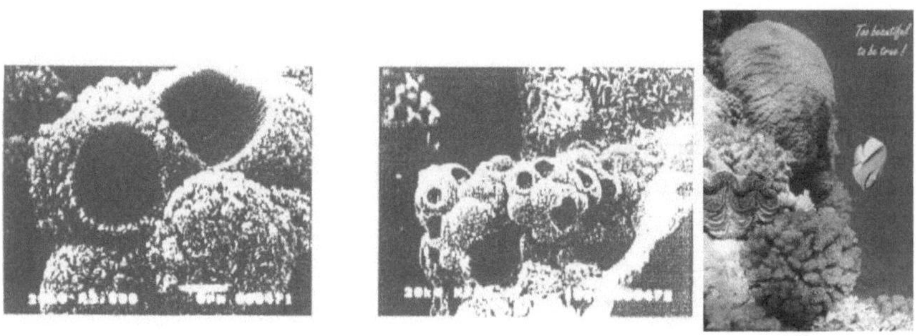

a b

Fig. 6.18 (a) SEM photo of rust (×3,500 and ×1,500; (b) coral reefs. (For a full color version of this figure, see the Color Section)

Many people enjoy Chinese pictures of the 16th century. Carbon steel coupons immersing in the aqueous solutions of peroxides are not less beautiful than the Chinese pictures (Figure 6.17).

The friable rust formed inside a gasoline storage tank is not a pleasant sight. But when we observe this rust by means of SEM, we see various beautiful biological formations suggesting "birds' nests", or wonderful "coral reefs" (Figure 6.18).

Fig. 6.19 (a) SEM photo (× 10,000) of rust formed in the heat exchanger of a overhead crude distillation column; (b) black sea urchins. (For a full color version of this figure, see the Color Section)

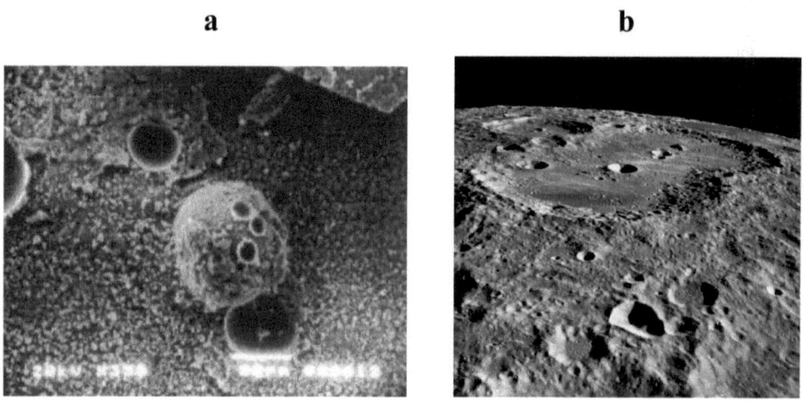

Fig. 6.20 (a) SEM photo (× 350) of rust after immersion of carbon steel strip in water-kerosene mixtures; (b) the Moon's surface. (For a full color version of this figure, see the Color Section)

Rust formed in the overhead of the atmospheric tower for crude oil distillation when magnified 10,000 times, resembles the black sea urchins or leaves of the pinetrees needles (Figure 6.19).

Rust formed on a steel surface in a water-kerosene mixture when magnified 350 times, resembles the Moon's surface with beautiful fantastic objects (Figure 6.20).

Humor is one of the very important factors in life. I would like to list several examples of how humor helps in corrosion education. When the Belgian physicochemist Marcel Pourbaix calculated and presented his thermodynamic "potential-pH" diagrams, the generations of students and engineers studying various zones in this diagram found it difficult to remember (Figure 6.21a). An interesting picture of this diagram was carried out by a painter Vladimir Frenkel (Figure 6.21b).

There are so many examples in nature and in life that resemble corrosion phenomena. *Fretting corrosion* – the accelerated deterioration at the interface between

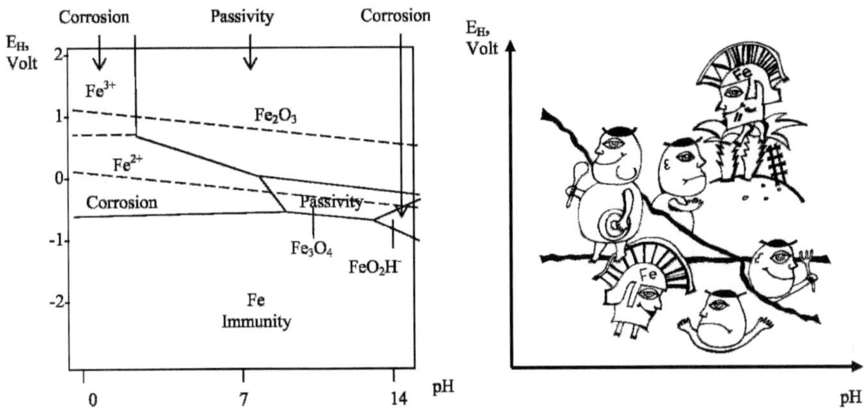

Fig. 6.21 (a) Diagram "potential (E_H) – pH" as Marcel Pourbaix suggested; (b) diagram "potential (E_H) – pH" as painter Vladimir Frenkel proposed.

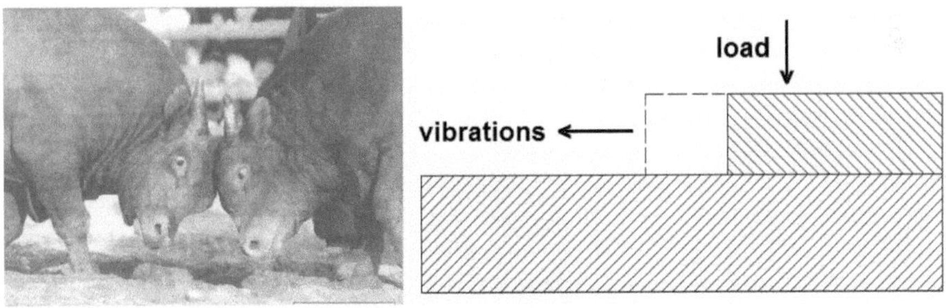

Fig. 6.22 "Fretting corrosion" – attack due to rubbing contact. (For a full color version of this figure, see the Color Section)

contacting surfaces as the result of corrosion and oscillatory movement between the two surfaces (Figure 6.22).

Impingement corrosion or *impingement attack* – a form of erosion-corrosion generally associated with the local impingement of a high-velocity substance (environment) moving against another solid surface. Usually, as we see, impingement is an unexpected, a sudden, surprise attack (Figure 6.23).

There are some phenomena in nature that even people are not able to create. Here is a carbon steel mast of the ship found not so far from Haifa Bay (Figure 6.24). This mast corroded under Mediterranean Sea weather into large holes. The birds built nests inside of the corrosion holes and live there with love.

An interesting definition exists: "Corrosion rate is a number inversely proportional to the corrosion engineer's remaining tenure on the job".

Fig. 6.23 "Impingement attack" – localized attack related to high velocity. (Right – impingement by escaping steam from cracked tube). (For a full color version of this figure, see the Color Section)

Fig. 6.24 "Corrosion led to contentment". Corroded mast of a ship (Haifa Bay). (For a full color version of this figure, see the Color Section)

6.3.1 Philosophy of Links between Corrosion and Art

Both *corrosion science* and *art* have much in common:

(a) Art has been developed from ancient times. Every historical period gives impulse to the development of new forms of art. It is impossible to complete art. Similar to art, corrosion science is in continuous evolution.
(b) Another similarity: *the system of attempts*. The artist creates several versions and chooses the best one. A similar process exists in corrosion science. Corrosion scientists and engineers carry out experiments with the aim of obtaining good useful results.
(c) Another common aspect is the use of organs of senses (sight, sound, smell, taste, touch, and intuition) in comprehending art and corrosion phenonena.

An engineer is listening to a pump in order to define a *cavitation* – a specific noise similar to the sounds which we hear during the movement of stones. A mechanical engineer can identify engine faults by listening to the "noise". We are able to define corrosion products according to their color (see Appendix I). We detect ammonia, hydrogen sulphide, hydrochloric acid, acetone, acetic acid according to their smell. We can define the presence of biofouling at a metal surface according to the specific slippery feel of slime. The black color of biofilm points out the presence of sulphate-reducing bacteria.

We may emphasize some difference between art and corrosion science in the sphere of use of organs of sense. Not the whole spectrum of electromagnetic waves is used in art. Unlike art, such methods as *acoustic emission* are used for the detection of cracks (materials "cry" when atomic bonds break), or *electrochemical noise measurements* are the "ears" of corrosion monitoring and are used for the detection of corrosion processes, especially localized attack. Why link *corrosion* and *art*?

(a) First, this comparison helps in the study and understanding of corrosion processes and phenomena. In addition, the subject studied acquires special "color", and the learning process becomes enjoyable. Aesthetic values enter corrosion science, exactly as in art. People understand and remember well when receiving explanations by means of a *system of analogies*.
(b) The second reply concerns the philosophical aspect. We look at *beauty* during the whole of our lives without even being conscious of it. If the aim of art is the creation of beauty, the main aim of corrosion science and engineering is to save the beauty of equipment and structures from deterioration.

We see various works of art in corrosion of materials. That shows that there is one indivisible world. The English poet and artist William Blake (1757–1827) wrote:

> To see a World in a Grain of Sand
> And a Heaven in a Wild Flower,
> Hold Infinity in the Palm of your hand
> And Eternity in an hour.

Recommended Literature

1. Smith, C.S., *From Art to Science*, 1982, MIT Press, 118 pp.
2. Smith, C.S., *Search for Structure: Selected Essays on Science, Art and History*, 1981, MIT Press, Cambridge, Massachusetts and London, England, 410 pp.
3. Smith, C.S., *History of Metallography*, 1988, MIT Press.
4. Groysman, A., Aesthetic, Philosophical and Historical Aspects in the Physical Chemistry Education. In: *Trends in Electrochemistry and Corrosion at the Beginning of the 21st Century*, Dedicated to Professor Dr. Josep M. Costa on the occasion of his 70th birthday, E. Brillas and P.-L. Cabot (Eds.), Publicacions Universitat de Barcelona, 2004, pp. 1203–1225.

Appendix A
Thermodynamics of Oxidation of Iron and Carbon Steels in Water

The Belgian scientist Marcel Pourbaix (1904–1998) was the first who showed the power of thermodynamics in the definition of the tendency corrosion of metals in pure water at 25°C [1]. Townsend [2] expanded the thermodynamics diagram of Marcel Pourbaix for the system Fe–H_2O up to 200°C and showed, by means of complicated calculations, that corrosion regions are increased at low and high pH. Let us show how simple thermodynamic calculations predict the influence of temperature and pressure on the corrosion reactions of iron with water and some components dissolved in aqueous solutions.

A.1 Reactions of Iron and Steels with Water

When there is no dissolved oxygen in water, iron reacts with water according to the following reactions:

$$Fe_{(s)} + H_2O_{(l)} \rightarrow FeO_{(s)} + H_{2(g)}, \tag{A.1}$$

$$\Delta G^\circ_{298} = -14.26\,\text{kJ/mol}, \quad \Delta S^\circ_{298} = +94.16\,\text{J/(mol}\cdot\text{K)},$$

$$Fe_{(s)} + 2H_2O_{(l)} \rightarrow Fe(OH)_{2(s)} + H_{2(g)}, \tag{A.2}$$

$$\Delta G^\circ_{298} = -15.72\,\text{kJ/mol}, \quad \Delta S^\circ_{298} = +51.36\,\text{J/(mol}\cdot\text{K)}.$$

ΔG°_{298} and ΔS°_{298} are changes of Gibbs energy and entropy for the reactions (A.1) and (A.2).

The data for the calculations of ΔG°_{298} and ΔS°_{298} were taken from [3].

$\Delta G^\circ_{298} < 0$ for the reactions (A.1) and (A.2). This means that they occur spontaneously at standard conditions ($T = 298$ K, $P = 1$ atm). "Spontaneous" does not mean "fast". The rate of these reactions may be very low. $\Delta S^\circ_{298} > 0$ for the reactions (A.1) and (A.2). This means that a rise in temperature results in an increase of the thermodynamic tendency of the reaction between Fe and H_2O. This derives from the thermodynamic equation (see [4]):

$$\Delta S^\circ_{298} = -\left(\frac{\partial \Delta G^\circ_{298}}{\partial T}\right)_P.$$

When $\Delta S^\circ_{298} > 0$, $(\partial \Delta G^\circ_{298}/\partial T)_P < 0$, and an increase of temperature will result in a decrease of ΔG°_{298}.

In addition to the reactions (A.1) and (A.2), the following reactions may occur:

$$Fe_{(s)} + 3H_2O_{(l)} \rightarrow Fe(OH)_{3(s)} + 3/2H_{2(g)}, \tag{A.3}$$

$$\Delta G^\circ_{298} = +6.42\ \text{kJ/mol}, \quad \Delta S^\circ_{298} = +63.45\ \text{J/(mol}\cdot\text{K)},$$

$$2Fe_{(s)} + 3H_2O_{(l)} \rightarrow Fe_2O_{3(s)} + 3H_{2(g)}, \tag{A.4}$$

$$\Delta G^\circ_{298} = -30.78\ \text{kJ/mol}, \quad \Delta S^\circ_{298} = +214.95\ \text{J/(mol}\cdot\text{K)}.$$

For reaction (A.3), $\Delta G^\circ_{298} > 0$; this means that this reaction is not spontaneous at 25°C. For reaction (A.4), $\Delta G^\circ_{298} < 0$; and this reaction can spontaneously occur at 25°C. For reaction (A.4), $\Delta S^\circ_{298} > 0$; this means that an increase of temperature results in an increase in the tendency of a reaction between Fe and H_2O with the formation of the haematite Fe_2O_3. Ferrous hydroxide (II) $Fe(OH)_2$ may convert to magnetite Fe_3O_4 according to the reaction

$$3Fe(OH)_{2(s)} \rightarrow Fe_3O_{4(s)} + 2H_2O_{(l)} + H_{2(g)}, \tag{A.5}$$

$$\Delta G^\circ_{298} = -19.68\ \text{kJ/mol}, \quad \Delta S^\circ_{298} = +152.15\ \text{J/(mol}\cdot\text{K)}.$$

As $\Delta S^\circ_{298} > 0$, the increase of temperature results in increasing the decomposition tendency of $Fe(OH)_2$. The most stable form for iron oxide is Fe_2O_3.

$$2Fe(OH)_{3(s)} \rightarrow Fe_2O_{3(s)} + 3H_2O_{(l)}, \tag{A.6}$$

$$\Delta G^\circ_{298} = -43.62\ \text{kJ/mol}, \quad \Delta S^\circ_{298} = +88.05\ \text{J/(mol}\cdot\text{K)}.$$

Ferric hydroxide (III) $Fe(OH)_3$ is not thermodynamically stable and, on transforming to Fe_2O_3, gives intermediate products characterized by partial dehydration and the polymeric structure of ferric hydroxides.

An increase of hydrogen gas pressure in reactions (A.1), (A.2), and (A.4) should decrease and even stop the corrosion of iron. For the determination of the minimum pressure of hydrogen gas in these reactions, in order to prevent corrosion of iron in water, we can use the equation [4]:

$$\Delta G(P_2) = \Delta G(P_1) + \Delta n RT \cdot \ln\left(\frac{P_2}{P_1}\right), \tag{A.7}$$

$\Delta G(P_2)$ – change of Gibbs energy for chemical reaction at pressure P_2; $\Delta G(P_1)$ – change of Gibbs energy for the same chemical reaction at pressure P_1 (usually P_1 is the standard pressure, 1 atm); Δn – change in moles of gaseous substances: $\Delta n = \Sigma n$ (gaseous products) – Σn (gaseous reactants); R – universal gaseous constant; T – temperature in Kelvin. If we know $\Delta G(P_1) = \Delta G$ (1 atm) $= -14.26$ kJ/mol for reaction (A.1), it is simple to calculate the minimum hydrogen pressure P_2 for reaction (A.1) in order to stop it. For this we have to equal $\Delta G(P_2) = 0$ (at the equilibrium), that is,

$$0 = -14.26\ \text{kJ/mol} + 1\ \text{mol} \cdot 8.314\ \text{J/(mol} \cdot \text{K)} \cdot 10^{-3}\ \text{(kJ/J)} \cdot 298\ \text{K} \cdot \ln\left(\frac{P_2}{1}\right). \quad \text{(A.8)}$$

Calculation according to (A.8) gives $P_2 = 316$ atm, that is, corrosion of iron according to reaction (A.1) will stop at a hydrogen pressure above 316 atm. Similar calculations for reactions (A.2) and (A.4) will give $P_2 = 570$ atm and $P_2 = 63$ atm, respectively. Such simple and enjoyable thermodynamic calculations result in very important conclusions:

1. Pure iron and iron in alloys can be oxidized with water (liquid or steam) to haematite Fe_2O_3 or magnetite Fe_3O_4 at ambient temperatures. The increase of temperature results in the growth of the tendency of reactions (A.1), (A.2) and (A.4) between iron (including steels) and water.
2. The increase of hydrogen pressure results in diminishing of the tendency of iron to react with water. Iron will be resistant to water at high hydrogen pressures. For reaction (A.1), iron will be resistant at $P_{H_2} > 316$ atm, for reaction (A.2) at $P_{H_2} > 570$ atm, and for reaction (A.4) at $P_{H_2} > 63$ atm, respectively.

A.2 Thermodynamics of Oxidation of Ferrous Ions with Oxygen in Aqueous Solutions

Ferrous cations Fe^{2+} can be oxidized with dissolved oxygen in water according to the following reactions:

$$Fe^{2+}_{(aq)} + O_{2(aq)} + H^{+}_{(aq)} \rightarrow Fe^{3+}_{(aq)} + HO_{2(aq)},\ \Delta G^{\circ}_{298} = +86.95\text{kJ/mol}, \quad \text{(A.9)}$$

$$2Fe^{2+}_{(aq)} + O_{2(aq)} + 2H^{+}_{(aq)} \rightarrow 2Fe^{3+}_{(aq)} + H_2O_{2(aq)},\ \Delta G^{\circ}_{298} = +12.8\text{kJ/mol}, \quad \text{(A.10)}$$

$$4Fe^{2+}_{(aq)} + O_{2(aq)} + 4H^{+}_{(aq)} \rightarrow 4Fe^{3+}_{(aq)} + 2H_2O_{(l)},\ \Delta G^{\circ}_{298} = -177.40\ \text{kJ/mol}. \quad \text{(A.11)}$$

Only the last reaction (A.11) may occur spontaneously.

A.3 Reaction of Pure Iron Fe with Ferric Cations Fe^{3+} in Aqueous Solutions

In order to understand the processes occurring in the system iron–water–oxygen, we have to know the thermodynamic character of the following reaction:

$$Fe_{(s)} + 2Fe^{3+}_{(aq)} \rightarrow 3Fe^{2+}_{(aq)}, \quad \Delta G^{\circ}_{298} = -233.5\,\text{kJ/mol},$$
$$\Delta S^{\circ}_{298} = +273\,\text{J/(mol}\cdot\text{K)}. \qquad \text{(A.12)}$$

Equilibrium constant K for this reaction:

$$\log K = \frac{-\Delta G^{\circ}_{298}}{2.3RT} = \frac{-233.5\cdot 10^3}{2.3\cdot 8.314\cdot 298} = 41,$$

$$K = \frac{a^3_{Fe^{2+}}}{a^2_{Fe^{3+}}} = 10^{41}.$$

$a_{Fe^{2+}}$ and $a_{Fe^{3+}}$ are the activities of ferrous and ferric cations, respectively, in a solution; $T = 298$ K; $R = 8.314$ J/(mol·K).

These data show that iron is not resistant in water in the presence of dissolved Fe^{3+} cations from the thermodynamic point of view, and if the concentration of Fe^{3+} cations in a solution is high, metallic iron will dissolve to full dissolution. An increase of temperature will decrease the iron resistance to ferric cations Fe^{3+}, as $\Delta S^{\circ}_{298} > 0$ for reaction (A.12).

References

1. Pourbaix, M., *Atlas of Electrochemical Equilibria in Aqueous Solutions*, Pergamon Press, Oxford, 1966, 644 pp.
2. Townsend, H.E., Potential–pH diagrams at Elevated Temperature for the System Fe–H_2O, *Corrosion Science* **10**, 1970, pp. 343–358.
3. Bard, A.J., Parsons, R. and Jordan, J. (Eds.), *Standard Potentials in Aqueous Solutions*, International Union of Pure and Applied Chemistry (IUPAC), Marcel Dekker, Inc., New York/Basel, 1985, 834 pp.
4. Castellan, G.W., *Physical Chemistry*, Third Edition, Benjamin/Cummings Publishing Company, USA, 1983, 943 pp.

Appendix B
Reversible and Irreversible Electrode Potential

We have read about different terms of electrode potentials on a metal surface: *reversible*, *equilibrium*, *irreversible*, *standard*, *stationary*, *corrosion*, *electrode*, "*mixed*", and "*rest*" potential. What is the difference between them? We shall begin with the reversible potential.

B.1 Reversible Potential

If pure iron is immersed in 1 M $FeCl_2$ solution, the following redox *reversible* reaction takes place on the iron surface:

$$Fe^{2+}_{(aq)} + 2e^- \leftrightarrow Fe_{(s)}. \tag{B.1}$$

The reversible electrode potential corresponding to the redox reaction (B.1) is set on an iron surface immersed in 1 M $FeCl_2$ solution. An aqueous solution of $FeCl_2$ must not contain dissolved oxygen and other oxidizers. Otherwise, a complicated system containing two cations, Fe^{2+} and Fe^{3+}, will be formed. In other words, *reversible potential* represents the equilibrium of redox reaction (B.1) of only one material, iron in this case, and is also called *equilibrium potential*. Reversible electrode potential is defined only by activity (concentration) of species of electrochemical reaction (B.1) occurring on the metal surface. Such an electrode is called reversible. If we do not bring an electric current to a metal, there are no changes on the metal surface. In the equilibrium state on the reversible electrode, the rates of anodic i_a (iron dissolution) and cathodic i_c (reduction of Fe^{2+}) reactions equal each other:

$$i_\circ = |i_a| = |i_c|. \tag{B.2}$$

The value $i_\circ$ is called an *exchange current density* and is defined as a velocity of exchange of electrons among two species of electrochemical individual reaction in the equilibrium state.

An exchange current density exists only at a reversible potential. The state of the equilibrium at a constant temperature is defined only by the nature of a metal and cations' activity of the same metal in a solution. A reversible potential E is defined by the Nernst equation:

$$E = E^\circ + \left(\frac{RT}{nF}\right) \ln a_{\mathrm{M}}^{z^+}. \tag{B.3}$$

E° is a standard reversible potential at $a_{\mathrm{M}}^{z^+} = 1$; $a_{\mathrm{M}}^{z^+}$ is the activity of cations M^{z+} in solution; in this particular case, $M^{z+} = Fe^{2+}$. Therefore, reversible (equilibrium) potential of a metal does not depend on the dimensions of an electrode, volume of solution, and remains constant with time. If a reversible potential on a metal surface is measured in relation to the standard hydrogen electrode, the reversible potential is called the *standard electrode potential.*

Reversible potentials exist for Fe in 1 M $FeCl_2$ (or $FeSO_4$), Cu in 1 M $CuSO_4$, Zn in 1 M $ZnSO_4$, Ag in 1 M $AgNO_3$, Sn in 1 M $SnCl_2$, Cd in 1 M $CdSO_4$, Hg in 1 M Hg_2Cl_2. Although it is not possible for all metals to measure the reversible potential, their standard potentials are given in Table 1.3. It is very difficult to carry out the equilibrium in a general form

$$\mathrm{M}^{z+}_{(\mathrm{aq})} + ze^- \leftrightarrow \mathrm{M}_{(\mathrm{s})}, \tag{B.4}$$

because of some disturbances:

(a) Alkali (Li, Na, K, Rb, Cs) and alkaline earth (Be, Mg, Ca, Sr, Ba) metals are not resistant to aqueous solutions because they decompose in water.
(b) Some metals (Pt, Pd, Ir, Os) are inert (noble) from the electrochemical point of view. Electric potentials on their surface are defined by redox processes which are not related to the existence of cations of noble metals in solutions.
(c) Many metals (Cr, Ni, Co, Ti, Al, Mg) may be in a passive state, as they contain oxide films on their surfaces which do not allow measuring the reversible potentials. For such metals, reversible potentials were calculated on the thermodynamic basis. Usually, electric potentials on surfaces of such metals are not influenced by changes of concentrations of their cations according to the Nernst equation (B.3). Electrode potentials of these metals depend on concentrations of anions and molecules containing in solution.

B.2 Corrosion Potential

If we put the same piece of pure iron in aerated potable water, two different reactions defining the corrosion process take place on the iron's surface (see Section 1.6):

$$\text{anodic:} \quad \mathrm{Fe}_{(\mathrm{s})} \rightarrow \mathrm{Fe}^{2+}_{(\mathrm{aq})} + 2e^-, \tag{B.5}$$

$$\text{cathodic:} \quad \mathrm{O}_{2(\mathrm{aq})} + 2\mathrm{H}_2\mathrm{O}_{(\mathrm{l})} + 4e^- \rightarrow 4\mathrm{OH}^-_{(\mathrm{aq})}. \tag{B.6}$$

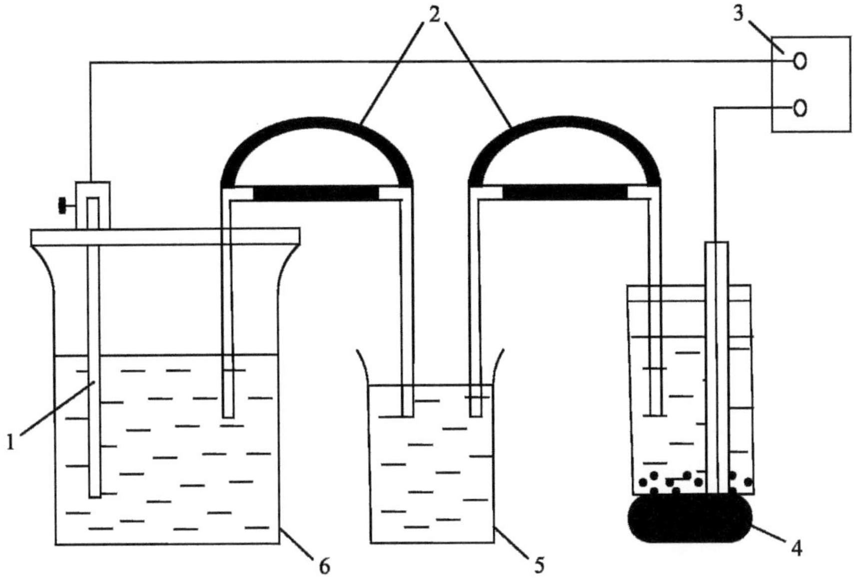

Fig. B.1 Scheme of measuring corrosion potential of metal in an aqueous solution of electrolyte. 1 – a metal; 2 – salt bridges; 3 – potentiometer; 4 – reference electrode; 5 – intermediate vessel with electrolyte; 6 – vessel with electrolyte. Anodic and cathodic reactions occur simultaneously on a metal surface.

An electric potential value is formed on an iron surface, when these two reactions take place simultaneously. This potential is called *irreversible*, *corrosion potential*, *electrode potential*, or *stationary potential*. The rate of anodic reaction (B.5) (anodic current i_a) equals the rate of cathodic reaction (B.6) (cathodic current i_c) at a corrosion potential. A similar situation occurs if we put Zn in H_2SO_4 aqueous solution. Two different reactions occur on the Zn surface:

$$Zn_{(s)} \rightarrow Zn^{2+}_{(aq)} + 2e^-, \tag{B.7}$$

$$2H_3O^+_{(aq)} + 2e^- \rightarrow H_{2(g)} + 2H_2O_{(l)}. \tag{B.8}$$

Polarization curves (Figure C.4, Appendix C) help us to understand the appearance of *corrosion potential*, and the German scientists C. Wagner and W. Traud were the first who explained this in 1938, and named it the "*mixed*" *potential*. Some scientists called it the "*rest*" *potential*. Corrosion potential can be measured by two ways. The first one is based on measuring the corrosion potential in relation to some reference electrode when there is no net electrical current flowing through the metal surface (see Figure B.1). This is corrosion potential at an open electrical circuit, or OCP (open circuit potential), or zero-current potential.

The second way of measuring corrosion potential is based on polarization curves (see Figure C.4, Appendix C).

If a metal is immersed in sulphuric acid (H_2SO_4) solution, two cathodic reactions can occur on the surface of a metal (B.8) and in the presence of dissolved oxygen in the solution:

$$O_{2(aq)} + 4H_3O^+_{(aq)} + 4e^- \rightarrow 6H_2O_{(l)}. \tag{B.9}$$

Thus, several cathodic reactions, for example, (B.8) and (B.9), and several anodic reactions (if alloy consists of two or more metals, for example, Fe–Cr) may take place simultaneously on a surface of alloy:

$$Fe_{(s)} \rightarrow Fe^{2+}_{(aq)} + 2e^-, \tag{B.10}$$

$$Cr_{(s)} \rightarrow Cr^{2+}_{(aq)} + 2e^-. \tag{B.11}$$

In this case, the sum of anodic currents must be equal to the sum of cathodic currents at the *corrosion potential*: $\Sigma|i_a| = \Sigma|i_c|$. Now we can sum up the distinction between reversible and corrosion electrode potential.

B.3 Distinction between Reversible and Corrosion Electrode Potential

1. *Reversible* (*equilibrium*) electrode potential represents the equilibrium redox reaction of one material, for example, reversible reaction (B.1) for the iron electrode. It is simply to verify if the potential is reversible. Its value must suit the Nernst equation (B.3), namely, linear graph $E - \ln a^{z+}_{\mathrm{M}}$ must exist.
   The *irreversible* potential is not suitable for the thermodynamic Nernst equation, that is, it is impossible to calculate irreversible potential. The latter may be defined only experimentally. *Corrosion* (*irreversible*, *stationary*, *electrode*, *mixed*, or *rest*) potential represents the equilibrium of the sum of chemical rates of oxidation and reduction for different systems, for example, (B.5) and (B.6) for iron immersed in water, and aqueous solutions of electrolytes, or (B.7) and (B.8) for zinc immersed in acidic solutions. Corrosion potential is defined by factors attributing to a metal (its type, physico-chemical conditions and structure, surface conditions, presence of oxides, adsorbed gases, mechanical stresses, and defects), and to a solution (its type, concentration of ions in solution, presence of gases and other non-electrolytes, temperature, pressure, and flow velocity).
2. The chemical composition of phases is constant when *reversible* potential exists and changes when *corrosion* potential exists. For example, if iron corrodes in water, the concentration of iron and hydroxide ions increases in water, and the concentration of dissolved oxygen decreases. The rates of chemical reactions depend on concentration of species in the solution, therefore corrosion potential must change in time, but this change is usually slow.
3. *Reversible* potential does not depend on the conditions of the metal surface, type of surface preparation, type of crystallographic border, even on the adsorption and presence of organic substances on the metal surface. All these factors influ-

ence, by a similar value, the rate of anodic and cathodic reaction (B.1) defining the reversible potential. Reversible potentials are highly reproducible. Corrosion potentials depend on all factors influencing the rates of electrochemical reactions. Therefore, corrosion potentials are not reproducible. Do not be surprised if you cannot obtain similar values of corrosion potentials of some metal in a solution. For this, you have to prepare metallic surface identically, but this is nearly impossible. Also, uncontrollable changes of concentrations of reactants occur at all times.

4. There are no changes of weight of metal electrode at a *reversible* potential. If an *irreversible* potential is formed on a metal electrode, corrosion occurs, that is metal losses its weight.

B.4 Influence of Temperature on Reversible Electrode Potential of Iron in Aqueous Solutions

For the reaction of iron oxidation by water

$$Fe_{(s)} + 6H_2O_{(l)} \rightarrow [Fe(H_2O)_6]^{2+}_{(aq)} + 2e^-, \tag{B.12}$$

the conjugated cathodic reaction will be the reduction of water

$$2H_2O_{(l)} + 2e^- \rightarrow H_{2(g)} + 2OH^-_{(aq)}. \tag{B.13}$$

We shall analyze the influence of temperature on the occurrence of two reactions (B.12) and (B.13). For reaction (B.12), the standard electric potential $E^\circ = -0.440$ V, and the standard temperature coefficient $\mathrm{d}E^\circ/\mathrm{d}T = +0.052$ mV/K [1].

The Nernst equation for reaction (B.12) can be written as

$$E = E^\circ + \frac{RT}{nF} \cdot \ln a_{Fe^{2+}}, \tag{B.14}$$

$a_{Fe^{2+}}$ is the activity of cations Fe^{2+} in solution.

The temperature coefficient $\mathrm{d}E/\mathrm{d}T$ of the electric potential of iron electrode is defined by the equation:

$$\frac{\mathrm{d}E}{\mathrm{d}T} = \frac{\mathrm{d}E^\circ}{\mathrm{d}T} + \left(\frac{R}{nF}\right) \cdot \ln a_{Fe^{2+}} = \frac{\mathrm{d}E^\circ}{\mathrm{d}T} + 4.35 \cdot 10^{-5} \ln a_{Fe^{2+}}. \tag{B.15}$$

In the first approach we took that the activity of Fe^{2+} cations was not influenced by the temperature $a_{Fe^{2+}} = \text{const.}$, $R = 8.314$ J/(mol·K), $n = 2$, $F = 96{,}500$ C/g-eqv. The results of calculations of temperature coefficients $\mathrm{d}E/\mathrm{d}T$ of iron electrodes at different ferrous cation activities $a_{Fe^{2+}}$ in a solution according to (B.15) are shown in Table B.1.

Table B.1 Influence of activity of ferrous cations (Fe^{2+}) on temperature coefficient dE/dT of electric potential of an iron electrode.

$a_{Fe^{2+}}$	10^{-10}	10^{-7}	10^{-4}	10^{-3}	10^{-2}	10^{-1}	1
$\frac{dE}{dT}$, mV/K	–0.940	–0.642	–0.345	–0.246	–0.146	–0.047	+0.052

When the activity of cations Fe^{2+} in a solution decreases from 1 to 10^{-10}, the temperature coefficient diminishes, and changes its sign from plus to minus. We can observe a significant influence of temperature on the reversible electrode potential of iron in diluted aqueous solutions containing Fe^{2+} cations. For reaction (B.13) occurring in neutral water (pH = 7) at 25°C, the electric potential is equal to -0.477 V; $\Delta S^{\circ}_{298} = +0.17$ J/(mol·K). This value of ΔS°_{298} is correlated with the temperature potential coefficient $dE/dT = +0.26$ mV/K. Therefore, an increase of temperature results in an increase of the reduction potential of water: electric potential is changed to positive values.

The conclusion from Table B.1 is that iron can be resistant from the thermodynamic point of view to deaerated water only at low temperatures when concentration of Fe^{2+} is high. In equation (B.15) we did not use the fact that the activity of Fe^{2+} cations depended on the temperature. We can take into account such dependence:

$$\frac{dE}{dT} = \frac{dE^{\circ}}{dT} + \left(\frac{R}{nF}\right) \cdot \ln a_{Fe^{2+}} + \left(\frac{RT}{nF}\right) \cdot \left(\frac{\partial \ln a_{Fe^{2+}}}{\partial T}\right). \tag{B.16}$$

Activity $a_{Fe^{2+}}$ is defined by molality m_+ and the coefficient of acitivity γ_+:

$$a_{Fe^{2+}} = m_+ \cdot \gamma_+, \tag{B.17}$$

m_+ – molality (value of moles of cations in 1000 gram of solvent), which does not depend on temperature; γ_+ – activity coefficient of Fe^{2+} cations, which depends on the temperature. Substituting (B.17) into (B.16), we obtain

$$\frac{dE}{dT} = \frac{dE^{\circ}}{dT} + \left(\frac{R}{nF}\right) \cdot \ln a_{Fe^{2+}} + \left(\frac{RT}{nF}\right) \cdot \left(\frac{\partial ln\gamma_+}{\partial T}\right). \tag{B.18}$$

We can use the equation of the Debye–Hückel theory for the rational coefficient of activity f_+ of cations Fe^{2+} in the interpretation of Güntelberg for aqueous solutions [2, 3]:

$$\log f_+ = -\frac{A(Z_+)^2\sqrt{I}}{1+\sqrt{I}}, \tag{B.19}$$

where

$$I = \frac{1}{2} \cdot \sum m_i (Z_i)^2$$

is the ionic strength of the solution; m_i is the molality of all ions in solution and Z_i presents the charges of these ions.

Parameter A is expressed by

$$A = \frac{e^3 \cdot (N_A)^{1/2}}{2.3 \cdot 4\pi \cdot \sqrt{2}(\varepsilon kT)^{3/2}},$$

e – electron charge; N_A – Avogadro number; ε – dielectric constant of solution; k – Boltzmann constant; T – temperature in Kelvins.

The rational coefficient of activity f_+ and molal activity coefficient γ_+ are related by the following equation [3]:

$$f_+ = \gamma_+(1 + 0.001\nu_+ m_+ M_\circ), \tag{B.20}$$

where ν_+ is the number of cations and $M_\circ$ the molar weight of the solvent.

Using (B.20) in (B.19) we receive:

$$\log \gamma_+ = \frac{-[A(Z_+)^2\sqrt{I}]}{(1 + \sqrt{I})} - \log(1 + 0.001\nu_+ m_+ M_\circ). \tag{B.21}$$

Only parameter A depends on the temperature in equation (B.21), therefore

$$\frac{\mathrm{d}\log \gamma_+}{\mathrm{d}T} = -\frac{(Z_+)^2\sqrt{I}}{1 + \sqrt{I}} \cdot \frac{\mathrm{d}A}{\mathrm{d}T}. \tag{B.22}$$

Using (B.17) in (B.18), we obtain

$$\frac{\mathrm{d}E}{\mathrm{d}T} = \frac{\mathrm{d}E^\circ}{\mathrm{d}T} + \left(\frac{2.3R}{nF}\right) \cdot \log m_+ + \left(\frac{2.3R}{nF}\right) \cdot \left[\log \gamma_+ + T\left(\frac{\partial \log \gamma_+}{\partial T}\right)\right]. \tag{B.23}$$

Using (B.21) and (B.22) in (B.23), we obtain

$$\frac{\mathrm{d}E}{\mathrm{d}T} = \frac{\mathrm{d}E^\circ}{\mathrm{d}T} + \left(\frac{2.3R}{nF}\right) \times \left\{\log\left[\frac{m_+}{(1 + 0.001\nu_+ m_+ M_\circ)}\right] - \left[\frac{(z_+)^2\sqrt{I}}{1 + \sqrt{I}}\right] \cdot \left(\frac{A + T \cdot \partial A}{\partial T}\right)\right\}. \tag{B.24}$$

In order to calculate $\mathrm{d}E/\mathrm{d}T$ for different m_+ and T, we have to know the influence of temperature on parameter A. The values of A for different T are taken in [2]. We can change $\mathrm{d}A/\mathrm{d}T$ for $\Delta A/\Delta T$. The calculated values of $\mathrm{d}E/\mathrm{d}T$ for the iron electrode in the solutions of $FeCl_2$ and $FeSO_4$ at their different concentrations and temperatures, are shown in Table B.2. We may conclude from this table, that temperature does not influence the temperature coefficient $\mathrm{d}E/\mathrm{d}T$, at low concentrations of $FeCl_2$ and $FeSO_4$ ($m < 10^{-2}$). At higher concentrations of these salts

Table B.2 The influence of the ionic strength I of $FeCl_2$ and $FeSO_4$ aqueous solutions and temperature T on the coefficient $\mathrm{d}E/\mathrm{d}T$ of iron electrode.

m	$\sqrt{I}$		$-\left(\frac{dE}{dT}\right)_{278}$ mV/K		$-\left(\frac{dE}{dT}\right)_{298}$ mV/K		$-\left(\frac{dE}{dT}\right)_{323}$ mV/K		$-\left(\frac{dE}{dT}\right)_{343}$ mV/K		$-\left(\frac{dE}{dT}\right)_{363}$ mV/K	
	$FeCl_2$	$FeSO_4$	$FeCl_2$	$FeSO_4$	$FeCl_2$	$FeSO_4$	$FeCl_2$	$FeSO_4$	$FeCl_2$	$FeSO_4$	$FeCl_2$	$FeSO_4$
10^{-6}	0.002	0.002	0.543	0.543	0.543	0.543	0.543	0.543	0.543	0.543	0.543	0.543
10^{-5}	0.005	0.006	0.445	0.446	0.445	0.446	0.445	0.446	0.446	0.446	0.446	0.446
10^{-4}	0.017	0.020	0.347	0.350	0.348	0.350	0.348	0.351	0.348	0.352	0.349	0.352
10^{-3}	0.055	0.063	0.253	0.260	0.254	0.262	0.255	0.264	0.256	0.267	0.257	0.269
10^{-2}	0.173	0.200	0.167	0.188	0.170	0.192	0.173	0.199	0.176	0.206	0.180	0.213
10^{-1}	0.548	0.630	0.097	0.146	0.103	0.158	0.111	0.174	0.119	0.190	0.127	0.206
1	1.732	2.000	0.037	0.126	0.047	0.146	0.062	0.176	0.076	0.204	0.090	0.233

($m > 10^{-2}$), increase of temperature causes growth of the temperature coefficient $\mathrm{d}E/\mathrm{d}T$, and this influence is stronger in solutions with a higher ionic strength I.

Comparison of the data in Tables B.1 and B.2 shows that when the concentrations of $FeCl_2$ and $FeSO_4$ are low, taking into consideration the influence of temperature on Fe^{2+} activity does not influence the $\mathrm{d}E/\mathrm{d}T$ coefficient of iron electrodes. This influence is significant at concentrations $m > 10^{-3}$. The type of electrolytes ($FeCl_2$ or $FeSO_4$) also influences the temperature coefficient at concentrations $m > 10^{-3}$.

References

1. Bard, A.J., Parsons, R. and Jordan, J. (Eds.), *Standard Potentials in Aqueous Solutions*, International Union of Pure and Applied Chemistry (IUPAC), Marcel Dekker, Inc., New York/Basel, 1985, 834 pp.
2. Dean, J.A., *Lange's Handbook of Chemistry*, 15th Edition, 1999, McGraw-Hill, Inc., New York, USA, pp. 8.2–8.5.
3. Damaskin, B.B. and Petri, O.A., *The Principles of Theoretical Electrochemistry*, The Higher School, Moscow, 1978, 239 pp.

Appendix C
Electrochemical Kinetics and Polarization Curves

Thermodynamics can define the fundamental possibility of corrosion reactions but cannot forecast their rate and mechanism, because it deals with processes in equilibrium. Time does not exist in thermodynamics. Only kinetic experiments can help to define the corrosion rate (see Section 1.4) and mechanism. An electrode potential formed on a metal surface in a solution of electrolytes after a long immersion time (~20 to 30 min) is related to the "thermodynamic" parameter, but an electric current relates to the kinetic one. How to connect them? Electrochemical kinetics, which relates to corrosion reactions in the solutions of electrolytes, can answer this question. Corrosion processes occurring according to the chemical mechanism (in gases and other non-electrolytes) are not covered in this appendix.

Electrochemical kinetics is the science that deals with rates of electrochemical reactions occurring at the border between metal and liquid electrolytes. From the engineering point of view, kinetics is a very important subject as it answers the questions, what is the corrosion rate and corrosion mechanism. As a result the "lifetime" of metallic constructions may be defined. Corrosion electrochemical process as any heterogeneous process consists of following continuous stages:

(a) Reactants (H_2O, H_3O^+, O_2, Fe^{3+}) must reach metal surface. This is a *diffusion* (*mass transport*) stage.
(b) *Electrochemical* stages are related to electrons or ions passing through the border (double electric layer) between phases of metal and liquid. This is a *redox* stage:

$$O_{2(aq)} + 2H_2O_{(l)} + 4e^- \rightarrow 4OH^-_{(aq)}, \tag{C.1}$$

$$2H_3O^+_{(aq)} + 2e^- \rightarrow H_{2(g)} + 2H_2O_{(l)}, \tag{C.2}$$

$$M_{(s)} \rightarrow M^{z+}_{(aq)} + ze^-, \tag{C.3}$$

where M is metal.
(c) Products of cathodic and anodic electrochemical reactions must leave a metal surface. This is a *diffusion* stage, in the opposite direction from the metal surface to the solution.

In addition to these stages, different chemical reactions in the volume of the liquid solution or on the metal surface may occur:

$$2Fe^{2+}_{(aq)} + 2OH^{-}_{(aq)} + O_{2(aq)} + 2H_2O_{(l)} \rightarrow 2Fe(OH)_{3(s)}, \tag{C.4}$$

as well as diffusion of atoms and molecules on the metal surface and its combining:

$$H_{(ads)} + H_{(ads)} \rightarrow H_{2(g)}. \tag{C.5}$$

$H_{(ads)}$ presents the adsorbed hydrogen atom on the metal surface.

Any of these stages may limit the general corrosion process and define the corrosion rate. The main task of electrochemical kinetics is defining the corrosion rate and limiting stages, as well as the corrosion mechanism, in order to influence the corrosion process: to predict, to control, and even to prevent it. The bases of electrochemical kinetics are the *polarization curves*, which describe the dependence between electric current and electrode potential on the metal surface. However, we will first discuss *polarization*.

C.1 Polarization

Let us put a zinc electrode in 1 M $ZnSO_4$ and an iron electrode in 1 M $FeSO_4$ aqueous solutions free of dissolved oxygen in an inert non-oxidized atmosphere in one vessel differentiating two solutions with a porous membrane (Figure C.1).

Two electrodes, made of zinc and iron, are connected through reostat R, voltmeter V, and ammeter A. If there is no electric current through these electrodes (electrical resistance $R > 10^6$ Ohm), standard electrode potentials are realized on the metal surfaces (see Table 1.3): $E^{\circ}_{Zn^{2+}/Zn} = -0.763$ V; $E^{\circ}_{Fe^{2+}/Fe} = -0.44$ V, and the voltage (potential difference between zinc and iron electrodes) equals $E^{\circ} = E^{\circ}_{Fe^{2+}/Fe} - E^{\circ}_{Zn^{2+}/Zn} = 0.323$ V. Zinc is the anode and iron is the cathode in this galvanic pair. The reversible potentials of zinc and iron electrodes $E^{\circ}_{Zn^{2+}/Zn}$ and $E^{\circ}_{Fe^{2+}/Fe}$ (E°_{Zn} and E°_{Fe} for the simplest description) are realized when the electrical outer circuit is open (R is very high), and are called *open circuit potentials* (OCP). Any reversible potential is OCP, but not any OCP is a reversible potential. Usually OCP is an electric potential of the electrode immersed in any electrolyte solution where the net electric current is zero (see Appendix B). Electrochemical corrosion occurs in this galvanic cell (zinc is the anode and iron is the cathode) according to the reaction

$$Zn_{(s)} + Fe^{2+}_{(aq)} \rightarrow Zn^{2+}_{(aq)} + Fe_{(s)}. \tag{C.6}$$

If we decrease the electrical resistance R, a small electric current I_1 will flow through two electrodes, iron and zinc, and the voltage between them will be less than 0.323 V, that is, the Ohm equation will be invalid:

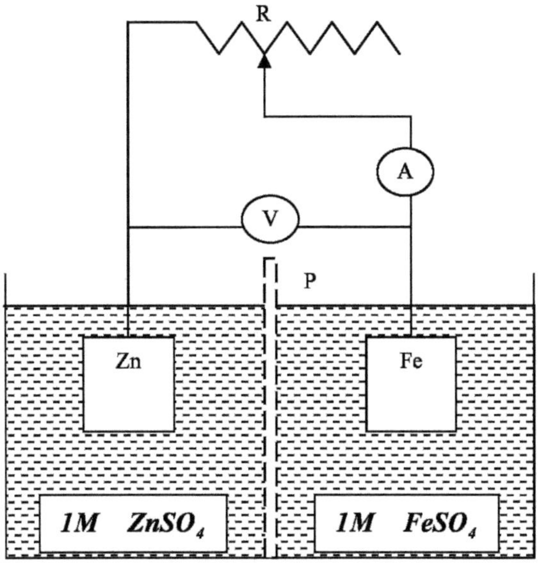

Fig. C.1 Galvanic (corrosion) cell for polarization measurements. R – Reostat; V – voltmeter; A – ammeter; P – porous membrane.

$$I_1 \neq \frac{E_c^\circ - E_a^\circ}{R_e + R_m}. \tag{C.7}$$

In this case, $E_c^\circ = E_{\text{Fe}}^\circ$ and $E_a^\circ = E_{\text{Zn}}^\circ$; R_e and R_M represent the electrical resistance of the liquid electrolyte solution and the metal, respectively. The difference between electrode potentials will be other than $E_c^\circ - E_a^\circ$, when electric current flows through two electrodes. If we measure the potentials of zinc and iron electrodes during the electric current flow, the potential of zinc (anode) will be more positive:

$$E_a = E_a^\circ + \Delta E_a \tag{C.8}$$

and the potential of iron (cathode) will be more negative:

$$E_c = E_c^\circ - \Delta E_c. \tag{C.9}$$

We can draw the dependence of change of the potential of zinc and iron electrodes as a function of electric current (Figure C.2).

The higher the electric current I, the more are the deviations of the electrode potentials from the open circuit potentials E_{Zn}° and E_{Fe}°. In other words, electrode potentials of anode and cathode draw together. If we short circuit two electrodes, zinc and iron, the difference $\Delta E = E_{\text{Fe}} - E_{\text{Zn}}$ will be nearly zero, and we obtain the maximum electric current I_{max}. Such a galvanic (corrosion) cell, including zinc and iron electrodes, is the analog of microgalvanic, or corrosion, cells which are

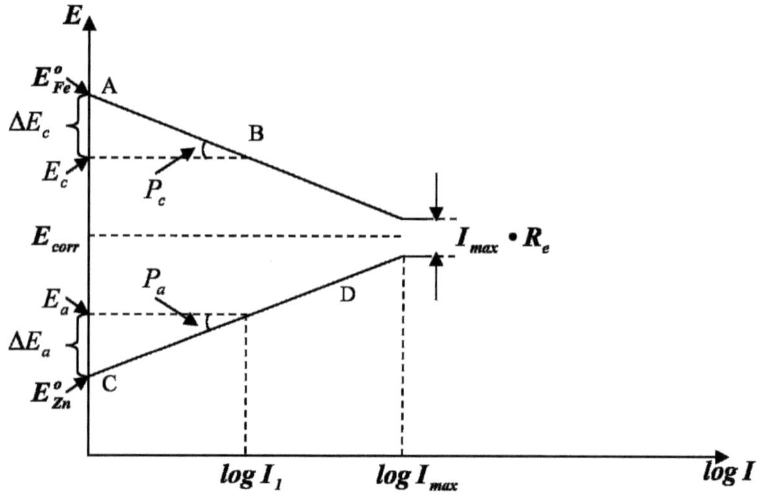

Fig. C.2 Changes of electrode potentials (E) of iron and zinc electrodes in a galvanic cell as a function of electric current (I) flow (polarization diagram). E_a and E_c are electrode potentials of anode and cathode when electric current I_1 flows. P_a and P_c are the slopes of anodic CD and cathodic AB polarization curves respectively. $I_{\max} = I_{\text{corr}}$ is the maximum (corrosion) current in the polarization diagram.

formed on the surface of corroding metal because of the heterogeneity of the metal surface and/or the environment.

The phenomenon of the influence of general electric current I on voltage between two electrodes ($E_{\text{Fe}} - E_{\text{Zn}}$) is called *polarization*. The values ΔE_a and ΔE_c are called the *polarizations* of anode and cathode, respectively. In other words, polarization is the deviation of electrode potential from the *reversible potential* (*equilibrium*, or *open circuit potential* in the general case) as a result of electric current flow between electrodes. The term polarization was suggested by E.Ch. Lentz in 1839 for electrolysis. We can calculate the values of anodic and cathodic polarization, P_a and P_c, as slopes for the two lines CD (anodic) and AB (cathodic) in Figure C.2:

$$P_a = \frac{\Delta E_a}{I}, \tag{C.10}$$

$$P_c = \frac{\Delta E_c}{I}. \tag{C.11}$$

We wrote I instead of I_1 for the general case. Using (C.8) and (C.9), we can write the values of electrode potentials at anode (E_a) and cathode (E_c) electrodes during the flow of electric current I:

$$E_a = E_a^\circ + \Delta E_a = E_a^\circ + I \cdot P_a, \tag{C.12}$$

$$E_c = E_c^\circ - \Delta E_c = E_c^\circ - I \cdot P_c. \tag{C.13}$$

The electric current I may be calculated according to Ohm's law:

$$I = \frac{E_c - E_a}{R_e} = \frac{E_c^\circ - I \cdot P_c - E_a^\circ - I \cdot P_a}{R_e}, \tag{C.14}$$

$$I = \frac{E_c^\circ - E_a^\circ}{R_e + P_c + P_a}. \tag{C.15}$$

Formula (C.15) shows that polarization of anode and cathode is equivalent to the addition of electrical resistance which is called *resistance to polarization*, or *polarization resistance*. The latter term was first used in 1951 by K.F. Bonhoeffer and W. Jena, following a suggestion by N.A. Lange to replace the previously used term *polarizability*. Existence of *polarization resistance* results in such a situation that even when Ohmic resistance of solution is very low ($R_e \rightarrow 0$), corrosion electric current cannot reach the infinite value. In this case, the maximum electric current I_{max} is called the corrosion current I_{corr} (see Figure C.2):

$$I_{\text{max}} = I_{\text{corr}} = \frac{E_c^\circ - E_a^\circ}{P_c + P_a}. \tag{C.16}$$

The difference of potentials of cathode and anode decreases to a minimum value $I_{\text{max}} \cdot R_e$, named "IR-drop". The less R_e, the lower is the "IR-drop". At corrosion current I_{corr}, electrode potentials of cathode and anode are equal and some "*mixed*" potential are present between E_c° and E_a°. Polarization occurs both in galvanic (corrosion) cells and in electrolytic baths (during electrolysis), namely, during a flow of constant electric current through electrodes. In the next section, we shall explain the causes of *polarization*.

C.2 The Causes of Polarization

Let us use the model of the *electric double layer* formed on a metal surface in a liquid electrolyte (see Section 1.5). This layer plays an important role in corrosion kinetics, as the corrosion process occurs at the border of metal–liquid electrolyte. The electric double layer is the obstacle for flow of electrons, ions, and molecules. This is the energetic barrier which must be overcome for the electrochemical process to occur. In order to understand how the electric double layer forming on cathode and anode influences the electric current flow in the outer electric circle and the difference between the electric potentials formed on these electrodes, let us differentiate all processes in several stages:

1. *Transition of electrons in the outer electric circle*. Before the connection of zinc and iron electrodes, there was a difference between electric charges on all of them. Zinc was more negatively charged than iron, because it had a greater tendency to be oxidized. Connection between zinc and iron causes a transition of electrons from Zn to Fe. As a result, Zn will be less negative and Fe will be more

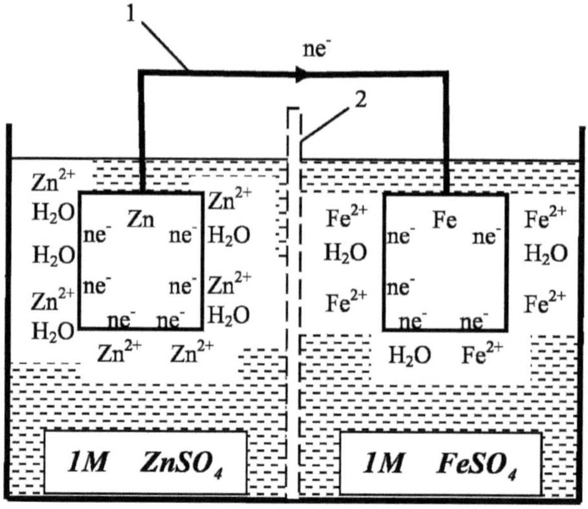

Fig. C.3 Scheme of anodic and cathodic polarization in "macro" galvanic cell: corrosion cell of Zn (anode) and Fe (cathode) in neutral aqueous solution of electrolytes. 1 – metallic conductor (outer electric circle); 2 – porous membrane. "Micro" galvanic cells are formed on one whole electrode.

negative (see Figure C.2). Transition of electrons in the metallic conductor (outer electric circle) is a very fast process.

2. *Transition of ions and molecules at the border between the electrodes and a liquid solution.* When zinc is less negative, more quantity of positive zinc cations must leave its surface. An electric double layer, which consists of zinc cations and polar water molecules, resists the new zinc cations leaving the surface of a zinc electrode (Figure C.3). This means that the anodic process of zinc dissolution drops behind the transition of electrons from zinc (anode) to iron (cathode), which in its turn causes the lessening of negative charge at the surface of the anode and results in more positive potential of the anode (see Figure C.2). What happens on the iron (cathode) surface? Ferrous cations (Fe^{2+}) or other species (dissolved O_2 and H_2O molecules if iron is immersed in neutral aqueous solutions) taking part in cathodic reactions, must pass through the electric double layer at the iron electrode (Figure C.3).
   This layer is the obstacle for species taking part in the cathodic process to pass. Thus, similar to the anodic process, the cathodic process also drops behind the transition of electrons from zinc (anode) to iron (cathode), which in its turn causes an increase of negative charge at a surface of the cathode and results in a more negative potential of the cathode (see Figure C.2). When electrochemical (corrosion) reactions are slow and define the velocity of all (general) corrosion process, this is called *activation polarization*, because slow kinetics in chemistry is associated with *activation energy*. In other words, the influence of surface processes is related to activation polarization, or activation control. This can be overcome by applying extra voltage, which is known as *activation overvoltage*.

3. *Diffusion* (*mass transport*) of species to cathode and taking part in cathodic processes, and products of anodic reaction from anode into a solution, may be slow compared with the electrochemical stages. The influence of concentrations of various species in a solution is related to *concentration polarization*, or *mass transport control.*

Polarization is the result of a lag in electrochemical anodic and cathodic processes or diffusion of species from the surface of an anode or to the surface of a cathode from the transition of electrons from anode to cathode in a galvanic (corrosion) cell. Such delay causes a decrease in difference between the electric potentials of two electrodes during the transition of electric current in the outer circle. To sum up, the causes of anodic polarization are:

1. Deceleration of anodic reaction on metal surface – *activation polarization.*
2. Slowing down the diffusion of metallic cations from the surface into the volume of solution causes *concentration polarization* of an anode.

The causes of cathodic polarization are:

1. Deceleration of cathodic reaction on metal surface – *activation polarization.*
2. Slowing down the diffusion of species (dissolved O_2, H_3O^+, Fe^{3+}, and others) taking part in cathodic process, causes *concentration polarization* of a cathode.

C.3 Polarization Curves

Kinetics of electrochemical corrosion process is adopted to show the *polarization curves*. The latter is a graphic representation of the dependence of electric potential (**E**) of an electrode on current density (**i**) (Figure C.4).

Polarization curves describe the behavior of a metal as an anode or cathode of a corrosion cell. When iron is immersed in a neutral or alkali aqueous solution of electrolyte, polarization curves reflect anodic dissolution of iron and cathodic reduction of dissolved oxygen. When iron is immersed in an acidic solution, polarization curves reflect the anodic dissolution of iron and cathodic reduction of H_3O^+ (alone or together with O_2). Two main parameters are defined from polarization curves: corrosion potential E_{corr} and corrosion current i_{corr} (corrosion rate). The latter is obtained from extrapolating the currents in the two Tafel regions (AB and CD) to the corrosion potential E_{corr}. This method is called Tafel extrapolation and is used in the laboratory and sometimes under industrial conditions. The slopes (called Tafel constants) of polarization curves define the large or small difficulty of anodic and cathodic processes. A degree of polarization is a measure of how the rates of anodic and cathodic reactions are influenced by various environmental (concentrations of dissolved oxygen, cations, anions, pH, temperature, agitation in solution) and surface processes (electrochemical reactions, adsorption, film formation) factors. The first defines *concentration polarization*. The latter defines *activation polarization*. Polarization curves are the basis for determining corrosion rates by the electrochemical

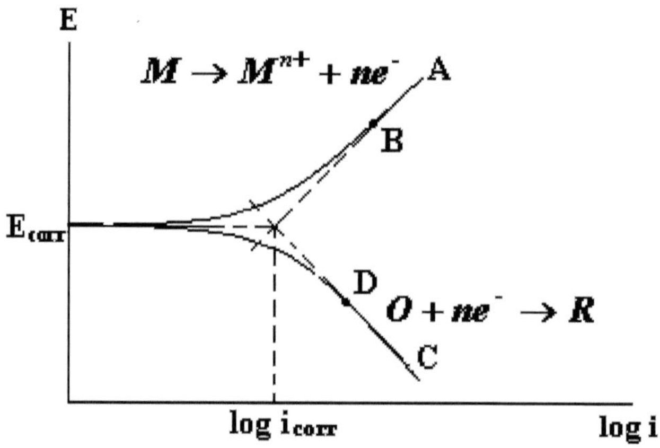

Fig. C.4 Polarization curves (Evans diagram, suggested by him in 1929) for corrosion process $M + O \rightarrow M^{n+} + R$. M – metal; O – Oxidizer (O_2, H_3O^+, Fe^{3+}); *R* – reduced species.

method, defining the efficiency and mechanism of corrosion inhibitors, influence of various factors on corrosion, and the possibility of *passivation* (see Appendix D).

References

1. Uhlig, H.H. and Revie, W.R., *Corrosion and Corrosion Control: An Introduction to Corrosion Science and Engineering*, Third Edition, John Wiley & Sons, USA, 1985, pp. 35–56.
2. Fontana, M.G., *Corrosion Engineering*, Third Edition, McGraw-Hill, Inc., USA, 1986, pp. 454–468, 499–502.
3. Korb, L.J. and Sprowls, D.O., *Metals Handbook, Vol. 13: Corrosion*, ASM International, USA, 1987, pp. 29–35.

Appendix D
Passivity

If we studied chemistry at school, we would probably remember the dependence of iron dissolution from concentration of sulphuric and nitric acids (Figure D.1).

How did the teacher explain these curves? An increase in the concentration of sulphuric and nitric acids causes an increased corrosion rate in iron. Then after the concentration of 60% (for H_2SO_4) and 40% (for HNO_3), the corrosion rate drastically decreases. The greater the concentration of acid, the less concentration of water, namely, less acid molecules capable of dissociating for H_3O^+, and as a result, the corrosion rate diminishes. The phenomenon of high resistance of iron in concentrated sulphuric and nitric acids is explained by *passivity*. Concentrated sulphuric and nitric acids are stored and transported in tanks made of carbon steel. You must not coat them inside and use stainless steel, but be careful. Iron dissolves in diluted H_2SO_4 and HNO_3 acids. Therefore, any dilution of these acids will cause a catastrophic acceleration of the corrosion rate of carbon steel. For example, vessels and tanks with concentrated nitric and sulphuric acids must be hermetically closed with valves to protect them from water penetration. Another point is very important: if iron is immersed in concentrated (~70%) HNO_3 and then is put into diluted (~30%) HNO_3, the passive state of iron may be maintained. Any change in the surface condition of iron (for example, scratch) or agitation may cause a disturbance of *passivity*.

Although passivity has been known for about 250 years, there has been no lucid explanation until now. *Passivity* is the phenomenon of relatively high corrosion resistance as a result of diminution of the anodic process. *Passivity* is the specific phenomenon occurring only with some metals and alloys in particular media.

The Russian scientist Michail Lomonosov (1711–1765), the English scientists James Keir (1735–1820), Michael Faraday (1791–1867) and the Swiss chemist Christian Friedrich Schönbein (1799–1868) studied this phenomenon. The latter coined *passive iron* and *passivity* in 1836 in a letter to Michael Faraday. The word *passive* came from the Latin word *passivus* which means "*indifferent, dispassionate, inactive*", and is the opposite of *active*. The difference in corrosion rate between passive and active states of metals may be one million and more. Iron, chromium, nickel, cobalt, molybdenum, titanium, aluminum, magnesium, and their alloys can

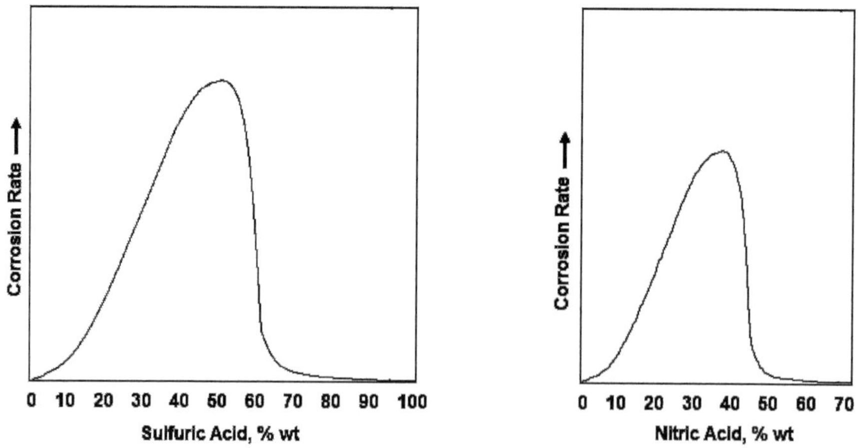

Fig. D.1 Corrosion rate of iron vs concentration of sulphuric and nitric acids.

possess passivity under particular conditions (in specific media). The following causes can result in corrosion resistance of metals and alloys: thermodynamic stability, difficulties in transportation of oxidizers to the surface of metals, retardation of the cathodic process, and inhibition of anodic process. The latter is called *passivation*, *passive state*, or *passivity*. There are several ways to transfer metals and alloys into a passive state.

(a) Anodic polarization from the outer source of direct electric current, or by means of connection to another metal which is a cathode, towards the metal to be polarized.
(b) Immersion of metals in specific media: iron in aqueous solutions of HNO_3 (>70%), H_2SO_4 (>96%), Na_2SO_4 (1 M), and others.
(c) The presence of oxidizers in solutions, for example, dissolved oxygen (O_2), peroxides (H_2O_2, KBO_3, $K_4P_2O_8$), and other substances (HNO_3, $NaNO_2$, $NaNO_3$, Na_2WO_4, K_2CrO_4, $K_2Cr_2O_7$).
(d) Addition of "noble" elements (Pt, Pd, Rh and Ir) to chromium, titanium, and their alloys (noble-metal alloying).

Figure D.2a shows the relationship between electric potential (**E**) and current density (**i**) for metals/alloys possessing a passive state. The polarization curve in Russian literature is usually depicted in mirror reflection (Figure D.2b). The correct presentation of the relationship between **E** and **i** is shown in Figure D.2c, because we measure electric current density (**i**) as a function of electric potential (**E**). In any case, this anodic polarization curve can be divided into several regions. AB – active, or Tafel region; a realm when an increase of potentials **E** causes a raise in corrosion rate (**i**). When electric potential **E** reaches the point B, metallic oxides are formed on a metal's surface, and electric current **i** (corrosion rate!) diminishes drastically.

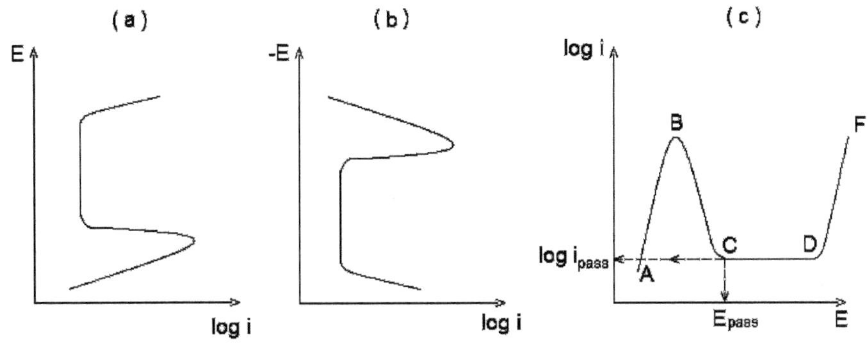

Fig. D.2 Anodic polarization curves of metals/alloys possessing passive state: (a) accepted in corrosion literature in Europe and America; (b) in Russia; (c) correct presentation (usually in the electrochemical literature). **E**, Volt; **i**, A·cm^{-2}.

Usually metal (M) is oxidized to M_xO_y at point B. The area BC is an active–passive transition, when a protective film is formed on the metal surface and the metal "moves" in the *passive* region. When electric current drops to point C, metal reaches the *passive* state. This potential ($\mathbf{E}_{pass}$ or $\mathbf{E}_{pp}$) is named *passivation potential*, or *Flade potential*, in honor of the scientist F. Flade who was the first to observe *passivity* of iron in sulphuric acid in 1911. Other names for the potential $\mathbf{E}_{pp}$ are the *primary passive potential* or the *potential of full passivity*. The electric current in point B is named *critical passivating current density*.

CD is a passive region. The electric current (corrosion rate) does not depend on the electric potential in this region. The corrosion rate of metal is not zero in a passive field, but is a very small and constant value ($\mathbf{i}_{pass}$).

We can detect different oxides (Fe_3O_4, γ-Fe_2O_3) or hydroxides (FeOOH) on an iron surface in *passive state*. These oxides and hydroxides grow directly into the metal and outside it, forming a film which effectively isolates the iron surface from the corrosive environment. Such films are so thin that it is impossible to see them with the naked eye. Contrary to rust, thin oxides are very dense, but are fragile. We can point out the characteristic features of *passivity*:

(a) The electric potential of metals changes in the positive direction, for example, from −0.5 V to +0.5–1.0 V for iron; from −0.6 V to +0.9 V for chromium.
(b) The corrosion rates of metals in the passive state are very low and are stable in time. Usually the transition from active to passive state results in the diminution of the corrosion rate ~1 million times.
(c) Passivity is a *metastable* state, from the Greek "*meta*" ("*after*"), movement or change to unstable condition, and is very simple to disturb by means of scratch, grinding, or changes to the outer conditions: temperature, concentration of aggressive variables, and agitation.

Passivity exists because of a thin delicate film of several nanometers thickness (3 to 5 nm) on the metal or alloy surface. Passivity is not an inert or static state, but a dynamic condition and belongs to a kinetic phenomenon in which there is continuous dissolution and repair of passive layers at discrete points on the surface. When metal reaches point D, the electric current increases to further enhance the electric potential. Region DF is called *transpassivity*. "*Trans*" shows the change, crossing, or passing the boundary of *passivity*. Several reasons exist for transpassivity:

(a) Molecular oxygen (O_2) can be formed on the outer metal surface, that is on oxides, according to the anodic reaction

$$4OH^-_{(aq)} \rightarrow O_{2(aq)} + 2H_2O_{(l)} + 4e^-. \tag{D.1}$$

Certainly, passive film on a metal surface must be electrically conductive. Even if the film is a semi-conductor, it is very thin, and reaction (D.1) can occur on the film's surface.

(b) Electric potential in point D is so large that metal oxide can be oxidized, for example:

$$Fe_2O_{3(s)} + 5H_2O_{(l)} \rightarrow 2FeO^{2-}_{4(aq)} + 5H_{2(g)}, \tag{D.2}$$

$$Fe_2O_{3(s)} + 5OH^-_{(aq)} \rightarrow 2FeO^{2-}_{4(aq)} + 5/2H_{2(g)}. \tag{D.3}$$

Passive film is destroyed on the metal surface at electric potentials above E_D. Thus, stability of the passive state is restricted by the concentration of oxidizers, or by the electric potential value during anodic polarization. Any damage of the passive film, such as fracture, crumbling, and disintegration, causes initiation of localized corrosion: pitting, cracking or crevice corrosion.

We can compare passivity with travelling by car from home to a place of work. If we go out before (earlier than 6 a.m.), or after heavy traffic (later than 9 a.m.), we quickly reach the working place. If we go out exactly during heavy traffic between 6 a.m. and 9 a.m. (when all people leave their home), we travel very slowly – "*passive movement*".

We see in passivity the dialectic principle of the *struggle of opposites* in philosophy. When electric potential increases, two processes are in competition on the metal surface: *accelerated corrosion* and *passivation*. The equilibrium between these two opposite processes is very delicate. Therefore, it is easy to disturb passivity. Activation is a transfer process of metal from passive into an active state as a result of outer conditions. Here are the ways to move metal from passivity to activity:

(a) Use of reducers: H_2, Na_2SO_3, $Na_2S_2O_3$.
(b) Use of sea water; aqueous solutions of NaCl (concentrations above 0.01 M); and ions (Cl^-, Br^-, I^-, SO_4^{2-}, and H_3O^+).
(c) Cathodic polarization: reduction of oxides on the metal surface by means of direct electric current from a power supply or, alternatively, by connecting passive metal to more cathodic metal.

Passivity is used in anodic protection.

References

1. Fontana, M.G., *Corrosion Engineering*, Third Edition, McGraw-Hill, Inc., USA, 1986, pp. 469–473.
2. Uhlig, H.H. and Revie, W.R., *Corrosion and Corrosion Control: An Introduction to Corrosion Science and Engineering*, Third Edition, John Wiley & Sons, USA, 1985, pp. 60–89.
3. Korb, L.J. and Sprowls, D.O., *Metals Handbook, Vol. 13: Corrosion*, ASM International, USA, 1987, pp. 35–36.
4. Shreir, L.L., Jarman, R.A. and Burstein, G.T. (Eds.), *Corrosion, Vols. 1 and 2*, Third Edition, Butterworth Heinemann, UK, 1994, pp. 1:118–1:150.
5. Macdonald, D.D., Passivity – The Key to Our Metals-Based Civilization, *Pure Appl. Chem.* **71**(6), 1999, pp. 951–978.
6. Revie, R.W. (Ed.), *Uhlig's Corrosion Handbook*, Second Edition, Wiley-Interscience, USA, 2006, pp. 165–171, 602.

Appendix E
Solubility of Oxygen in Water and Aqueous Solutions of Electrolytes

Dissolved oxygen takes part in the cathodic processes on a metal surface (see Section 1.5). Electrode potential and corrosion kinetics depend on the oxygen concentration in solutions.

In spite of polarity of water molecules, non-polar oxygen molecules dissolve in the voids between water molecules in the tetrahedral arrangement, reaching about 8 ppm (2.5×10^{-4} mol O_2 in 1 liter H_2O) at 25°C. Such a small oxygen concentration is enough for corrosion and, by the way, for many biological processes occurring in water. Solubility of non-polar oxygen molecules in water and other polar solvents is less than that in non-polar solvents (such hydrocarbons as pentane, hexane, toluene, fuels, etc.). Solubility of oxygen in water depends on temperature, pressure, and presence of electrolytes.

E.1 Influence of Temperature on Solubility of Oxygen in Water

Increase of temperature from 0 to 100°C results in a decrease of oxygen solubility to nearly zero and a subsequent increase of temperature above 100°C results in increase of oxygen solubility (Table E.1). The following increase above 100°C is so considerable that the solubility at 200°C (473 K) is equal to the oxygen solubility at 25°C. This fact is very important for corrosion at power stations as processes occur at high temperatures and pressures.

E.2 Influence of Pressure on Solubility of Oxygen in Water

The dependence of oxygen solubility in water on pressure is described with Henry's Law:

$$X_{O_2} = K_H \cdot P_{O_2}, \tag{E.1}$$

Table E.1 Solubility ($X_{O_2} \times 10^5$) of oxygen in water at 273 to 573 K ($P_{O_2} = 101{,}325$ Pa) [1].

T K	$X_{O_2} \times 10^5$	T K	$X_{O_2} \times 10^5$	T K	$X_{O_2} \times 10^5$	T K	$X_{O_2} \times 10^5$
273	3.953	313	1.870	353	1.478	453	2.038
278	3.462	318	1.775	358	1.462	473	2.388
283	3.072	323	1.697	363	1.450	493	2.852
288	2.758	328	1.633	368	1.445	513	3.460
293	2.504	333	1.580	373	1.444	533	4.250
298	2.297	338	1.539	393	1.487	553	5.276
303	2.126	343	1.507	413	1.599	573	6.603
308	1.986	348	1.483	433	1.780		

Note: $X_{O_2} = n(O_2)/(n(O_2)+n(H_2O))$ is the mole fraction solubility of oxygen in water; $n(O_2)$ and $n(H_2O)$ represent the number of moles of oxygen and water, respectively.

or, more precisely,

$$a_{O_2} = K_H \cdot f_{O_2}, \tag{E.2}$$

where K_H is Henry's Law constant; P_{O_2} is the partial pressure of oxygen; a_{O_2} is the activity of oxygen in water; f_{O_2} is the fugacity of oxygen in water.

There is a straight line between oxygen solubility and its pressure at low pressures (below 3 MPa), but deviation from Henry's Law occurs at high pressures (>3 MPa) of oxygen gas.

E.3 Influence of Type and Concentration of Electrolyte on Oxygen Solubility in Water

Addition of salt or any other electrolyte to water usually decreases oxygen solubility. This phenomenon is called *salting out*. We can observe the salting out if we add some table salt to beer. Many bubbles of CO_2 and air escape from the beer. The polar salt molecules ionize and capture the "voids" in the beer instead of CO_2 and air molecules, and the latter escape. The greater the salt or other electrolyte concentration in water, the lesser oxygen concentration is in it. The quantitative value for the salting out phenomenon is the Setschenov coefficient:

$$K_{c\alpha} = \left(\frac{1}{C}\right) \log\left(\frac{\alpha_o}{\alpha}\right) \quad [\text{L/mol}], \tag{E.3}$$

where C is the the concentration of electrolyte, mol/L; α_o, α is the solubility of oxygen (Bunsen coefficient) in pure water (α_o) and in a solution of electrolyte (α), L = liter.

The Russian physiologist Iliya Setschenov (1889) was the first to study the influence of salts on the solubility of gases in liquids and found the empirical law (E.3).

Table E.2 The Setschenov coefficients ($K_{c\alpha}$) for oxygen in 1 M HCl, H_2SO_4 and HNO_3 solutions [1].

Acid	T K	P_{O2}, MPa	$K_{c\alpha}$, L/mol
HCl	288.15	0.1	0.044
	298.15	0.1	0.031
	310.2	0.1	0.027
H_2SO_4	288.15	0.1	0.105
	298.15	0.1	0.087
	323.15	5	0.10
HNO_3	283.15	1 to 6	0.044
	293.15	1 to 6	0.0255
	303.15	1 to 6	0.0175
	310.2	0.1	0.010
	313.15	1 to 6	0.0115
	323.15	1 to 6	0.0090
	333.15	1 to 6	0.0050

Table E.3 The Setschenov coefficients ($K_{c\alpha}$) for oxygen in alkaline solutions [1].

Alkali	298.15 K		310.2 K	
	C^*, mol/L	$K_{c\alpha}$, L/mol	C^*, mol/L	$K_{c\alpha}$, L/mol
LiOH	0 to 0.07	0.332	0 to 4.058	0.137
	0.15 to 1.2	0.091	-	-
NaOH	0 to 1.2	0.181	0 to 4.071	0.169
KOH	0 to 0.8	0.176	0 to 4.871	0.160
RbOH	0 to 0.5	0.168	0 to 3.187	0.161
CsOH	0 to 0.5	0.158	0 to 2.325	0.153

Notes: C^* is the concentration range of electrolyte.
$K_{c\alpha} = 0.006$ L/mol for NH_4OH (0 to 8 M) at 323 to 423 K.

The Bunsen coefficient is defined as the volume of gas absorbed by unit volume of solvent (at the temperature of measurement) under a gas partial pressure of 1 atm and is calculated according to

$$\alpha = \left(\frac{V_2}{V_1}\right) \cdot \left(\frac{273.15}{T}\right) \quad [\mathrm{cm}_\mathrm{g}^3/\mathrm{cm}_\mathrm{liq}^3], \tag{E.4}$$

where V_1 is the volume of liquid (water); V_2 is the volume of gas (oxygen) corrected to 273.15 K and 101,325 Pa (1 atm) pressure.

The Setschenov coefficient $K_{c\alpha}$ (*salting out coefficient*) shows the ability of electrolyte to "salt out" the gas (oxygen) from the solution (Tables E.2–E.5). Usually the Setschenov coefficients depend on the type of cations and anions, temperature,

Table E.4 The Setschenov coefficients ($K_{c\alpha}$) for oxygen in solutions of halogenide of alkali metals [1] ($T = 298$ K, $P_{O_2} = 0.1$ MPa).

Ions	F^-	Cl^-	Br^-	I^-
Li^+	0.725	0.112	0.252	0.163
	-0.385	0.070	0.084	
Na^+	0.284	0.145	0.131	0.120
K^+	0.203	0.129	0.117	0.093
Rb^+	0.146	0.120	0.104	0.086
Cs^+	0.114	0.098	0.078	0.062

Note: The concentration range for salts of Na, K and Li are 0 to 1.2 M; salts of Rb and Cs are 0 to 0.5 M. For LiF, the first value $K_{c\alpha}$ was measured in the range 0 to 0.01 M, the second value is in the range 0.04 to 0.1 M. For LiCl and LiBr, the first value $K_{c\alpha}$ was measured in the range 0 to 0.07 M, for the second value it was in the range 0.15 to 1.2 M.

Table E.5 The Setschenov coefficients ($K_{c\alpha}$) for oxygen in salt solutions [1] ($T = 298$ K, $P_{O_2} = 0.1$ MPa).

Ions	ClO_4^-	NO_3^-	SO_4^{2-}	CO_3^{2-}	PO_4^{3-}
Li^+	0.218 (0 to 0.07 M)	0.196 (0 to 0.07 M)	0.244	0.254	–
	0.066 (0.15 to 1.2 M)	- 0.040 (0.15 to 1.2M)	–	–	–
Na^+	0.160	0.124	0.376	0.464	0.652 (0 to 0.6 M)
K^+	0.150 (0 to 0.12 M)	0.105	0.297 (0 to 0.6 M)	0.379	0.507 (0 to 0.6 M)
Rb^+	0.138 (0 to 0.1 M)	0.096 (0 to 0.5 M)	0.290 (0 to 0.3 M)	–	–
Cs^+	–	0.066 (0 to 0.5 M)	0.255 (0 to 0.45 M)	–	–

Note: In the parenthesis concentrations were given for which $K_{c\alpha}$ were calculated; in other cases concentrations were 0 to 1.2 M.

pressure, and sometimes on the electrolyte concentration. The higher the charge and dimensions of ions, the higher the Setschenov coefficient.

References

1. Groysman, A. and Khomutov, N., Solubility of Oxygen in Aqueous Solutions of Electrolytes, *Uspekhi Chimii* **59**(8), 1990, pp. 1217–1250 [in Russian].
2. Battino, R. (Ed.), *Solubility Data Series. Oxygen and Ozone, Vol. 7*, Pergamon Press, Oxford, 1981, 519 pp.

Appendix F
Chemical Compositions of Alloys

Table F.1 Chemical compositions (wt%) of stainless steels (see Tables 2.2 and 2.3).

Alloy		Cr	Ni	Mo	C	Fe	Other elements
Name	UNS No.						
302	S30200	17-19	8-10		0.15	Bal.	
304	S30400	18-20	8-10		0.08	Bal.	
304L	S30403	18-20	8-12		0.03	Bal.	
316	S31600	16-18	10-14	2-3	0.08	Bal.	
316L	S31603	16-18	10-14	2-3	0.03	Bal.	
317L	S31703	18-20	11-15	3-4	0.03	Bal.	
317LM	S31725	18-20	12-16	4-4.5	0.03	Bal.	
317LMN	S31726	17-20	13.5-17.5	18	0.03	Bal.	0.1-0.2N
321	S32100	17-19	9-12		0.08	Bal.	Ti 5 x C_{min}
347	S34700	17-19	9-13		0.08	Bal.	Nb+Ta10x C_{min}
410	S41000	11.5-13.5			0.15	Bal.	
416	S41600	12-14			0.15	Bal.	
430	S43000	14-18			0.12	Bal.	
439	S43035	17-19	0.5		0.07	Bal.	0.2Ti
Duplex steel 2205	S31803	21-23	4.5-6.5	2.5-3.5	0.03	Bal.	0.08-0.2N
AL-6XN	N083687	20-22	23.5-25.5	6-7	0.03	Bal.	0.18-0.25N
SEA-CURE	S44660	25-28	1-3.5	3-4	0.03	Bal.	1 Ti/Nb

Table F.2 Chemical compositions (wt%) of some light alloys.

Alloy		Al	Zn	Mn	Si	Cu	Fe	Ni	Cr	Ti	Mg
Name	UNS No.										
Aluminum 6063	A96063	Bal.	0.1	0.1	0.2-0.6	0.1	0.35		0.1	0.1	0.45-0.9
Magnesium AZ61	M11610	2.6-3.5	1.4-4	0.2-1	0-0.3	0.05	0.01	0-0.001			Bal.

Table F.3 Chemical compositions (wt%) of some carbon steels.

Alloy		C	Si	P	S	Mn	Al	Fe
Name	UNS No.							
Carbon steel C1018	G10180	0.21	0.38	0.09	0.05	0.05	0.01	Bal.
Carbon steel C1010	G10100	0.10	0.38	0.09	0.05	0.05	0.01	Bal.

Table F.4 Chemical compositions (wt%) of alloys (see Table 2.3).

Alloy		Fe	Ni	Cr	Mo	Cu	Zn	Al	Sn	Co	W	Mn	Other
Name	UNS No.												
Alloy C	N10002	6.0	Bal.	15.5	17.0					2.5	4.0		
Alloy G	N06007	19.5	Bal.	22.0	6.5	2.0							
Alloy 825	N08825	30.0	Bal.	21.5	3.0	2.2							
Alloy 20		31-44	32-38	19-21	2-3	3-4						2	Nb (8xC)
Alloy 400 (Monel)	N04400	1.2	Bal.			31.5						1.1	
Alloy K-500	N05500	1.0	Bal.			29.5		2.7					0.6Ti
Alloy 600	N06600	8.0	Bal.	15.5									
Silver-bronze alloys						Bal.							Copper alloy with 2-6Ag
Nickel-aluminum bronze	C95800	4.0	4.5			79		9.0				1.2	
Manganese bronze	C67500	0.8-2.0				57-60	Bal.	0.25	0.5-1.5			0.05-0.5	
Silicon bronze	C65500	0.8	0.6			94.8	1.5					0.5-1.3	2.8-3.8Si
Tin bronzes	C90700	0.15	0.5			88-90	0.5	0.005	10-12				0.2 Sb
Admiralty brass	C44300					70-73	Bal.		0.9-1.2				0.02-0.1As
Aluminum brass	C68700					76-79	Bal.	1.8-2.5					0.02-0.1As
Naval brass	C46400					59-62	Bal.		0.5-1.0				
Yellow brass	C27000					63-68.5	Bal.						
Red brass	C23000					84-86	Bal.						

Appendix G
Biocides Used in Industry

References

1. Korb, L.J. and Sprowls, D.O., *Metals Handbook, Vol. 13: Corrosion*, ASM International, USA, 1987, p. 493.
2. Flemming, H.C., Biofouling in Water Treatment. In: *Biofouling and Biocorrosion in Industrial Water Systems*, H.C. Flemming and G.G. Geesey (Eds.), Springer-Verlag, Berlin, 1991, p. 47.

Table G.1 Biocides using in industry [1, 2].

Biocide	Effectiveness			Advantages	Disadvantages	Notes
	Bacteria	Fungi	Algae			
Chlorine (Cl_2)	E	S	E	Broad spectrum of activity, residual effect, active in low concentrations, effective at neutral pH.	Toxic byproducts (reacts with –NH_2 groups), losses efficiency at high pH, corrosive.	Oxidizing. Destroys biofilm matrix and supports detachment. Concentration: 0.1-0.2 ppm.
Chlorine dioxide (ClO_2)	E	G	G	Activity less pH dependent, less sensitivity against hydrocarbons, can be used in presence of –NH_2 groups, effective in low concentrations.	Toxic byproducts, explosive gas (safety problems), corrosive.	Oxidizing. Concentration: 0.1-1 ppm.
Hypo-chlorite (NaOCl)	E	S	S	Cheap, easy to handle.	Rapid microbial after-growth is observed, toxic byproducts, corrosive.	Oxidizing. Destabilizes and detaches the biofilm matrix.
Chloramine (RNH_2Cl) R – radical (R=CH_3-, CH_3CH_2-, etc.)				Good penetration through biofilm, less toxic byproducts.	Less effective than chlorine to suspended bacteria	
Bromine (Br_2)	E	S	E	Effective over broad pH range and against broad microbial spectrum. Effective in low concentration.	Toxic byproducts. Corrosive.	Oxidizing. Substitute for Cl_2 and ClO_2. Use concentration: 0.05-0.1 ppm.

Table G.1 Continued

Biocide	Effectiveness			Advantages	Disadvantages	Notes
	Bacteria	Fungi	Algae			
Ozone (O_3)				No residues. Similar efficiency as chlorine.	Reacts with many organic substances. Short life time. Corrosive.	Oxidizing. Weakens biofilm matrix.
Hydrogen peroxide (H_2O_2)				Cheap. Relatively non-toxic. Can easily be generated in situ.	High concentrations (>3%) necessary. Corrosive.	Oxidizing. Weakens biofilm matrix.
Peroxy-acetic acid				Broad spectrum, kills spores, no toxic byproducts known.	Not very stable. Corrosive.	Oxidizing. Penetrates biofilm.
Form-aldehyde				Broad spectrum, low cost, stability, easy application.	Toxic, suspected to promote cancer, resistant in some organisms. Legal restrictions.	Non-oxidizing. Reacts with protein fixing biofilms on surfaces.
Glutar-aldehyde	E	E	E	Cheap. Effective over broad pH range. Noncorrosive.	Does not penetrate well through biofilm. Degrades to formic acid. Deactivated by $-NH_2$ groups.	Non-oxidizing. Use concentration: 10-75 ppm.
Isothia-zolones	E	G	E	Broad antibiotic spectrum. pH insensitive.	Deactivated by -HS and $-NH_2$ groups. Noncompatible with some water ingredients.	Non-oxidized. Use concentration: 0.9-13 ppm.

Table G.1 Continued

Biocide	Effectiveness			Advantages	Disadvantages	Notes
	Bacteria	Fungi	Algae			
Quaternary ammonium salts	E	G	E	Surface active, adsorbs to surface and prevent biofilm growth. Relatively non-toxic.	Ineffective in highly oil- or organic-fouled systems. Inactivation by low pH, Ca^{2+} and Mg^{2+} cations. Tendency to foam.	Non-oxidized. Use concentration: 8-35 ppm.
Organo-bromide (DBNPA – 2,2-dibromo-3-nitrilopropionamide)	E	NA	S	Effective at pH=6-8.5.		Non-oxidizing. Use concentration: 0.5-24 ppm.
Methylene bis-thiocyanate	E	S	S		Hydrolyzes above pH=8.	Non-oxidizing. Use concentration: 1.5-8 ppm.
Organo-tin/ quaternary ammonium salts	E	G	E	Functions best at high pH.	Tendency to foam.	Non-oxidizing. Use concentration: 7-50 ppm.

Notes: E – excellent, G – good, S – slight, NA – Not Applicable.

Appendix H
Physico-Chemical Properties of Crude Oil, Petroleum Distillates/Fuels, Naphthenic, and Some Aliphatic Acids

Table H.1 Hydrocarbons containing in crude oil and petroleum distillates/fuels.

Generic Name		Chemical Formula or Structure	State	Corrosiveness or aggressiveness to	
General name	Typical Representative			metals and alloys	polymers and coats
Alkanes (paraffins) C_nH_{2n+2}	Methane	CH_4	gas	No	No
	Ethane	C_2H_6	gas		
	Propane	C_3H_8	gas		
	Butane	C_4H_{10}	gas		
	Pentane	C_5H_{12}	liquid		
	Hexane	C_6H_{14}	liquid		
		C_7H_{16} to $C_{17}H_{36}$	liquid		
		$C_{18}H_{38}$ etc.	solid		
Alkenes C_nH_{2n}	Ethylene	C_2H_4	gas	No	No
	Propylene	C_3H_6	gas		
	Butylene	C_4H_8	gas		
	Pentene	C_5H_{10}	liquid		
	etc.	C_nH_{2n}	liquid		
Cyclo-paraffines	Cyclopentane	C_5H_{10}	liquid	No	Unknown
	Cyclohexane	C_6H_{12}	liquid		
Aromatics	Benzene	C_6H_6	liquid	No	Aggressive
	Toluene	$C_6H_5-CH_3$	liquid		
	Xylene	$C_6H_4-(CH_3)_2$	liquid		
Naphthenic Acids	Cyclopentane carboxylic acid, Cyclohexane carboxylic acid, etc.	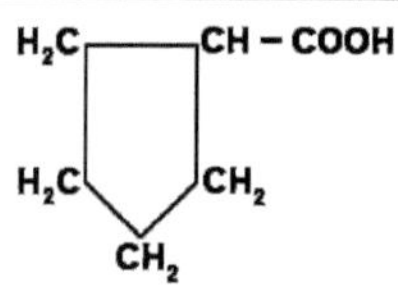	liquid	Corro-sive at 190 to 400°C	Unknown
Sulfur containing compounds	Sulfur	S_8	solid	Depends on temperature	
	Hydrogen Sulfide	H_2S	gas	Yes	Yes
	Mercaptans	R-S-H	gas-liquid	Yes	Unknown
	Sulfides	R-S-R_1			
	Disulfides	R-S-S-R_1			
	Polysulfides	R-S_n-S-R_1			
	Thiophenes	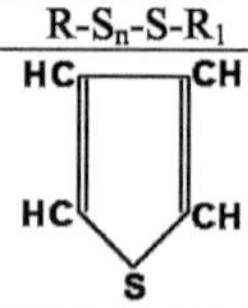	liquid		

Table H.1 Continued.

Generic Name		Chemical Formula or Structure	State	Corrosiveness or aggressiveness to	
General name	Typical Representative			metals and alloys	polymers and coats
	Sulfones	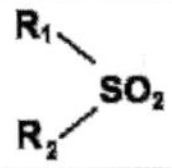	solid		
Nitrogen containing compounds	Pyridine	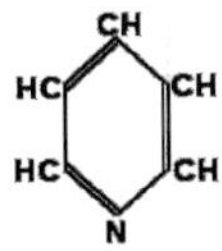	liquid	Corrosion Inhibitor	
	Quinoline	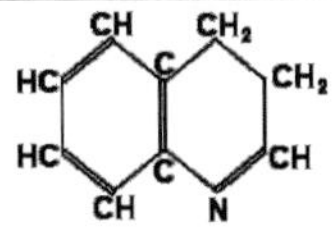	liquid		

Note: The state is done for standard conditions (298 K, 1 atm)

Table H.2 Chemical content of crude oils and petroleum distillates/fuels.

Element	Weight %
C	83 to 87
H	11 to 14
O	0.3 to 1.7
S	0.03 to 5
N	0 to 0.3

Table H.3 Fuels obtained by distillation from crude oils.

Petroleum distillate	Number of C atoms	Molecular weight
Gas	C to C_4	16 to 58
Gasoline	C_5 to C_{12}	72 to 170
Kerosene	C_{10} to C_{16}	156 to 226
Gas oil (diesel)	C_{15} to C_{22}	212 to 294
Fuel oil	above C_{22}	above 300

Table H.4 Chemical content (volume %) of the petroleum distillates/fuels.

Chemical substance	Gasoline	Naphtha	Kerosene (jet fuel)	Diesel oil
Paraffins	30	65 to 85	45	50 to 80
Naphthenes	5	30	35	–
Aromatics	35	5	20	20 to 50
Olefins	25	–	–	–
MTBE[a]	10 to 15 (weight %)	–	–	–

[a]MTBE – Methyl Tertiary Butyl Ether.

Table H.5 Physical properties of some NA and low molecular weight aliphatic (fatty) acids.

NA (organic acid) type	Formula	MW, g/mol	T_b, °C
Cyclopentane carboxylic acid	C_5H_9COOH	114	216
Cyclohexane carboxylic acid	$C_6H_{11}COOH$	128	232
Cyclopentane acetic acid	$C_5H_9CH_2COOH$	128	234
Cyclohexane acetic acid	$C_6H_{11}CH_2COOH$	142	242
Cyclopentane propionic acid	$C_5H_9(CH_2)_2COOH$	142	252
Cyclohexane propionic acid	$C_6H_{11}(CH_2)_2COOH$	156	265
Cyclohexane butyric acid	$C_6H_{11}(CH_2)_3COOH$	170	270
Formic acid	$HCOOH$	46	102
Acetic acid	CH_3COOH	60	113
Propionic (propanoic) acid	C_2H_5COOH	74	141
Butanoic (butyric) acid	C_3H_7COOH	88	164

Notes: NA = Naphthenic Acids; MW = Molecular Weight; T_b, °C = boiling point.

Appendix I
Identification of Corrosion Products According to Their Colors

Table I.1 Identification of corrosion products according to their colors. (For a full color version of this figure, see the Color Section)

Metal	Corrosion product (Deposit)	Chemical Name	Color	Corrosive species causing corrosion/ deposit
Iron (Fe)	$Fe(OH)_2$	Iron (II) hydroxide	Blue - green	H_2O
	$Fe(OH)_3$, or FeOOH $Fe_2O_3 \cdot nH_2O$	Iron (III) hydroxide	Red – brick - brown	H_2O, O_2
	FeO	Iron (II) oxide, Ferrous oxide, Maghemite	Gray - black	O_2 H_2O
	Fe_2O_3	Iron (III) oxide, Ferric oxide, Haematite	Orange - red	O_2 H_2O
	Fe_3O_4	Iron (II, III) oxide, Magnetite	Black - grey	O_2
	FeS	Iron (II) Sulfide (troilite), $Fe_{1-X}S$ (pyrrhotite), $Fe_{1+X}S$ (mackinawite)	Black	H_2S, S_8
	Fe_2S_3	Iron (III) Sulfide	Black	H_2S, S_8
	FeS_2	Iron disulfide, or Pyrite	Black	H_2S, S_8
	$FePO_4$	Ferric Phosphate	Yellow – Brown	H_3PO_4
	$Fe_3(PO_4)_2$	Ferrous Phosphate	Brown	H_3PO_4
	$FeCO_3$	Ferrous Carbonate, Siderite	Black	H_2CO_3
	$FeSiO_3$	Ferrous Silicate	Black (shiny-lustre)	SiO_3^{2-}
	$Fe_2(SiO_3)_3$	Ferric Silicate		SiO_3^{2-}
	$Fe(NO_3)_2 \cdot H_2O$	Ferrous Nitrate Hydrate	Green	HNO_3
	$FeSO_4 \cdot 7H_2O$	Ferrous Sulfate Hydrate	Green	H_2SO_4

Table I.1 Continued.

Metal	Corrosion product (Deposit)	Chemical Name	Color	Corrosive species causing corrosion/ deposit
	$FeBr_2 \cdot 6H_2O$	Ferrous Bromide Hydrate	Green	Br_2, HBr
	$(NH_4)_2Fe(SO_4)_2 \cdot 6H_2O$	Ferrous Ammonium Sulfate Hexa-Hydrate, Salt of Mohr	Green	$(NH_4)_2SO_4$
Aluminum (Al)	$Al(OH)_3$, or AlOOH, or $Al_2O_3 \cdot H_2O$	Aluminum Hydroxide	White or Gray - white	H_2O, O_2, OH^-
	$AlPO_4$	Aluminum Phosphate	White	H_3PO_4
Zinc (Zn)	$Zn(OH)_2$	Zinc hydroxide	White	H_2O
	ZnO	Zinc oxide	White	O_2
	$Zn(OH)_2 \cdot ZnCO_3$	Zinc hydroxide – carbonate	White	H_2O, O_2, CO_2
	$Zn_3(PO_4)_2$	Zinc phosphate	White	H_3PO_4
	ZnS	Zinc sulfide	White	H_2S, S_8
Copper (Cu)	Cu_2O	Copper (I) oxide, Cuprite	Red	H_2O, O_2
	CuO	Copper (II) oxide	Black	H_2O, O_2
	Cu_2S	Copper (I) sulfide	Black	H_2S, S_8
	CuS	Copper (II) sulfide	Black	H_2S, S_8
	$Cu(OH)_2 \cdot CuCO_3$	Copper (II) hydroxide – copper (II) carbonate, Malachite (Patina)	Green	H_2O, O_2, CO_2
	$Cu(OH)_2 \cdot CuSO_4$	Copper (II) hydroxide – copper (II) sulfate (Patina)	Blue – green	H_2O, O_2, SO_2
	$Cu(OH)_2$	Copper (II) hydroxide	Blue	H_2O, O_2
	$Cu_3(PO_4)_2$	Copper (II) phosphate	Blue	H_3PO_4, O_2,

Table I.1 Continued.

Metal	*Corrosion product (Deposit)*	*Chemical Name*	*Color*	*Corrosive species causing corrosion/ deposit*
	$Cu(NO_3)_2 \cdot 6H_2O$	Copper (II) nitrate hexa-hydrate	Blue	HNO_3
	$CuSO_4 \cdot 5H_2O$	Copper (II) sulfate pentahydrate, chalcanthite	Blue	H_2O, O_2, SO_2
Calcium (Ca)	$CaCO_3$	Calcium carbonate	White	$Ca^{2+}_{(aq)}$, $CO_3^{2-}{}_{(aq)}$
	$Ca(OH)_2$	Calcium hydroxide	White	$Ca^{2+}_{(aq)}$, $OH^-_{(aq)}$
	$CaHPO_4$	Calcium hydro-phosphate	White	$Ca^{2+}_{(aq)}$, $PO_4^{3-}{}_{(aq)}$
	CaC_2O_4	Calcium oxalate	White	$Ca^{2+}_{(aq)}$, $C_2O_4^{2-}{}_{(aq)}$
Magnesium (Mg)	$Mg(OH)_2$	Magnesium hydroxide	White	H_2O
	$MgCO_3$	Magnesium carbonate	White	CO_2, H_2O
	$MgCO_3 \cdot Mg(OH)_2$	Carbonate hydroxy magnesium, Dolomite	White	CO_2, H_2O
	$MgHPO_4$	Magnesium hydro-phosphate	White	H_3PO_4
Silver (Ag)	Ag_2O	Silver oxide	Brown	O_2
	Ag_2S	Silver sulfide	Black	H_2S, S_8
	$AgCl$	Silver chloride	White	Cl_2, HCl, H_2O, O_2
	$AgBr$	Silver bromide	Yellow	Br_2, HBr
	AgI	Silver iodide	Yellow	I_2, HI, H_2O, O_2
	Ag_3PO_4	Silver phosphate	Yellow	$Ag^+_{(aq)}$, $PO_4^{3-}{}_{(aq)}$
	Ag_2CO_3	Silver carbonate	White	CO_2, O_2
	Ag_2SO_4	Silver sulfate	White	$H_2SO_{4(conc.)}$, H_2O, O_2, SO_2
	Ag_2CrO_4	Silver chromate	Red	CrO_4^{2-}

Table I.1 Continued.

Metal	*Corrosion product (Deposit)*	*Chemical Name*	*Color*	*Corrosive species causing corrosion/ deposit*
Lead (Pb)	$PbSO_4$	Lead sulfate	White	H_2SO_4
	$PbCO_3$	Lead carbonate	White	CO_2, O_2
	$Pb_3(PO_4)_2$	Lead phosphate	White	H_3PO_4
	Pb_3O_4	Lead Oxide, Minium	Red	O_2
	PbS	Lead sulfide	Black	H_2S, S_8
	$PbCl_2$	Lead chloride	White	HCl
	PbI_2	Lead iodide	Yellow	HI
Manganese (Mn)	$Mn(OH)_2$	Manganese (II) hydroxide	White	$OH^-_{(aq)}$, O_2
	$MnCO_3$	Manganese (II) carbonate	White	CO_2, H_2O, O_2
	MnO	Manganese (II) oxide	Grey - green	O_2
	MnO_2	Manganese (IV) oxide, Pyroluzite	Black-brown	O_2
	MnS	Manganese (II) sulfide	Black	H_2S, S_8
	$Mn(SO_4)_2$	Manganese (IV) sulfate	Black	H_2SO_4, O_2
	$Mn_3(PO_4)_2$	Manganese (II) phosphate	Black	H_3PO_4
Nickel (Ni)	$Ni(OH)_2$	Nickel hydroxide	Green (as apple)	$OH^-_{(aq)}$, O_2
	NiO	Nickel oxide	Green	O_2
	NiS	Nickel sulfide	Black	H_2S, S_8
	NiI_2	Nickel iodide	Black	HI, I_2
	$NiCl_2$	Nickel chloride	Yellow of various tint	HCl, Cl_2
	NiF_2	Nickel fluoride		HF, F_2
	$NiSO_4$	Nickel sulfate		H_2SO_4, O_2
	$Ni(NO_3)_2$	Nickel nitrate		HNO_3
	$Ni(CN)_2$	Nickel cyanide		HCN
Tin (Sn)	SnO	Tin (II) oxide	Black	O_2
	SnO_2	Tin (IV) oxide, Cassiterite	Black	O_2
	SnS_2	Tin (IV) sulfide	Yellow	H_2S, S_8
	SnS	Tin (II) sulfide	Brown	H_2S, S_8
	SnI_4	Tin (IV) iodide	Yellow	I_2
	SnI_2	Tin (II) iodide	Red	HI, I_2

Table I.1 Continued.

Metal	*Corrosion product (Deposit)*	*Chemical Name*	*Color*	*Corrosive species causing corrosion/ deposit*
Titanium (Ti)	TiO_2	Titanium (IV) oxide, Rutile	White	O_2
	Ti_2O_3	Titanium (III) oxide	White	O_2
	TiO	Titanium (II) oxide	White	O_2
Molybdenum (Mo)	MoO_3	Molybdenum (VI) oxide	White	O_2
	$MoCl_2$	Molybdenum (II) chloride	Yellow	HCl
	$MoCl_3$	Molybdenum (III) chloride	Red-dark	Cl_2
	MoI_3	Molybdenum (III) iodide	Black	I_2
	$MoBr_3$	Molybdenum (III) bromide	Black	Br_2
	MoF_3	Molybdenum (III) fluoride	Yellow	F_2
Chromium (Cr)	$Cr_2O_3^*$ (crystalline)	Chromium oxide	Black with metallic lustre	O_2
	Cr_2O_3 (amorphous, powder)	Chromium oxide	Green	O_2
	CrO_3 (poison!)	Chromium (VI) oxide	Deep-red	O_3, peroxides
	$Cr(OH)_2$	Chromium (II) hydroxide	Yellow	$OH^-_{(aq)}$
	$Cr(OH)_3$	Chromium (III) hydroxide	Grey-blue	$OH^-_{(aq)}$, O_2
	$Cr_2O_3 \cdot nH_2O$	Chromium (II) oxide hydrate	Grey-blue	O_2, H_2O
	$CrCl_3$	Chromium (III) chloride	Red-violet	Cl_2
	$CrCl_3 \cdot 6H_2O$	Chromium (III) chloride hexahydrate	Green-dark	Cl_2, H_2O
	CrF_3	Chromium (III) fluoride	Green	F_2
	$CrBr_3$	Chromium (III) bromide	Black	Br_2
	CrI_3	Chromium (III) iodide	Black	I_2

Table I.1 Continued.

Metal	*Corrosion product (Deposit)*	*Chemical Name*	*Color*	*Corrosive species causing corrosion/ deposit*
	Cr_2S_3	Chromium (III) sulfide	Black	H_2S, S_8
	$Cr_2(SO_4)_3$	Chromium (III) sulfate	Pink	H_2SO_4
	$Cr_2(SO_4)_3 \cdot 18H_2O$	Chromium (III) sulfate hydrate	Violet	H_2SO_4, H_2O
Cobalt (Co)	CoO	Cobalt (II) oxide	Grey-green	O_2
	$Co(OH)_2$	Cobalt (II) hydroxide	Blue (ambient temperature) Pink (heating)	$OH^-_{(aq)}$, O_2
	Co_3O_4	Cobalt (II, III) oxide	Brown	O_2
	$CoCl_2 \cdot 6H_2O$	Cobalt (II) chloride hexahydrate	Pink	HCl

*Thin film Cr_2O_3 on stainless steel surface has no color.

Postscript

Is there an end for *science*, *technology*, and *art*? They are integral parts of people's existence. People cannot survive without usage of their achievements even if they do not belong to scientific, technological, and artistic fields of society. Life is based on paradoxes and contradictions. A tourist is happy twice: when he leaves and returns home. I came to the saddest point: we have to separate. We went together this interesting way. I began this way many years ago, and I cannot stop it, because I find over again new and original fascinating faces of corrosion, especially in connection with humanitarian aspects of life. Discussion between lyrics and physics (Plato and Aristotle) failed. This is supported by the work of famous metallurgist Cyril Stanley Smith (1904–1992) who said that "*the objects here displayed under the twin banners of Art and Science are to be viewed as reflections of the inner nature of man*". Both sides of life, our existence, have mutual influence, and they enrich each other. I hope you found some useful, interesting, and fascinating aspects in the behavior of metals and alloys, in their corrosion products, in corrosion phenomena, in the methods of corrosion control and corrosion monitoring. Corrosion of metals is an eternal process, as thermodynamics points out. Our efforts to overcome thermodynamic tendency and struggle to continue the healthy life of metals are based on the importance of *corrosion study* from the following points of view: economical, safety, quality of the environment, reliability of substances (chemicals, medicines, food, etc.), and preservation (restrictions) of sources of metals in the crust of the Earth.

Kozma Prutkov (the fictional author invented by Aleksey Konstantinovich Tolstoy and three Zhemchuzhnikov brothers who lived during Russia's authoritarian reign of Nicholas I) said: "One cannot embrace the unembraceable". I did not write an encyclopedia. I wanted to show the *corrosion world* from unusual, philosophical, enjoyable, and beautiful points. In other words, I wanted to show that even complicated or simple at first glance corrosion phenomena can be explained in an enjoyable way. This is a new approach to understanding and teaching such an old and nasty phenomenon as *corrosion*. Our planet, the Earth, is a ball. All phenomena are interrelated. Maybe our universe is also a ball, or some other geometrical form without beginning and end. There is no end to corrosion, as well as to art, to universe, to time … But I finish this book with the certainty that somebody will continue this interesting and enjoyable work of connecting corrosion with all aspects of life for everybody.

Alec Groysman

Color Section

Fig. 1.12 Corroded anchor made of carbon steel.

a b

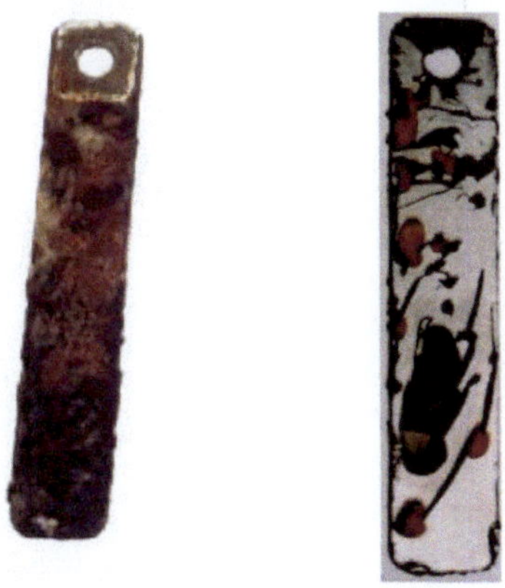

Fig. 1.17 Carbon steel after 7 days immersion in cooling water (a) and in water with the addition of sodium peroxoborate ($NaBO_3$) (b). The "beautiful" rust is formed on the steel's surface in a $NaBO_3$ solution.

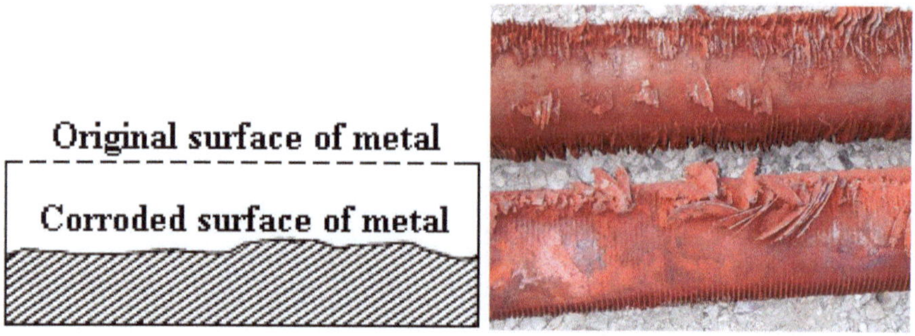

a b

Fig. 2.1 Uniform, or general, corrosion. (a) Scheme of corroded surface of metal; (b) carbon steel tubes after uniform corrosion in the furnace.

a b c

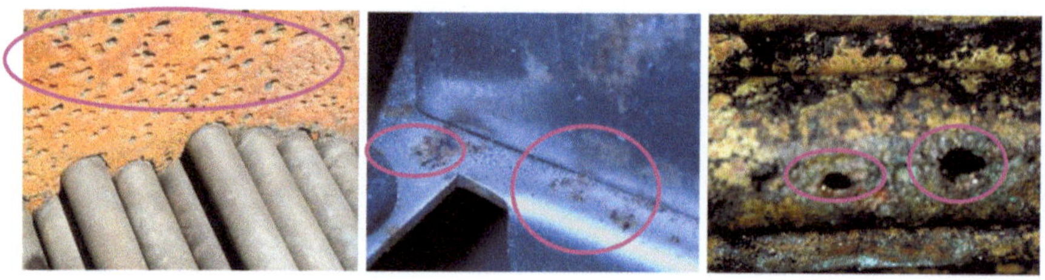

Fig. 2.3 Pits on the carbon steel tube-sheet of a heat exchanger (a), stainless steel 316 bath (b) and Admiralty brass tubes (c). Pits are encircled.

Fig. 2.6 Heat exchanger: tubes made of titanium and baffles made of carbon steel (cooling water, 4 years). Deposits on the titanium tubes are the salts (scale) formed from the cooling water, and not corrosion products.

Fig. 2.9 Galvanic corrosion of coated carbon steel tube sheet because of high ratio between cathode (Admiralty brass tubes) and anode area (carbon steel tube sheet coated with epoxy). Gray – failed epoxy coating; yellow – rusted tube sheet.

Fig. 2.10 MIC on a metal surface compared with pitting, crevice, differential aeration cell, or galvanic corrosion. The bottom of a crude oil storage tank after a microbial attack.

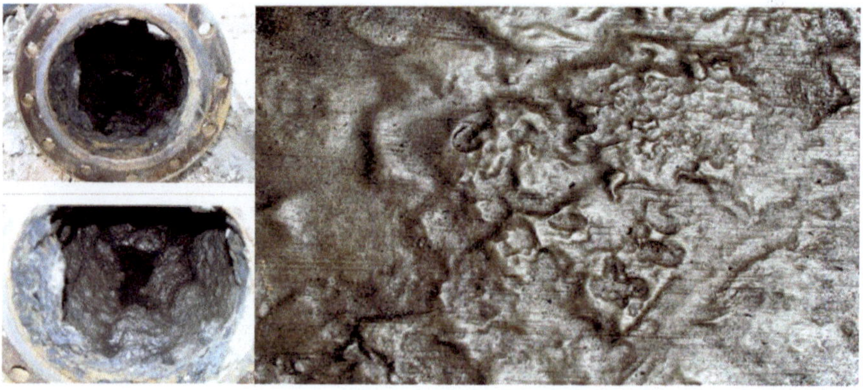

a b

Fig. 2.14 (a) SRB inside cooling water carbon steel pipes (after four years); the thickness of biofilm with SRB was about 6 to 10 cm. (b) Corroded carbon steel bottom of the crude oil aboveground storage tank after 18 years of service as a result of SRB activity.

Fig. 2.16 Erosion-corrosion inside a liquefied petroleum gas (LPG) carbon steel tube. "Grooves" are formed as a result of erosion.

Fig. 3.1 Carbon steel tubes used in contact with water.

Fig. 3.2 Scale (calcium and magnesium carbonates) formed inside of heat exchanger with cooling water after 4 years.

Fig. 3.3 Metallic structures and equipment in the atmosphere.

Fig. 3.5 Natural patina formed on a bronze statue.

Fig. 3.7 Sculpture "Sun – Man" (1998) made of *weathering steel* (CORTEN) by Jorge Vieira (Lisbon, Portugal).

Fig. 3.8 The exhaust system of a vehicle after dewpoint corrosion.

Fig. 3.9 Corrosion of a heat exchanger under thermal insulation.

Fig. 3.11 Corrosion "balls" formed on a carbon steel surface (strip coupons) after 3 days' immersion in kerosene with water (1 wt %) added.

Fig. 3.15 Corroded roof of the gas oil AST after 5 years of service.

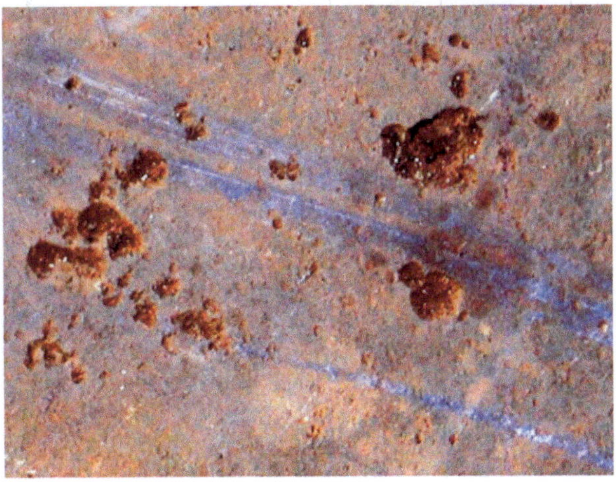

Fig. 3.16 Corroded bottoms of the fuel oil AST after 10 years of service.

Fig. 3.17 Corroded bottom of a crude oil AST. The morphology of the surface and microbiological examination showed MIC.

Fig. 4.8 Aluminum sacrificial anodes (A) installed on the inside of heat exchangers coated with epoxy paint (green). The cooling water flows inside the tubes.

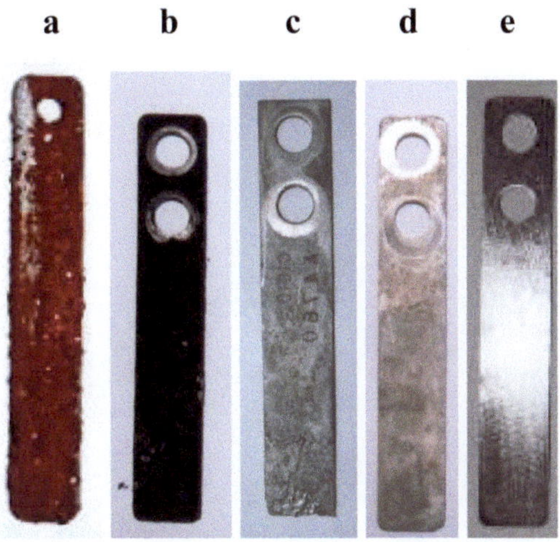

Fig. 5.2 The coupons after being exposed to cooling water (a, c) and to the process stream in the overhead of the crude distillation column (b, d): a, b – coupons with corrosion products; c, d – coupons after cleansing of corrosion products; e – original coupon.

Fig. 6.1 Left – Gold mask of Pharaoh Tutankhamun (14th century B.C.); Right – silver coins (Israel, 69 A.D.).

Fig. 6.3 Miniature from the alchemy book. Seven metals are depicted as gods – planets' symbols sheltered in the cave.

Fig. 6.4 Sasha Okun "Woman". (Reprinted with permission)

Fig. 6.5 Noam Ben-Jacov "Motion". (Reprinted with permission)

Fig. 6.6 Beauty of corroded Raschig rings in an oil refinery.

Fig. 6.8 Cathedral of Christ the Savior (Moscow), stainless steel with decorative and protective titanium nitride (TiN) coating.

Fig. 6.9 "Bone structure V", carbon steel, treated with vinegar and natural pigments, Natasha Carsberg, England. (Reprinted with permission)

Fig. 6.10 "Come on! Let us fight against corrosion!" Bronze, Israel.

Fig. 6.11 Alberto Giacometti (1901–1966) "Tall Figure" (1949, painted bronze), resembles uniform corrosion. © Photo Scala, Florence, The Museum of Modern Art, New York/Scala Florence, 2009.

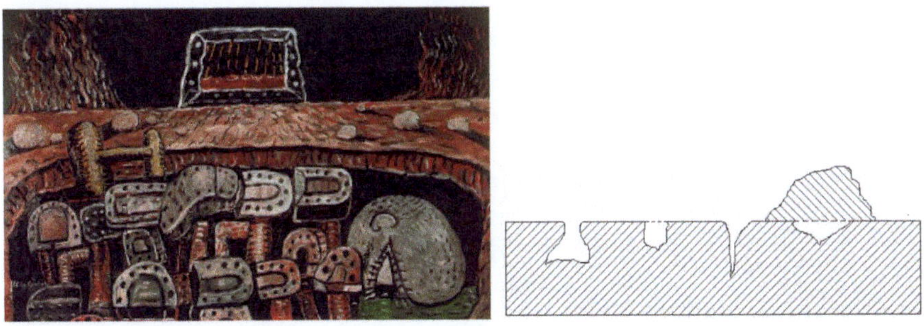

Fig. 6.12 Philip Guston "Pit" (1976) National Gallery of Australia, Canberra. © The Estate of Philip Guston. Pitting corrosion – severe local corrosion. Reprinted with permission.

a b c

Fig. 6.13 Umberto Boccioni (1882–1916) "Unique Forms of Continuity in Space" (1913) or "Erosion". (a, c) Erosion inside brass tube; (b) the sculpture of Umberto Boccioni. © Photo Scala, Florence, The Museum of Modern Art, New York/Scala, Florence, 2009.

Fig. 6.14 Hieronymus Bosch "The Garden of Earthly Delights" (1510–1515) and three periods of car's life: new car, beginning of corrosion, and destruction. (Reprinted with permission, © Museo Nacional del Prado, Madrid, Spain.)

Fig. 6.15 Anti-corrosion protection as well as decoration of a storage tank (Israel).

Fig. 6.16 Russian version of the "The Wizard of Oz" (1939): "Iron Woodman". Dorothy brings an oil-can against atmospheric corrosion.

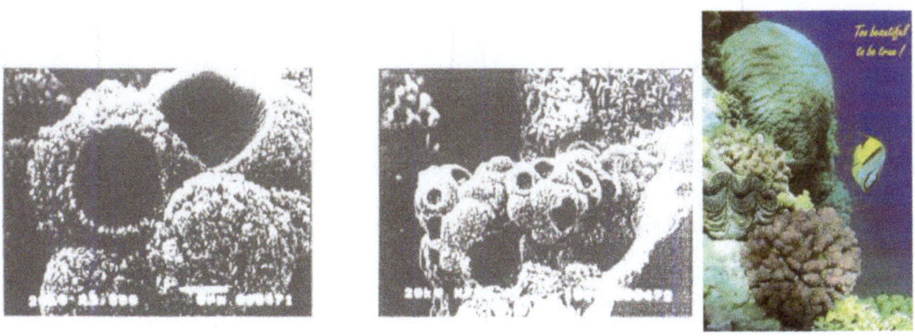

a b

Fig. 6.18 (a) SEM photo of rust (×3,500 and ×1,500; (b) coral reefs.

a b

Fig. 6.17 (a) "The Old Trees and Waterfall" (China, 1549); (b) carbon steel strips after immersion in sodium peroxocarbonate aqueous solutions.

a b

Fig. 6.19 (a) SEM photo (× 10,000) of rust formed in the heat exchanger of a overhead crude distillation column; (b) black sea urchins.

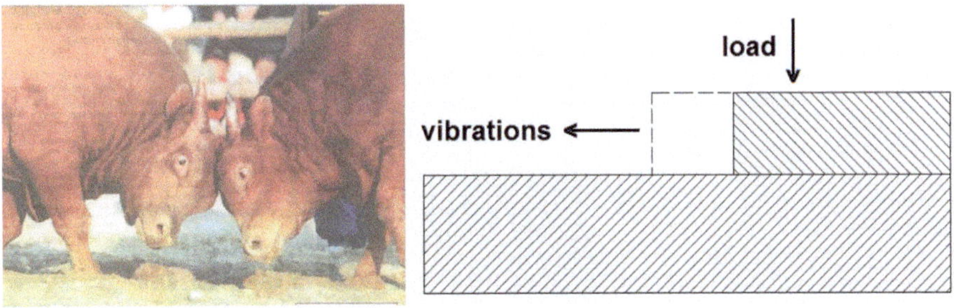

Fig. 6.22 "Fretting corrosion" – attack due to rubbing contact.

Fig. 6.23 "Impingement attack" – localized attack related to high velocity. (Right – impingement by escaping steam from cracked tube).

Fig. 6.24 "Corrosion led to contentment". Corroded mast of a ship (Haifa Bay).

Table I.1 Identification of corrosion products according to their colors.

Metal	Corrosion product (Deposit)	Chemical Name	Color	Corrosive species causing corrosion/ deposit
Iron (Fe)	$Fe(OH)_2$	Iron (II) hydroxide	Blue - green	H_2O
	$Fe(OH)_3$, or FeOOH $Fe_2O_3{\cdot}nH_2O$	Iron (III) hydroxide	Red – brick - brown	H_2O, O_2
	FeO	Iron (II) oxide, Ferrous oxide, Maghemite	Gray - black	O_2 H_2O
	Fe_2O_3	Iron (III) oxide, Ferric oxide, Haematite	Orange - red	O_2 H_2O
	Fe_3O_4	Iron (II, III) oxide, Magnetite	Black - grey	O_2
	FeS	Iron (II) Sulfide (troilite), $Fe_{1\text{-}X}S$ (pyrrhotite), $Fe_{1+X}S$ (mackinawite)	Black	H_2S, S_8
	Fe_2S_3	Iron (III) Sulfide	Black	H_2S, S_8
	FeS_2	Iron disulfide, or Pyrite	Black	H_2S, S_8
	$FePO_4$	Ferric Phosphate	Yellow – Brown	H_3PO_4
	$Fe_3(PO_4)_2$	Ferrous Phosphate	Brown	H_3PO_4
	$FeCO_3$	Ferrous Carbonate, Siderite	Black	H_2CO_3
	$FeSiO_3$	Ferrous Silicate	Black (shiny-lustre)	SiO_3^{2-}
	$Fe_2(SiO_3)_3$	Ferric Silicate		SiO_3^{2-}
	$Fe(NO_3)_2 \cdot H_2O$	Ferrous Nitrate Hydrate	Green	HNO_3
	$FeSO_4 \cdot 7H_2O$	Ferrous Sulfate Hydrate	Green	H_2SO_4

Table I.1 Continued.

Metal	Corrosion product (Deposit)	Chemical Name	Color	Corrosive species causing corrosion/ deposit
	$FeBr_2 \cdot 6H_2O$	Ferrous Bromide Hydrate	Green	Br_2, HBr
	$(NH_4)_2Fe(SO_4)_2 \cdot 6H_2O$	Ferrous Ammonium Sulfate Hexa-Hydrate, Salt of Mohr	Green	$(NH_4)_2SO_4$
Aluminum (Al)	$Al(OH)_3$, or AlOOH, or $Al_2O_3 \cdot H_2O$	Aluminum Hydroxide	White or Gray - white	H_2O, O_2, OH^-
	$AlPO_4$	Aluminum Phosphate	White	H_3PO_4
Zinc (Zn)	$Zn(OH)_2$	Zinc hydroxide	White	H_2O
	ZnO	Zinc oxide	White	O_2
	$Zn(OH)_2 \cdot ZnCO_3$	Zinc hydroxide – carbonate	White	H_2O, O_2, CO_2
	$Zn_3(PO_4)_2$	Zinc phosphate	White	H_3PO_4
	ZnS	Zinc sulfide	White	H_2S, S_8
Copper (Cu)	Cu_2O	Copper (I) oxide, Cuprite	Red	H_2O, O_2
	CuO	Copper (II) oxide	Black	H_2O, O_2
	Cu_2S	Copper (I) sulfide	Black	H_2S, S_8
	CuS	Copper (II) sulfide	Black	H_2S, S_8
	$Cu(OH)_2 \cdot CuCO_3$	Copper (II) hydroxide – copper (II) carbonate, Malachite (Patina)	Green	H_2O, O_2, CO_2
	$Cu(OH)_2 \cdot CuSO_4$	Copper (II) hydroxide – copper (II) sulfate (Patina)	Blue – green	H_2O, O_2, SO_2
	$Cu(OH)_2$	Copper (II) hydroxide	Blue	H_2O, O_2
	$Cu_3(PO_4)_2$	Copper (II) phosphate	Blue	H_3PO_4, O_2,

Table I.1 Continued.

Metal	*Corrosion product (Deposit)*	*Chemical Name*	*Color*	*Corrosive species causing corrosion/ deposit*
	$Cu(NO_3)_2 \cdot 6H_2O$	Copper (II) nitrate hexa-hydrate	Blue	HNO_3
	$CuSO_4 \cdot 5H_2O$	Copper (II) sulfate pentahydrate, chalcanthite	Blue	H_2O, O_2, SO_2
Calcium (Ca)	$CaCO_3$	Calcium carbonate	White	$Ca^{2+}_{(aq)}$, $CO_3^{2-}{}_{(aq)}$
	$Ca(OH)_2$	Calcium hydroxide	White	$Ca^{2+}_{(aq)}$, $OH^-_{(aq)}$
	$CaHPO_4$	Calcium hydro-phosphate	White	$Ca^{2+}_{(aq)}$, $PO_4^{3-}{}_{(aq)}$
	CaC_2O_4	Calcium oxalate	White	$Ca^{2+}_{(aq)}$, $C_2O_4^{2-}{}_{(aq)}$
Magnesium (Mg)	$Mg(OH)_2$	Magnesium hydroxide	White	H_2O
	$MgCO_3$	Magnesium carbonate	White	CO_2, H_2O
	$MgCO_3 \cdot Mg(OH)_2$	Carbonate hydroxy magnesium, Dolomite	White	CO_2, H_2O
	$MgHPO_4$	Magnesium hydro-phosphate	White	H_3PO_4
Silver (Ag)	Ag_2O	Silver oxide	Brown	O_2
	Ag_2S	Silver sulfide	Black	H_2S, S_8
	AgCl	Silver chloride	White	Cl_2, HCl, H_2O, O_2
	AgBr	Silver bromide	Yellow	Br_2, HBr
	AgI	Silver iodide	Yellow	I_2, HI, H_2O, O_2
	Ag_3PO_4	Silver phosphate	Yellow	$Ag^+_{(aq)}$, $PO_4^{3-}{}_{(aq)}$
	Ag_2CO_3	Silver carbonate	White	CO_2, O_2
	Ag_2SO_4	Silver sulfate	White	$H_2SO_{4(conc.)}$, H_2O, O_2, SO_2
	Ag_2CrO_4	Silver chromate	Red	CrO_4^{2-}

Table I.1 Continued.

Metal	*Corrosion product (Deposit)*	*Chemical Name*	*Color*	*Corrosive species causing corrosion/ deposit*
Lead (Pb)	$PbSO_4$	Lead sulfate	White	H_2SO_4
	$PbCO_3$	Lead carbonate	White	CO_2, O_2
	$Pb_3(PO_4)_2$	Lead phosphate	White	H_3PO_4
	Pb_3O_4	Lead Oxide, Minium	Red	O_2
	PbS	Lead sulfide	Black	H_2S, S_8
	$PbCl_2$	Lead chloride	White	HCl
	PbI_2	Lead iodide	Yellow	HI
Manganese (Mn)	$Mn(OH)_2$	Manganese (II) hydroxide	White	$OH^-_{(aq)}$, O_2
	$MnCO_3$	Manganese (II) carbonate	White	CO_2, H_2O, O_2
	MnO	Manganese (II) oxide	Grey - green	O_2
	MnO_2	Manganese (IV) oxide, Pyroluzite	Black-brown	O_2
	MnS	Manganese (II) sulfide	Black	H_2S, S_8
	$Mn(SO_4)_2$	Manganese (IV) sulfate	Black	H_2SO_4, O_2
	$Mn_3(PO_4)_2$	Manganese (II) phosphate	Black	H_3PO_4
Nickel (Ni)	$Ni(OH)_2$	Nickel hydroxide	Green (as apple)	$OH^-_{(aq)}$, O_2
	NiO	Nickel oxide	Green	O_2
	NiS	Nickel sulfide	Black	H_2S, S_8
	NiI_2	Nickel iodide	Black	HI, I_2
	$NiCl_2$	Nickel chloride	Yellow of various tint	HCl, Cl_2
	NiF_2	Nickel fluoride		HF, F_2
	$NiSO_4$	Nickel sulfate		H_2SO_4, O_2
	$Ni(NO_3)_2$	Nickel nitrate		HNO_3
	$Ni(CN)_2$	Nickel cyanide		HCN
Tin (Sn)	SnO	Tin (II) oxide	Black	O_2
	SnO_2	Tin (IV) oxide, Cassiterite	Black	O_2
	SnS_2	Tin (IV) sulfide	Yellow	H_2S, S_8
	SnS	Tin (II) sulfide	Brown	H_2S, S_8
	SnI_4	Tin (IV) iodide	Yellow	I_2
	SnI_2	Tin (II) iodide	Red	HI, I_2

Table I.1 Continued.

Metal	*Corrosion product (Deposit)*	*Chemical Name*	*Color*	*Corrosive species causing corrosion/ deposit*
Titanium (Ti)	TiO_2	Titanium (IV) oxide, Rutile	White	O_2
	Ti_2O_3	Titanium (III) oxide	White	O_2
	TiO	Titanium (II) oxide	White	O_2
Molybdenum (Mo)	MoO_3	Molybdenum (VI) oxide	White	O_2
	$MoCl_2$	Molybdenum (II) chloride	Yellow	HCl
	$MoCl_3$	Molybdenum (III) chloride	Red-dark	Cl_2
	MoI_3	Molybdenum (III) iodide	Black	I_2
	$MoBr_3$	Molybdenum (III) bromide	Black	Br_2
	MoF_3	Molybdenum (III) fluoride	Yellow	F_2
Chromium (Cr)	$Cr_2O_3^*$ (crystalline)	Chromium oxide	Black with metallic lustre	O_2
	Cr_2O_3 (amorphous, powder)	Chromium oxide	Green	O_2
	CrO_3 (poison!)	Chromium (VI) oxide	Deep-red	O_3, peroxides
	$Cr(OH)_2$	Chromium (II) hydroxide	Yellow	$OH^-_{(aq)}$
	$Cr(OH)_3$	Chromium (III) hydroxide	Grey-blue	$OH^-_{(aq)}$, O_2
	$Cr_2O_3 \cdot nH_2O$	Chromium (II) oxide hydrate	Grey-blue	O_2, H_2O
	$CrCl_3$	Chromium (III) chloride	Red-violet	Cl_2
	$CrCl_3 \cdot 6H_2O$	Chromium (III) chloride hexahydrate	Green-dark	Cl_2, H_2O
	CrF_3	Chromium (III) fluoride	Green	F_2
	$CrBr_3$	Chromium (III) bromide	Black	Br_2
	CrI_3	Chromium (III) iodide	Black	I_2

Table I.1 Continued.

Metal	*Corrosion product (Deposit)*	*Chemical Name*	*Color*	*Corrosive species causing corrosion/ deposit*
	Cr_2S_3	Chromium (III) sulfide	Black	H_2S, S_8
	$Cr_2(SO_4)_3$	Chromium (III) sulfate	Pink	H_2SO_4
	$Cr_2(SO_4)_3 \cdot 18H_2O$	Chromium (III) sulfate hydrate	Violet	H_2SO_4, H_2O
Cobalt (Co)	CoO	Cobalt (II) oxide	Grey-green	O_2
	$Co(OH)_2$	Cobalt (II) hydroxide	Blue (ambient temperature) Pink (heating)	$OH^-_{(aq)}$, O_2
	Co_3O_4	Cobalt (II, III) oxide	Brown	O_2
	$CoCl_2 \cdot 6H_2O$	Cobalt (II) chloride hexahydrate	Pink	HCl

*Thin film Cr_2O_3 on stainless steel surface has no color.

Index

MIX
Papier aus verantwortungsvollen Quellen
Paper from responsible sources
FSC® C105338

If you have any concerns about our products,
you can contact us on
ProductSafety@springernature.com

In case Publisher is established outside the EU,
the EU authorized representative is:
Springer Nature Customer Service Center GmbH
Europaplatz 3, 69115 Heidelberg, Germany

Printed by Libri Plureos GmbH
in Hamburg, Germany